MEGADESIGN and MegaOpt – German Initiatives for Aerodynamic Simulation and Optimization in Aircraft Design

Results of the Closing Symposium of the MEGADESIGN and MegaOpt Projects, Braunschweig, Germany, 23–24 May, 2007

Norbert Kroll
Dieter Schwamborn
Klaus Becker
Herbert Rieger
Frank Thiele
(Editors)

Springer

Prof. Dr. Norbert Kroll
German Aerospace Center (DLR)
Institute of Aerodynamics and Flow Technology
Lilienthalplatz 7
38108 Braunschweig
Germany
E-mail: norbert. kroll@dlr.de

Dr. Dieter Schwamborn
German Aerospace Center (DLR)
Institute of Aerodynamics and Flow Technology
Bunsenstraße 10
37073 Göttingen
Germany
E-mail: dieter.schwamborn@dlr.de

Dr. Klaus Becker
Airbus Deutschland GmbH
Airbusallee 1
28199 Bremen
Germany
E-mail: klaus.becker@airbus.com

Dr. Herbert Rieger
EADS Defence and Security
P.O. Box 80 1 160
81663 München
Germany
E-mail: herbert.rieger@eads.com

Prof. Dr. Frank Thiele
Hermann-Föttinger-Institut für
Strömungsmechanik
Müller-Breslau-Str. 8
10623 Berlin
Germany
E-mail: thiele@pi.tu-berlin.de

ISBN 978-3-642-26062-9 e-ISBN 978-3-642-04093-1

DOI 10.1007/978-3-642-04093-1

Notes on Numerical Fluid Mechanics
and Multidisciplinary Design ISSN 1612-2909

Typeset & Cover Design: Scientific Publishing Services Pvt. Ltd., Chennai, India.

Printed in acid-free paper

5 4 3 2 1 0

springer.com

NNFM Editor Addresses

Prof. Dr. Wolfgang Schröder
(General Editor)
RWTH Aachen
Lehrstuhl für Strömungslehre und
Aerodynamisches Institut
Wüllnerstr. zw. 5 u. 7
52062 Aachen
Germany
E-mail: office@aia.rwth-aachen.de

Prof. Dr. Kozo Fujii
Space Transportation Research Division
The Institute of Space
and Astronautical Science
3-1-1, Yoshinodai, Sagamihara
Kanagawa, 229-8510
Japan
E-mail: fujii@flab.eng.isas.jaxa.jp

Dr. Werner Haase
Höhenkirchener Str. 19d
D-85662 Hohenbrunn
Germany
E-mail: office@haa.se

Prof. Dr. Ernst Heinrich Hirschel
(Former General Editor)
Herzog-Heinrich-Weg 6
D-85604 Zorneding
Germany
E-mail: e.h.hirschel@t-online.de

Prof. Dr. Bram van Leer
Department of Aerospace Engineering
The University of Michigan
Ann Arbor, MI 48109-2140
USA
E-mail: bram@engin.umich.edu

Prof. Dr. Michael A. Leschziner
Imperial College of Science
Technology and Medicine
Aeronautics Department
Prince Consort Road
London SW7 2BY
U.K.
E-mail: mike.leschziner@ic.ac.uk

Prof. Dr. Maurizio Pandolfi
Politecnico di Torino
Dipartimento di Ingegneria
Aeronautica e Spaziale
Corso Duca degli Abruzzi, 24
I-10129 Torino
Italy
E-mail: pandolfi@polito.it

Prof. Dr. Jacques Periaux
38, Boulevard de Reuilly
F-75012 Paris
France
E-mail: jperiaux@free.fr

Prof. Dr. Arthur Rizzi
Department of Aeronautics
KTH Royal Institute of Technology
Teknikringen 8
S-10044 Stockholm
Sweden
E-mail: rizzi@aero.kth.se

Dr. Bernard Roux
L3M – IMT La Jetée
Technopole de Chateau-Gombert
F-13451 Marseille Cedex 20
France
E-mail: broux@l3m.univ-mrs.fr

Prof. Dr. Yurii I. Shokin
Siberian Branch of the
Russian Academy of Sciences
Institute of Computational
Technologies
Ac. Lavrentyeva Ave. 6
630090 Novosibirsk
Russia
E-mail: shokin@ict.nsc.ru

Preface

Over the last decade, Computational Fluid Dynamics (CFD) has become a mature technology for the development of new products in aeronautical industry. Aerodynamic design engineers have progressively taken advantage of the possibilities offered by the numerical solution of the Reynolds averaged Navier-Stokes (RANS) equations. Significant improvements in physical modeling and solution algorithms as well as the enormous increase of computer power enable high-fidelity numerical simulations in all stages of aircraft development.

In Germany, the national CFD project MEGAFLOW furthered the development and availability of RANS solvers for the prediction of complex flow problems significantly. MEGAFLOW was initiated by the first aviation research program of the Federal Government in 1995 under the leadership of the DLR (see Kroll, N., Fassbender, J.K. (Eds).: *MEGAFLOW – Numerical Flow Simulation for Aircraft Design*; Notes on Numerical Fluid Mechanics and Multidisciplinary Design, Volume 89, Springer, 2005). A network from aircraft industry, DLR and several universities was created with the goal to focus and direct development activities for numerical flow simulation towards a common aerodynamic simulation system providing both a block-structured (FLOWer-Code) and a hybrid (TAU-Code) parallel flow prediction capability. Today, both codes have reached a high level of maturity and reliability. They are routinely used at DLR and German aeronautic industry for a wide range of aerodynamic applications. For many universities the MEGAFLOW software represents a platform for the improvement of physical models and for the investigation of complex flow problems. The network was established as an efficient group of very closely co-operating partners with supplementing expertises and experience. Focusing on common software, the process of transferring latest research and technology results into production codes used at industry has been considerably accelerated.

Despite the progress made in CFD, future demands of aircraft industry with respect to more environmentally friendly, safer and more economical aircraft require further improvement of simulation capabilities. The need to achieve reliable results at a high level of accuracy for complex configurations within short turn-around time places severe constrains on the application of CFD for aerodynamic data production and the integration of RANS methods into multidisciplinary simulation and optimization procedures. Consequently, enhanced CFD capabilities for reducing design cycle time and cost are indispensable for the industry.

In order to meet future requirements of German aircraft industry, two MEGAFLOW follow-on projects were set up, MEGADESIGN within the third aviation program of the Federal Government mid 2003 and MegaOpt as an internal DLR-project linked to MEGADESIGN. Based on the MEGAFLOW software, the main objectives of these four-year projects were to ensure the prediction accuracy with a guaranteed error bandwidth for certain aircraft configurations at design conditions, to reduce the simulation turn-around time for large-scale applications significantly, to improve the reliability of the flow solvers for full aircraft configurations in the complete flight regime, to extend the flow solvers to allow for multidisciplinary simulations and to establish numerical shape optimization as a vital tool within the aircraft design process. Partners of the MEGADESIGN consortium were DLR (Institute of Aerodynamics and Flow Technology), Airbus, EADS Military Air Systems, Synaps Ingenieur-Gesellschaft mbH, FastOpt, HPCC Space GmbH, RWTH Aachen University (Department of Mechanics), Berlin Technical University (Institute of Fluid Mechanics and Technical Acoustics), Braunschweig Technical University (Institute for Fluid Mechanics), Darmstadt University of Technology (Institute of Fluid Mechanics and Aerodynamics), Trier University (Department of Mathematics). The project was coordinated by DLR. Both the Institute of Aerodynamics and Flow Technology and the Institute of Aeroelasticity were involved in the complementary DLR-project MegaOpt.

This volume contains results presented during the closing symposium of the MEGADSIGN project which took place at DLR Braunschweig, Germany, on May 23rd and 24th 2007 and was jointly held with contributions from the MegaOpt project which finished at the end of 2007. Selected papers give an overview of the main activities and results achieved within both projects. Improvements and enhancements of the flow solvers are described, followed by new developments with respect to aerodynamic shape optimization and multidisciplinary optimization. Improved numerical simulation capabilities are demonstrated by several industrial applications.

Thanks are due to all partners and colleagues who have contributed in an open and collaborative manner. The knowledge and engagement of each individual contributed to the success and world wide appreciation of the MEGADSIGN project.

The funding of partial activities through the German Government in the framework of the air transport research program is gratefully acknowledged. The editors would also like to express gratitude to M. Wagler and F. Prill for technical support in compiling this book. Finally, the editors are grateful to Prof. Dr. W. Schröder as the general editor of the Springer series "Notes on Numerical Fluid Mechanics and Multidisciplinary Design" and also to the staff of the Springer for the opportunity to publish the technical results of the German CFD projects in this series.

Braunschweig / Göttingen
May 2009

Norbert Kroll
Dieter Schwamborn

Contents

Part I

Reduction of Simulation Time

Recent Developments of TAU Adaptation Capability

T. Alrutz[1] and D. Vollmer[2]

[1] German Aerospace Center (DLR) Göttingen
Institute of Aerodynamics and Flow Technology – Numerical Methods Branch
Bunsenstrasse 10, 37073 Göttingen, Germany
thomas.alrutz@dlr.de

[2] German Aerospace Center (DLR) Braunschweig
Institute of Aerodynamics and Flow Technology – Numerical Methods Branch
Lilienthalplatz 7, 38108 Braunschweig, Germany
daniel.vollmer@dlr.de

Summary. We present an overview of the mesh adaptation facility of the DLR TAU code as well as details of improvements made to it in the recent MEGADESIGN project, in particular focusing on advances made in the core of the adaptation module (for example parallel (de-)refinement and other efficiency improvements) as well as a relatively new type of adaptation indicator based on the adjoint solution (among other things) for a goal-oriented mesh adaption.

These improvements to the existing algorithms already available in the TAU code allow us to produce improved computational meshes in a more distributed manner, which provide more accurate predictions for selected functionals of the flow solution such as drag or lift – in fact, for any for which the necessary adjoint solution can be computed.

1 Introduction

Today's computational fluid dynamics (CFD) solvers are used to solve problems of ever increasing size and complexity, with many industrial partners relying more and more on the predictions of their computations and less on actual testing until later development stages are reached.

A big improvement for complex configurations – although not the "panacea" it may once have been touted as – was the introduction of unstructured methods, which allow the cumbersome and time-consuming process of mesh generation for such computations to be largely automated. This was thought to reduce the reliance on experienced users for setting up such complicated computations. Unfortunately, the quality of the computational mesh is still directly related to the accuracy of the result.[1] The quality of the mesh is in turn often dependent on the experience of the user with similar flow configurations as it is usually not clear *a priori* where dominant flow features will occur, and even whether they will have an impact on the outcome of the computation.

[1] Or in some non-trivial cases, whether a result could be obtained at all, as a low quality mesh has a high impact on the robustness of the solver.

N. Kroll et al. (Eds.): MEGADESIGN and MegaOpt, NNFM 107, pp. 3–19.
springerlink.com

To remedy this and produce meshes that resolve each of the flow phenomena of interest, local mesh adaption was introduced (for examples in the DLR TAU code see [5]). This procedure takes an existing computational mesh and a flow solution thereon[2] to produce a new mesh that will in some sense yield a better result for the observed flow topology, usually by inserting additional points in "regions of interest" and sometimes removing points where they are not needed or moving existing points to improve element quality.

In order to isolate these regions, a so-called *adaption indicator* or *sensor* is computed for each point in the mesh to obtain a measure of the local mesh quality[3]. This is a non-trivial process as mesh quality is not a purely local phenomenon because it depends on how the flow solver itself operates – for example the size of the stencil of the scheme, the exact nature of its gradient computations, or the flux functions used all influence how accurate a solver can be on a particular mesh. For this reason, almost all mesh adaption indicators resort to computing gradients of the solution variables with the notion being that mesh regions where large changes in the solution occur (e.g. shocks/discontinuities) have a strong impact on the quality of the result. In a sense, this is certainly true as by Godunov's theorem such regions can only exhibit 1st order accuracy and as such, increased spatial resolution will improve the resolution of those discontinuities.

Unfortunately, the existence of such strong shocks can lead to a problem where most – if not all – of the added points during mesh adaption are spent over-resolving shocks all the while neglecting either weaker shocks or under-resolving smooth solution areas which nevertheless can have a larger impact on the accuracy of the computation, depending on how the "accuracy" is measured. This can result in solutions whose inaccuracy is *amplified* by the repeated application of local mesh refinement based on such a gradient-based indicator – a clear contradiction to the common opinion that using more points will always give a better solution.

2 TAU-Code Adaptation Overview

The adaptation module of the TAU-code consists of three different components for various grid manipulations to adapt a given grid to the solved flow field:

1. y^+ based grid adaptation to adjust the first wall distance over turbulent surfaces in hybrid grids,
2. hierarchical grid refinement and derefinement to introduce new grid points on a given egde-indicator function without producing hanging nodes,
3. surface approximation and reconstruction for curved surfaces after introduction of new grid points.

[2] As opposed to more dynamic methods that modify the mesh *during* the solver iteration itself and not as a separate post-process, for example [23].

[3] The quality of a mesh is not only dependent on the density of points, but also on the elements that are constructed out of them, but we presuppose that a good mesh adaption will produce adequate elements for the solver.

In this section we will give a brief description of the of grid refinement algorithm and of the implemented egde-indicator functions to detect regions of refinement and derefinement.

A detailed description of the capabilities of the y^+ based grid adaptation can be found in [1]. In [2] we have presented the hierarchical grid refinement and derefinement algorithm along with the requirements for a parallelization of the adaptation module on distributed memory machines. A brief overview of the algorithms used for the surface approximation may be found in [4].

2.1 Grid Refinement Algorithm

The basic concept of the local grid refinement and derefinement is similar to the red/green refinement in [7,16]. The main requirements of a local refinement strategy are the detection of grid areas which will be refined and a method of element subdivisions, which results from insertion of new points to these areas. A basic algorithm for the grid refinement reads as follows:

I. Build edge list and element to edge reference.
II. Evaluate edge indicators I_e.
III. Refine edge list considering:
  a. the edge indicators,
  b. the target point number and
  c. the grid conformity.
IV. Calculate coordinates of new points.
V. Construct new elements.
VI. Interpolate solution to new points.

The refinement module (2.) uses an edge based approach. The refinement indicators are therefore evaluated for all edges in the grid and new points are inserted at the edge mid points. The element subdivisions can then be determined from the configuration of refined edges [5]. Therefore, the edge list and the element to edge reference has to be build up first (I.). In stage II. all edges (N_e) are evaluated by the use of one of the sensor functions described in the next section.

2.2 Edge-Indicator Sensor Functions

The definition of a useful edge-indicator for local refinement depends on the investigated problem. Some approaches use a residual-based indicator or an adjoint approach (see for details Sect. 3) while other make use of gradients or differences of any suitable flow variable. The letter is the default setting for the refinement module (2.) of the TAU-code adaptation.

The approximated gradient $G_{(\Phi)}$ of a variable Φ in discrete form is $\Delta\Phi/h$ with $\Delta\Phi = \Phi_{p_1} - \Phi_{p_2}$, i.e. the difference between the point values of the two points p_1 and p_2 connected by one edge, where h is the length of the edge. We write the indicator function as

$$I_e = \Delta\Phi h^\alpha \, . \tag{1}$$

A widely used formulation is $\alpha = 1$, i.e.: $G_{(\Phi)}h^2$. The advantage of scaling the indicator with a positive value of α is that the refinement stops automatically in the corresponding area after several cycles.

Our choice of $\Delta\Phi$ for the indicator function is

$$\Delta\Phi_{\mathrm{e}} = \max_{i=0,\ldots N_\phi} \left(c_{\phi_i} \frac{(\Delta\phi_i)_{\mathrm{e}}}{(\Delta\phi_i)_{\max}} \right) \tag{2}$$

with N_ϕ being the number of different flow variables considered and e the edges in the grid.

The weights c_{ϕ_i} are scaling parameters which enable the choice of different combinations of the single parts of the indicator (to be set to zero in order to turn off ϕ_i).

The reference values $(\Delta\phi_i)_{\max}$ are for a balanced scaling of each part of the indicator function with

$$(\Delta\phi_i)_{\max} = \max_{\mathrm{e}=0,1,\ldots N_{\mathrm{e}}} \left((\Delta\phi_i)_{\mathrm{e}} \right) , \tag{3}$$

for all edges e in the grid. For the standard usage we have implemented three sensors functions for the edge-indicator

A. The differences (Δ_{d}) of the flow values

$$(\Delta_{\mathrm{d}}\phi_i)_{\mathrm{e}} = |\phi_i(x_{p_1}) - \phi_i(x_{p_2})| \, . \tag{4}$$

B. The differences of the gradients (Δ_{g}) of the flow values

$$(\Delta_{\mathrm{g}}\phi_i)_{\mathrm{e}} = |\partial(\phi_i(x_{p_1})) - \partial(\phi_i(x_{p_2}))| \, . \tag{5}$$

C. The differences of the reconstructed flow values (Δ_{r}) to the edge midfaces

$$\begin{aligned}(\Delta_{\mathrm{r}}\phi_i)_{\mathrm{e}} = |(\phi_i(x_{p_1}) + \tfrac{x_{\mathrm{e}}}{2}\partial(\phi_i(x_{p_1}))) - \\ (\phi_i(x_{p_2}) - \tfrac{x_{\mathrm{e}}}{2}\partial(\phi_i(x_{p_2})))| \, .\end{aligned} \tag{6}$$

with ϕ_i the flow value, p_1 and p_2 the two edgepoints of edge e and $x_{\mathrm{e}} = x_{p_1} - x_{p_2}$.

2.3 Target Point Number Iteration

In stage III. of the basic refinement algorithm we calculate an initial limit

$$L_0 = c\,(\max_{I_{\mathrm{e}}} + \min_{I_{\mathrm{e}}}) \; , \tag{7}$$

with $\max_{I_{\mathrm{e}}} = \max_{e=0,1,\ldots N_{\mathrm{e}}}(I_{\mathrm{e}})$ and $\min_{I_{\mathrm{e}}} = \min_{e=0,1,\ldots N_{\mathrm{e}}}(I_{\mathrm{e}})$ for the indicator I_{e} which results in the target number of new points if all edges with

$$I_{\mathrm{e}} > L_i \; , \; i = 0, 1, \ldots \tag{8}$$

are marked for refinement. The variable c depends on the distribution of I_{e} and the desired target point number. This data is used to modify the default value of $c = \frac{1}{2}$ in order to speed up the target point number iteration.

An addditional consistency loop over all elements in the grid marks additional edges to get a valid subdivision case for each element. This loop has to be repeated until no more edges are marked by the consistency test. The set of implemented subdivision cases of the supported element types can be found in [2].

At the end of stage III. we test if the actual number of new points ($\bar{N}_p$) will match the given target number of new points ($\acute{N}_p$). In case of a match ($0.95\acute{N}_p \leq \bar{N}_p \leq 1.05\acute{N}_p$) the algorithm will continue with stage IV. In case of a mismatch, we have to calculate a new limit

$$L_{i+1} = \begin{cases} \frac{1}{2}\left(L_i + \min_{I_e}\right), \ \max_{I_e} = L_i \ ; \ \bar{N}_p < 0.95\acute{N}_p \\ \frac{1}{2}\left(L_i + \max_{I_e}\right), \ \min_{I_e} = L_i \ ; \ \bar{N}_p > 1.05\acute{N}_p \end{cases}, \ i = 0, 1, \dots \qquad (9)$$

and start again on stage III. The extension of this loop for the derefinement is quite easy because the same threshold ($I_e < L_i$) can be used to detect edges that have been refined in a previous adaptation step [2].

The stages IV. to VI. of the basic refinement algorithm are straight forward and a description can be found in [4].

2.4 Recent Algorithmic Devolpments

Within the frame of the MEGADESIGN and the associated DLR project MegaOpt continuated progress was made in the major parts of the adaptation module. The detailed description of all the achievements is out of the scope for this paper, so we refer only to the important improvements:

- The parallelization of the refinement and derefinement algorithm [2].
- The extension of the TAU-code primary grid partitioner to support distributed grid hierarchy for parallel dynamic repartioning of adapted grids [2].
- The parallelization of the y^+ based grid adaptation [1].
- The support for 3D unstructured hexahedra elements for the y^+ adaptation [1].
- The implementation of vortex sensor functions for the edge-indicator (see Sect. 2.2) to detect regions of vortical flow [22].

From all of the improvements the parallelization has the strongest effect on the turn around time of a complex simulation. This holds not only for the adaptation tool but also for the primary grid partioning as well.

2.5 Results

In order to show the capabilities of the parallelized version, we consider a scenario like the one presented in [3], where we used the adaptation to ajust the grid of a pitching delta wing in the regions of vortical flow. Here we started from an initial grid with approximately 1×10^6 grid points and 2.5×10^6 volume elements. It consists of an inner region of 1.9×10^6 prisms in 20 layers for the boundary layer and an outer region of 650000 tetrahedra. After several steps of adaptation we achieved a total number of 4.47×10^6 grid points, around 15×10^6

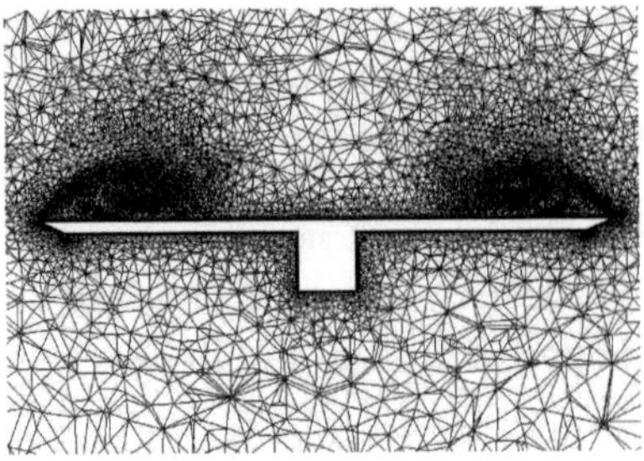

(a) adapted

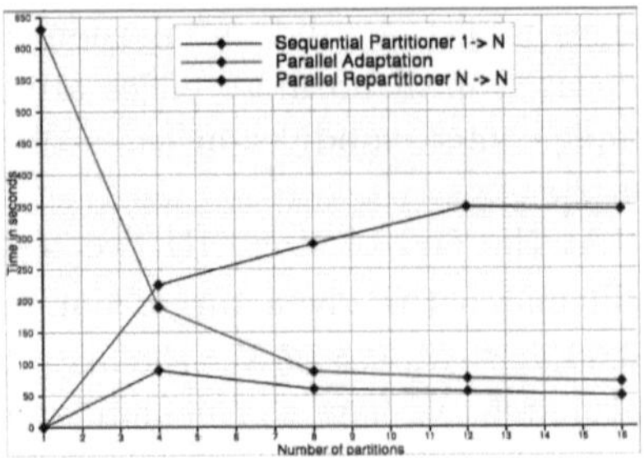

(b) performance evaluation

Fig. 1. Cut of the adapted grid at $x = 200$ (a), performance on a Linux Opteron cluster with a Gigabit interconnect (b)

volume elements and 4.17×10^6 parents (see Fig. 1(a)). This grid was then used for a performance evaluation of the parallel adaptation method. It can be seen in Fig. 1(b), that the parallel adaptation and the partitioner scale quite well. Even the overall time from the initial sequential grid partitioning, the parallel adaptation and the parallel repartitioning is smaller then the time needed for the sequential adaptation. In this case we could achive a speedup of 4.4 for 8 CPUs and 5.4 for 16 CPUs.

Table 1. Comparison of the performance of the parallel and sequential adaptation with an AIRBUS A380 starting configuration (performed on a Linux Opteron cluster with Infiniband interconnect)

	Adaptation steps					
Steps	1	2	3	4	5	6
# Mill. Points	7.5	9.3	10.2	11.9	11.5	12
1 CPU	-/-	484.37 s	-/-	635.85 s	-/-	769.37 s
16 CPU[4]	54.53 s	54.97 s	55.16 s	57.07 s	55.08 s	83.05 s
Speedup	-/-	8.81	-/-	11.52	-/-	9.26

Another example for the achieved speedup by the use of the parallelized version can be seen in Table 1. In this case we used the adaptation to refine the region of the jet blast of an AIRBUS A380 starting configuration [17]. It can clearly be seen that the parallel version not only saves the time for the adaptation process itself, but also saves the time needed for an additional (sequential) partitioning step in order to use the parallel TAU solver and preprocessing.

The presented method has worked quite well for a lot of typical applications in aerodynamic research [3, 13, 17, 20], but it has some restrictions with respect to solution accuracy as we will discuss in the following section.

[4] The timings for the parallel adaptation includes repartitioning.

3 Target Functional-Based Mesh Adaption

Most of the time it is not necessary for a flow solution to contain equally accurate results over the whole domain (which is also one of the reasons why constructing an accurate *and* efficient mesh is hard) as the main interest in many computations is the prediction of key performance functionals, such as the drag (C_{D}) or lift (C_{L}) coefficient of transport aircraft configurations for example. But not each point of the grid contributes equally to the evaluation of these functionals, and that can be used to our advantage. By using the so-called adjoint equation to obtain information about where discretization errors in the solution have an impact on the actual functional of interest, one can reliably identify regions of the mesh for adaption for a given target functional. Obtaining an estimate of the discretization error itself is a challenge of its own because if it *were* known we would know the exact solution to the problem.

Another advantage of the described approach to mesh adaption based on a target functional is that it – by its very construction – allows for an estimate of the remaining error in the functional for the current solution. This is very useful for measuring the reliability or conversely the uncertainty of a computation (as investigated among others in the NODESIM-CFD or MUNA projects), and thus can used to terminate a solver/mesh adaption cycle once the desired accuracy in the target functional is reached.

3.1 Mathematical Background

The approach to goal-based adaption for finite-volume methods described herein stems from the error *correction* approach to integral functionals with the dual (or adjoint) problem as shown by Pierce and Giles [12]. While they mention uses of the correction term as an adaptation indicator in passing, their focus lies in achieving higher-order convergence as measured by the error in the functional of choice.

Venditti and Darmofal [21] then develop a correction *and* mesh refinement strategy for unstructured finite-volume methods using a discrete adjoint solution. Their correction term is analogous to Pierce and Giles' adjoint correction term [19], but introduced by a slightly different argument. It is derived via a Taylor series expansion on a coarse mesh to an embedded fine mesh where then terms are approximated to avoid computations on the fine mesh altogether – except for a residual evaluation.

We will give a compact derivation of the procedure as described by Venditti and Darmofal [21] and as such will mostly adhere to their naming conventions, these being

- U – (Converged) flow solution of the Euler or Reynolds-averaged Navier–Stokes equations.
- R – Residual operator, i.e. to which degree the argument fails to solve the discretized equations.
- I – Target functional (of the solution U), such as aerodynamic drag C_{D} or lift C_{L}.

- Ψ – (Converged) solution of the dual (or adjoint) problem, with boundary conditions as imposed by the choice of target functional I.
- H, h – Used as super- and subscripts and refer to meshes on which particular operations have been carried out. H is a "coarse" mesh and h is an embedded "fine" mesh.
- L, J – Interpolation or prolongation operators for moving values between meshes, often used implicitly when both mesh-subscripts are used (e.g. $X_{\mathrm{h}}^{\mathrm{H}} = L_{\mathrm{h}}^{\mathrm{H}} X_{\mathrm{H}}$).

We depart from a Taylor-series expansion of the functional I on the fine grid based on the coarse-grid solution:

$$I_{\mathrm{h}}(U_{\mathrm{h}}) = I_{\mathrm{h}}(U_{\mathrm{h}}^{\mathrm{H}}) + \left.\frac{\partial I_{\mathrm{h}}}{\partial U_{\mathrm{h}}}\right|_{U_{\mathrm{h}}^{\mathrm{H}}} (U_{\mathrm{h}} - U_{\mathrm{h}}^{\mathrm{H}}) + \cdots . \tag{10}$$

Similarly, we can expand the residual on the fine mesh

$$R_{\mathrm{h}}(U_{\mathrm{h}}) = R_{\mathrm{h}}(U_{\mathrm{h}}^{\mathrm{H}}) + \left.\frac{\partial R_{\mathrm{h}}}{\partial U_{\mathrm{h}}}\right|_{U_{\mathrm{h}}^{\mathrm{H}}} (U_{\mathrm{h}} - U_{\mathrm{h}}^{\mathrm{H}}) + \cdots . \tag{11}$$

If we now truncate the above series and isolate terms, we obtain

$$\left(\left.\frac{\partial R_{\mathrm{h}}}{\partial U_{\mathrm{h}}}\right|_{U_{\mathrm{h}}^{\mathrm{H}}} \right)^{-1} (R_{\mathrm{h}}(U_{\mathrm{h}}) - R_{\mathrm{h}}(U_{\mathrm{h}}^{\mathrm{H}})) \approx (U_{\mathrm{h}} - U_{\mathrm{h}}^{\mathrm{H}}) . \tag{12}$$

By its very definition and the postulate that we deal with converged solutions $R_{\mathrm{h}}(U_{\mathrm{h}}) = 0$ for a given U_{h}.[5] If we use this fact, then the previous expansion of the residual reduces to

$$-\left(\left.\frac{\partial R_{\mathrm{h}}}{\partial U_{\mathrm{h}}}\right|_{U_{\mathrm{h}}^{\mathrm{H}}} \right)^{-1} R_{\mathrm{h}}(U_{\mathrm{h}}^{\mathrm{H}}) \approx (U_{\mathrm{h}} - U_{\mathrm{h}}^{\mathrm{H}}) . \tag{13}$$

If we now substitute (13) into (10) one obtains

$$I_{\mathrm{h}}(U_{\mathrm{h}}) \approx I_{\mathrm{h}}(U_{\mathrm{h}}^{\mathrm{H}}) - \left.\frac{\partial I_{\mathrm{h}}}{\partial U_{\mathrm{h}}}\right|_{U_{\mathrm{h}}^{\mathrm{H}}} \left(\left.\frac{\partial R_{\mathrm{h}}}{\partial U_{\mathrm{h}}}\right|_{U_{\mathrm{h}}^{\mathrm{H}}} \right)^{-1} R_{\mathrm{h}}(U_{\mathrm{h}}^{\mathrm{H}}) . \tag{14}$$

At this point, we introduce the adjoint variable Ψ with a definition as follows

$$\left(\Psi_{\mathrm{h}}|_{U_{\mathrm{h}}^{\mathrm{H}}} \right)^{\mathrm{T}} = \left.\frac{\partial I_{\mathrm{h}}}{\partial U_{\mathrm{h}}}\right|_{U_{\mathrm{h}}^{\mathrm{H}}} \left(\left.\frac{\partial R_{\mathrm{h}}}{\partial U_{\mathrm{h}}}\right|_{U_{\mathrm{h}}^{\mathrm{H}}} \right)^{-1} \tag{15}$$

---

[5] As an aside, for a given converged coarse-grid solution U_{H} we have $R_{\mathrm{H}}(U_{\mathrm{H}}) = 0$, but $R_{\mathrm{h}}(L_{\mathrm{h}}^{\mathrm{H}} U_{\mathrm{H}}) \neq 0$. Instead of using a finer mesh to create a residual of higher accuracy, one could also use a consistent residual operator with a higher order of accuracy than the one used to solve for U to gage this discretization error.

$$\Leftrightarrow \left(\Psi_{\mathrm{h}}\big|_{U_{\mathrm{h}}^{\mathrm{H}}}\right)^{\mathrm{T}} \left(\left.\frac{\partial R_{\mathrm{h}}}{\partial U_{\mathrm{h}}}\right|_{U_{\mathrm{h}}^{\mathrm{H}}}\right) = \left.\frac{\partial I_{\mathrm{h}}}{\partial U_{\mathrm{h}}}\right|_{U_{\mathrm{h}}^{\mathrm{H}}} \tag{16}$$

$$\Leftrightarrow \left(\left.\frac{\partial R_{\mathrm{h}}}{\partial U_{\mathrm{h}}}\right|_{U_{\mathrm{h}}^{\mathrm{H}}}\right)^{\mathrm{T}} \Psi_{\mathrm{h}}\big|_{U_{\mathrm{h}}^{\mathrm{H}}} = \left(\left.\frac{\partial I_{\mathrm{h}}}{\partial U_{\mathrm{h}}}\right|_{U_{\mathrm{h}}^{\mathrm{H}}}\right)^{\mathrm{T}} \tag{17}$$

based $U_{\mathrm{h}}^{\mathrm{H}}$, i.e. the primal coarse grid solution interpolated to the fine mesh, with the adjoint computation taking place on the *fine* level. This definition of the dual problem is analogous to the one used in the optimization context (see for example Gauger [10]) – except for the particular form of the primal solution (usually U instead of the $U_{\mathrm{h}}^{\mathrm{H}}$ used here) – and can be solved with exactly the same solver and boundary conditions that have been developed in the optimization context.

So as long as (17) holds true, we have

$$I_{\mathrm{h}}(U_{\mathrm{h}}) \approx I_{\mathrm{h}}(U_{\mathrm{h}}^{\mathrm{H}}) - \left(\Psi_{\mathrm{h}}\big|_{U_{\mathrm{h}}^{\mathrm{H}}}\right)^{\mathrm{T}} R_{\mathrm{h}}(U_{\mathrm{h}}^{\mathrm{H}}) \,, \tag{18}$$

which is an approximation of the sought-after functional evaluation from a fine-mesh solution based on a coarse-mesh solution *interpolated to the fine grid* minus a correction term. As written above, this correction term incorporates a solution for Ψ on the fine mesh, which we would like to avoid as the solution of the adjoint problem is very roughly about half as expensive as the solution U of the original problem so we could just have solved U on the fine mesh to start with. Thus we used following *ansatz* of replacing the fine-mesh adjoint with an interpolated adjoint from the coarse grid:

$$\Psi_{\mathrm{h}}\big|_{U_{\mathrm{h}}^{\mathrm{H}}} \approx L_{\mathrm{h}}^{\mathrm{H}}\, \Psi_{\mathrm{H}}\big|_{U_{\mathrm{H}}} = \Psi_{\mathrm{h}}^{\mathrm{H}} \,, \tag{19}$$

where $L_{\mathrm{h}}^{\mathrm{H}}$ is a prolongation operator from the coarse grid H to the fine grid h. This is a sensible approximation because the adjoint solution is smooth where the primal solution exhibits discontinuities (Giles [11]). The smooth regions in the adjoint solution are well-predicted by the interpolation operator, and these regions are also the ones that coincide with large residuals (e.g. across discontinuities in the primal solution), which in turn make a significant contribution to the correction term. When we now apply this approximation to (18) and write $U_{\mathrm{h}}^{\mathrm{H}}$ as $J_{\mathrm{h}}^{\mathrm{H}} U_{\mathrm{H}}$ (with $J_{\mathrm{h}}^{\mathrm{H}}$ being a similar prolongation operator), we arrive at

$$I_{\mathrm{h}}(U_{\mathrm{h}}) \approx I_{\mathrm{h}}^{\mathrm{H}}(J_{\mathrm{h}}^{\mathrm{H}} U_{\mathrm{H}}) - (L_{\mathrm{h}}^{\mathrm{H}} \Psi_{\mathrm{H}})^{\mathrm{T}} R_{\mathrm{h}}(J_{\mathrm{h}}^{\mathrm{H}} U_{\mathrm{H}}) \tag{20}$$

for which Ψ_{H} has to satisfy the traditional adjoint equation of

$$\left(\frac{\partial R_{\mathrm{H}}}{\partial U_{\mathrm{H}}}\right)^{\mathrm{T}} \Psi_{\mathrm{H}} = \left(\frac{\partial I_{\mathrm{H}}}{\partial U_{\mathrm{H}}}\right)^{\mathrm{T}} . \tag{21}$$

The correction term in (20) forms the basis of our mesh adaption indicator. It is made up of many individual contributions, one from each of the points in the computational mesh (of the form shown in (22) and as such we can use these point-wise contributions to identify regions in the grid where refinement would have a non-negligible impact on the target functional, while the sum over all these individual terms for all mesh points yields the correction term itself.

Another way of looking at the correction term in a more colloquial manner concentrates on the two components that make up each point-wise contribution. For the solution at a given mesh point i (expressed in primitive variables) this contribution boils down to

$$\Psi|_i \cdot R|_i = \Psi_\rho|_i \, R_\rho|_i + \Psi_u|_i \, R_u|_i + \Psi_v|_i \, R_v|_i + \cdots . \tag{22}$$

Individually, the residual R obtained from the fine mesh can be seen as a measure of the discretization error on the original mesh (i.e. where a fine mesh would result in a large update to the solution) whereas the adjoint solution Ψ relates that to the sensitivity of the target functional I to changes in the solution. This viewpoint led to the name of "dual-weighted residual (DWR)" method in the Finite Element (e.g. Becker and Rannacher [6]) and Discontinuous Galerkin (e.g. Hartmann [14]) community, where constructing residual operators of varying order is comparatively easy.

3.2 Implementation in the DLR TAU Code

The DLR TAU code [9] is a second order space and time[6] accurate cell-vertex finite-volume code working on hybrid unstructured meshes. The overall process-chain that has been implemented works as follows:

1. We compute a (fully converged) solution of the original configuration. For the purposes of this paper, the central scheme with an implicit LUSGS solver [8] was used, but the procedure is completely independent of the solver itself so any of the existing options in the TAU code can be used with no modification to the adaption (as long as the same solver is used for the residual evaluation).
2. Based on this solution, we obtain the matching discrete adjoint solution. Again, the procedure is not tied a particular adjoint implementation and could equally well use the continuous adjoint implementation that is available in the code for Euler computations.
3. In order to obtain the fine-mesh residual needed, the original mesh is globally refined and the solution from the first step is linearly interpolated onto the new mesh.
4. The (non-zero) residual of the coarse solution interpolated onto the fine mesh is calculated. This is essentially a single iteration of the solver-loop to sum all the fluxes across each of the faces and yields information on where a finer mesh would modify the original flow solution.
5. The volume-weighted residuals are summed from the fine mesh back to the original mesh, as that is where the adaption indicator is needed. This means that the residual contributions from the dual cells that were added for the

[6] Although the time accuracy can also be of third order if requested.

embedded fine mesh are distributed back to their coarse parent cell (see Fig. 3). A positive side effect of evaluating the indicator on the coarse mesh is that the adjoint solution does not need to be interpolated to the fine mesh. Also, no residual smoothing as proposed by Müller and Giles [18] was done as the residuals did not exhibit a checkerboard pattern, most likely due the different nature of the residual evaluation.
6. The actual adaption indicator ε_i whose magnitude is used to select elements for refinement is calculated point-wise as $\Psi^T \cdot R$, where Ψ is the adjoint solution vector and R is the residual vector at the point. This is currently performed in an external Python script with a thin layer called `netcdfdic` on top of the NetCDF-interface used by the TAU code. The script performs the sensor evaluation given the primal, adjoint and residual `pval`-files as arguments.
7. Finally, we use TAU's mesh adaption facility (as described in Sect. 2) to refine the edges where either endpoints' $|\varepsilon_i| \geq \sigma\, {}^{\varepsilon_t}/_{|\varepsilon_g|}$, where σ is the standard deviation of the local indicators/errors ε_i, ε_t is the prescribed uncertainty one wants to achieve in the functional and ε_g is the "global error estimate" which is simply $\sum \varepsilon_i$; similar to the limits used by Kim and Nakahashi [15] with $a = 1$.

The relevant processes and their inputs, outputs and dependencies are shown in Fig. 2. A feature-based adaption only needs the left-most components from the diagram, i.e. an initial grid with a flow solution is used with the adaption to produce an adapted grid. While the overhead in terms of added complexity is large, most of it uses existing – and thus well-tested – functionality.

The global refinement of the mesh that is used to produce the residual (i.e. how much a finer mesh would change each point in the solution – a measure of the

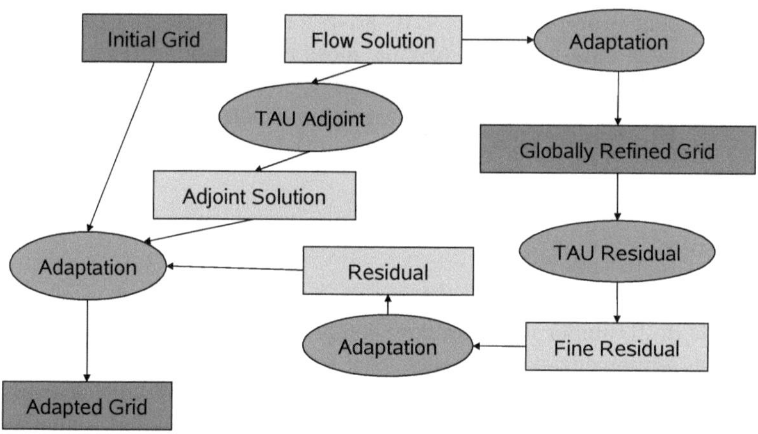

Fig. 2. Flow chart of steps currently involved in the proposed adaption mechanism. Note that most of the steps are necessary to produce a fine-mesh (and thus non-zero) residual of the coarse grid solution for the estimation of the discretization error.

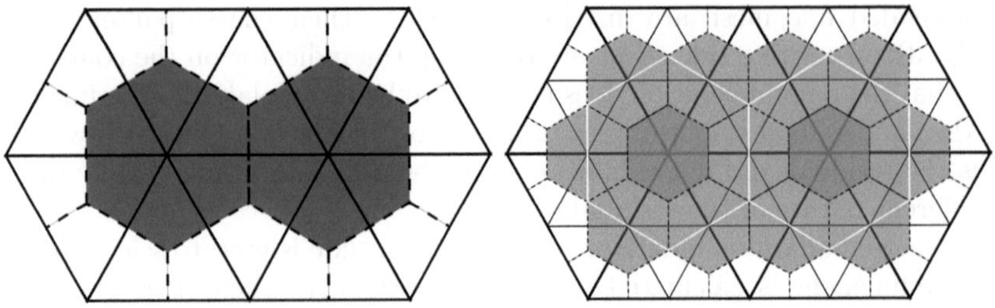

(a) Original Mesh (b) Globally Refined Mesh

Fig. 3. The left-hand side shows two dual cells cells and the accompanying primal mesh. On the right-hand side, the primal mesh has been globally refined, and one can see that by distributing 1/2 of each newly added vertex' dual volume to the endpoints of the parent edge that it bisects, we account completely for the original volume as shown in white.

discretization error) uses a slightly modified version of the normal TAU adaption module, where all edges are forcibly bisected. Unfortunately, the TAU adaption can not currently refine hexahedral sublayers (which are usually used for resolving the boundary layer in Navier–Stokes computations). This stems from the fact that the TAU adaption employs automatic y^+ adaption to redistribute the points there. Nevertheless, work is under way to remove this restriction. Simple trials on such hybrid Navier–Stokes grids seem to suggest that the sensor computation procedure itself works when one ignores the structured parts of the mesh – although this of course makes the correction term incorrect ...

For facilitating the evaluation of the sensor a new data-structure was introduced that specifically tracks edge bisections with parent-child relationships through the refinement and derefinement process. This allows us to use the normal derefinement procedure of the TAU adaption tool to restrict the sensor from the globally refined mesh h to the original mesh H in a mode where for each point that exists in h but not in H we distribute 1/2 of its value to each of the parent edge's endpoints. That this holds true for volume-weighted values can be seen in Fig. 3 as all of the dual cell's volume is accounted for and redistributed correctly back to the original mesh points.

3.3 Results

We have used the above-mentioned process for grid refinement studies on various airfoils in 2D under differing flow conditions. Comparisons are made with global mesh refinement, which is also used to estimate the "exact" value of the functional via Richardson extrapolation, as well as the default feature-based adaption already available in TAU. Some data-points for global refinement and the default feature-based adaptation were taken from an earlier mesh refinement study conducted by R. Dwight.

The feature-based adaptation was set up to introduce approximately 30% more points into the mesh at each adaptation step, whereas the goal-oriented procedure flagged as many elements of the mesh for refinement as it deemed necessary to achieve the prescribed error tolerance. This amount was usually well in excess of that produced by the feature-based adaptation, but as the goal-based adaption is more costly in terms of computational effort (as the adjoint solution has to be computed as well), this need not be a disadvantage as long as the mesh that is produced yields more accurate results.

Subsonic Flow (Euler)

The first test-case was the symmetric flow around the NACA0012 profile at $\mathrm{Ma}_\infty = 0.5$ and an angle of attack of $\alpha = 0.0°$ and it used drag (C_D) as the target functional for the adjoint computation (and thus the functional correction/mesh adaption procedure). The exact value of the drag is 0, due to completely subsonic inviscid flow.

For this case, essentially any refinement will reduce the error in the functional compared to the exact value. As expected, the different error thresholds for the goal-oriented refinement strategy modify the number of elements (or more accurately edges – as the TAU adaption tool is based on edge splitting) introduced during each adaption cycle, where a lower error threshold (i.e. a higher required accuracy) results in more refinement during a single adaption step.

The goal-oriented strategy does not improve on the already good results the feature-based adaption provides, in fact it even performs very slightly worse; although one has to remember that all of this take place in the space of a single drag count. This state of affairs can be improved significantly when we use the correction term as introduced beforehand, which comes at no extra cost provided we are already computing the goal-oriented adaption indicator. Using this correction factor, the new adaption strategy provides consistently better results than the default gradient sensor. For the largest given error threshold of $\varepsilon_\mathrm{t} = 0.0001$, a single adaption cycle is enough to achieve a result that is accurate enough – which the procedure realizes after evaluating and applying the correction term on the second mesh – and self-terminates.

Transonic Flow (Euler)

The left side of Fig. 5 shows a series of mesh refinements for the NACA0012 airfoil at $\mathrm{Ma}_\infty = 0.85$ and $\alpha = 2.0°$. The horizontal black line represents the "correct" value of the lift coefficient as obtained from the globally refined meshes by Richardson extrapolation. This transonic case exhibits two shocks, a stronger one on the upper side and a weaker one on the lower side. The goal-oriented adaptation is performed to produce a mesh that increases accuracy of the lift coefficient C_L of the profile for different error tolerances ε_t. This tolerance essentially increases or decreases the sensitivity of the adaption process and thus somewhat dictates the amount of the new points introduced. If one wants to achieve a very accurate end-result, even regions with a relatively small adaption indicator magnitude will be refined in order to meet the required tolerance whereas as these regions will be left as they are when only lower accuracy is desired.

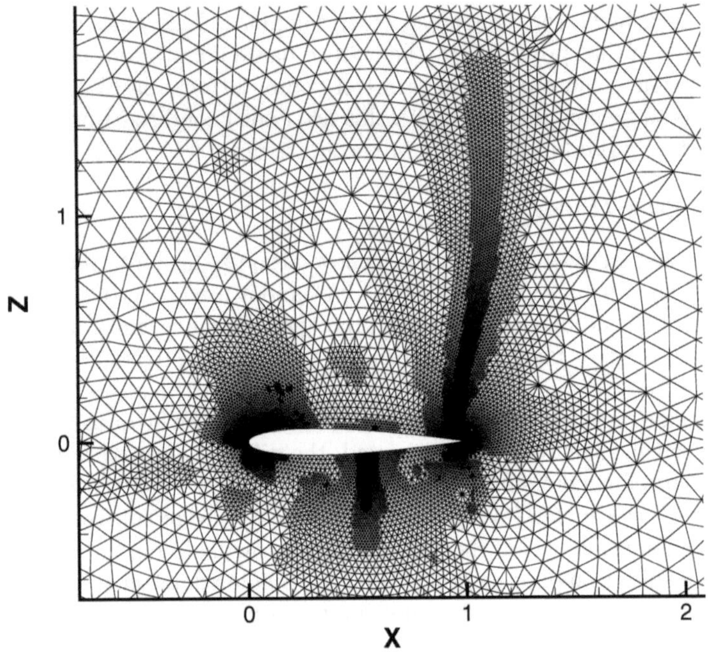

Fig. 4. Grid after the three goal-oriented adaptions with $\varepsilon_t = 0.001$ (NACA0012 with $Ma_\infty = 0.85$ and $\alpha = 2.0°$) in the transonic Euler case

In this particular case, the default feature-based adaptation produces rather erratic results and even seems to diverge from the "exact" value of C_L. We suspect this is the case due to the large gradients across the upper shock, which make the grid adaption spend every last bit of its allocated point budget there and thus neglecting other flow features such as the weaker shock and the trailing edge that also have an impact on the lift.

The goal-oriented adaptation gives very good results for all tolerances (although none of them has achieved a mesh on which $\varepsilon_g < \varepsilon_t$ yet). After three adaptation steps all meshes provide a lift coefficient that is within 0.0003 of the exact value, whereas as the overall best result[7] from 10 feature-based adaptations is off by 0.002 and getting worse.

The final mesh after the third adaptation for $\varepsilon_t = 0.001$ can be seen in Fig. 4. Easily visible are the two well-resolved shocks but the whole nose region as well as the trailing edge are also refined to provide a more accurate lift coefficient.

Transonic Flow (Navier–Stokes)

For the RAE2822 airfoil at $Ma_\infty = 0.734$, $\alpha = 2.29°$ with a Reynolds number of 6.5 million and the SAE turbulence model, the goal-oriented indicator is also much better than the default gradient sensor as it remains much closer to the

[7] Which is usually not known as the exact value is not available ...

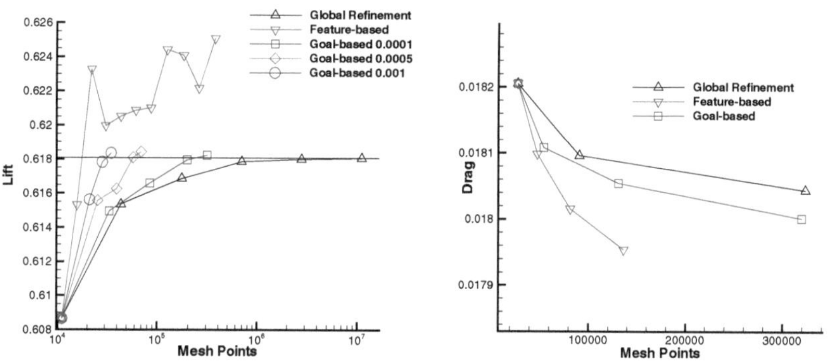

(a) Lift convergence (Euler) (b) Drag convergence (Navier–Stokes)

Fig. 5. Mesh convergence of different sensors for the transonic Euler (left) and Navier–Stokes (right) case

result of the "global" refinement (see right of Fig. 5), but certain restrictions of the TAU adaptation on hybrid grids remain in all cases, namely its inability to insert additional structured layers tangential to the surface or only in parts of the boundary layer. This also invalidates any possibility of using the correction term, as the residual is not produced on a truly globally refined grid.

4 Conclusion

We have described the recent developments of the TAU adaptation module, in particular as these pertain to the MEGADESIGN project. The improvements to the parallel refinement and derefinement along with the parallel repartioning have shown a significant increase of the overall performance. For most of the applications we could achieve a parallel efficiency well above 50%, which is necessary for a scalable code. The implementation of the vortical based flow sensors gives the adaptation tool the possibility to detect and refine regions of vortical flow automatically as long as the required data is precomputed by the solver. This feature is of major interest for all applications where vortical flow dominates and the standard set of gradient based sensors will fail.

For the goal-oriented mesh adaption approach encouraging results have been presented for inviscid 2D flows around the NACA0012 profile for a variety of flow conditions on triangular meshes and viscous flow for RAE2822 Case 9 on a hybrid 2D mesh.

The advantages of the approach become evident for more complex flows such as the transonic test-case presented, where the feature-based adaption process diverges from the correct lift value. This is a very real problem for users, who expect mesh adaption to *improve* their results and not make them worse, especially as the real or correct value is not known beforehand so that any prediction is as good or as bad as any other – but with more weight given to any result produced on a mesh with more points. This fallacy is clearly shown here as well as how the adaption based on a target functional alleviates the problem.

Nevertheless, complexity issues remain – it is questionable whether an adaptation indicator whose operation still requires a global refinement of the mesh (even if it is just for a single residual evaluation) is viable in an industrial context. Also, limitations of the adaptation module place inconvenient restrictions on the refinement of structured layers in hybrid grids. Solutions to these problems are sought after as part of future research.

References

1. Alrutz, T., Knopp, T.: Near-wall grid adaptation for turbulent flows. International Journal of Computing Science and Mathematics (2007) (accepted)
2. Alrutz, T., Orlt, M.: Parallel dynamic grid refinement for industrial applications. In: Wesseling, P., Onate, E., Periaux, J. (eds.) Proceedings ECCOMAS 2006, Egmond aan Zee, The Netherlands, September 5-8 (2006)
3. Alrutz, T., Rütten, M.: Investigation of Vortex Breakdown over a Pitching Delta Wing applying the DLR TAU-Code with Full, Automatic grid adaptation. In: 35th AIAA Fluid Dynamics Conference, Toronto, Canada, pp. 2005–5162 (2005)
4. Alrutz, T.: Hybrid Grid Adaptation in TAU. In: MEGAFLOW – Numerical Flow Simulation for Aircraft Design Results of the second phase of the German CFD initiative MEGAFLOW presented during its closing symposium at DLR, Braunschweig, Germany, December 10-11, 2002. Notes on Numerical Fluid Mechanics and Multidisciplinary Design, vol. 89, pp. 109–116. Springer, Berlin (2005)
5. Alrutz, T.: Erzeugung von unstrukturierten Netzen und deren Verfeinerung anhand des Adaptationsmoduls des DLR-TAU-Codes. Diplomarbeit Universität Göttingen IB 224-2002 A 10, DLR, Göttingen (2002)
6. Becker, R., Ranancher, R.: An optimal control approach to a posteriori error estimation in finite element methods. Institut für Angewandte Mathematik, Universität Heidelberg (2001)
7. Bey, J.: Tetrahedral grid refinement. Computing 55, 355–378 (1995)
8. Dwight, R.: Time-Accurate Navier–Stokes Calculations with Approximately Factored Implicit Schemes. In: Proceedings of the ICCFD3 Conference, Toronto. Springer, Heidelberg (2004)
9. Galle, M., Gerhold, T., Evans, J.: Technical Documentation of the DLR TAU-Code DLR IB 233-97/A43 (1997)
10. Gauger, N.: Das Adjungiertenverfahren in der aerodynamischen Formoptimierung. DLR Report 2003-05, DLR Braunschweig (2003) ISSN 1434-8454
11. Giles, M., Pierce, N.A.: On the properties of solutions of the adjoint Euler equations. Numerical Methods for Fluid Dynamics VI, ICFD (1998)
12. Giles, M., Pierce, N.A.: Improved lift and drag estimates using adjoint Euler equations. AIAA Paper 1999-3293 (1999)
13. Fritz, W.: RANS Solutions for the CAWAPI F-16XL in Solution Adapted Hybrid Grids. In: 45th AIAA Aerospace Sciences Meeting and Exhibit, Reno, NV 2007-492 (2007)
14. Hartmann, R.: Adaptive Finite Element Methods for the Compressible Euler Equations. PhD Thesis, Universität Heidelberg (2002)

15. Kim, H.J., Nakahashi, K.: Output-Based Error Estimation and Adaptive Mesh Refinement Using Viscous Adjoint Method. In: AIAA 44th Aerospace Sciences Meeting and Exhibit, Reno, NV 2006-1395 (2006)
16. Mavriplis, D.: Adaptive Meshing Techniques for Viscous Flow Calculation on Mixed-Element Unstructured Meshes. Paper 97-0857, AIAA (1997)
17. Melber-Wilkending, S.: Aerodynamic Analysis of Jet-Blast using CFD considering as example a Hangar and an AIRBUS A380 configuration. Technical report, STAB 11 (2004)
18. Müller, J., Giles, M.: Solution Adaptive Mesh Refinement Using Adjoint Error Analysis. AIAA Paper 2001-2550 (2001)
19. Pierce, N.A., Giles, M.: Adjoint and defect error bounding and correction for functional estimates. Journal of Computational Physics 200(2), 769–794 (2004)
20. Richter, K., Rosemann, H.: Numerical Investigation on the Aerodynamic Effect of Mini-TEDs on the AWIATOR Aircraft at Cruise Conditions. In: 25th Congress on International Council of the Aeronautical Sciences, ICAS 2006 (2006)
21. Venditti, D., Darmofal, L.: Grid Adaption for Functional Outputs: Application to Two-Dimensional Inviscid Flows. Journal of Computational Physics 176, 40–69 (2002)
22. Widhalm, M., Schütte, A., Alrutz, T., Orlt, M.: Improvement of the automatic grid adaptation for vortex dominated flows using advanced vortex indicators with the DLR-TAU code. In: STAB, 15th DGLR-Symposium of STAB (November 2006)
23. Vollmer, D.: Adaptive Mesh Refinement using Subdivision of Unstructured Elements for Conservation Laws. MSc Thesis, University of Reading, UK (2003)

15. [illegible]
16. [illegible] Mixed [illegible]
17. [illegible]
[illegible]

Adaptive Wall Function for the Prediction of Turbulent Flows

T. Schmidt[1], C. Mockett[2], and F. Thiele[3]

[1] Institute of Fluid Mechanics and Engineering Acoustics
tobias.schmidt@cfd.tu-berlin.de
[2] Institute of Fluid Mechanics and Engineering Acoustics
charles.mockett@cfd.tu-berlin.de
[3] Institute of Fluid Mechanics and Engineering Acoustics
frank.thiele@cfd.tu-berlin.de

1 Motivation

Conventionally, two types of wall boundary condition are available for the solution of turbulence transport equations in CFD. These exhibit very different requirements on the wall normal distance of the first grid point and any violation of these requirements leads to a drastic degeneration in the solution quality. This places a very high level of importance on the design of the numerical grid, and contributes to the excessive human resources typically spent on this task. Furthermore, these criteria depend strongly on the local flow field quantities, which means that prior knowledge of the solution is required for correct grid design. In practice this often means that an iterative grid design process is required, causing further grid generation expenditure.

The existence of two distinct wall boundary conditions has its origins in the physical nature of turbulent boundary layer flow, in which distinct zones exist at various distances to the wall. Normalisation of the flow variables and geometric quantities with respect to the friction velocity $U_\tau = \sqrt{\tau_w/\varrho}$ (where τ_w is the wall shear stress) allows the description of universal boundary layer profiles [7]. Such a profile of the normalised mean velocity $U^+ = U/U_\tau$ and normalised wall distance $Y^+ = U_\tau n/\nu$ (where n is the wall distance and ν is the kinematic viscosity) is given in Fig. 1. The region nearest to the wall where turbulent stresses are damped to negligible levels is termed the *viscous sublayer*, where $U^+ = Y^+$ and $Y^+ < 5$. "Low-Reynolds" boundary conditions target the full resolution of this region within the solution, and hence require the first grid point to be placed at $Y^+ \leq 1$. Further away from the wall where turbulence effects dominate at $Y^+ > 30$, the velocity profile is desribed by a logarithmic dependency on wall distance, hence the terminology *log-law region*. The transition region between the viscous sublayer and log-law region is known as the *buffer layer*. "High-Reynolds" boundary conditions (often referred to as "wall functions") target the positioning of the first point within the log-law region, hence bridging of the viscous sublayer and buffer layer through a parametric description that is valid only for a subset of physical circumstances (e.g. turbulent local equilibrium). Best

N. Kroll et al. (Eds.): MEGADESIGN and MegaOpt, NNFM 107, pp. 21–33.
springerlink.com

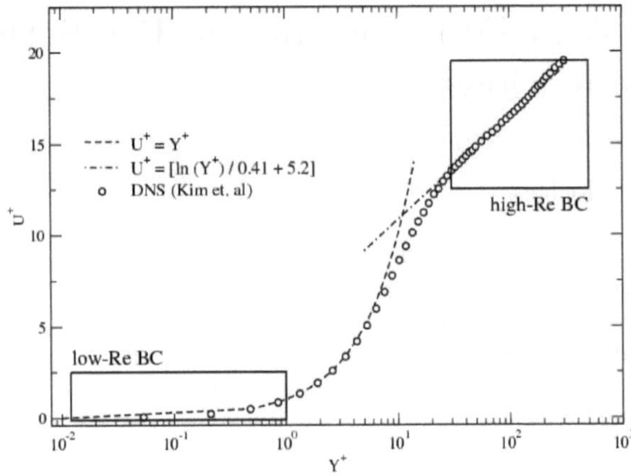

Fig. 1. Mean velocity profile from the direct numerical simulation of turbulent channel flow at $\mathrm{Re}_\tau = 395$ [4]

practice guidelines [2] stipulate a placement of the first grid point at $Y^+ > 20$ for high-Re boundary conditions.

A further problem is therefore apparent with the conventional boundary conditions: a gap is present in the region $1 < Y^+ < 20$ where neither is valid. A simple switch between the two formulations depending on the local Y^+ value is therefore not possible. Another practical problem can emerge when combining the use of low-Re boundary conditions with multi-grid convergence acceleration techniques, as is the case for the FLOWer solver – it is very likely that the auxiliary solutions on coarser grid levels will result in violation of the boundary condition requirements.

This state of affairs is therefore highly undesirable for industrial CFD, and a more general unified boundary condition is required that will deliver robust and reliable solutions for arbitrary near-wall grid spacing, i.e. the first cell placement at any value of Y^+. Previously, only low-Re boundary formulations have been implemented in FLOWer[1]. Such a generic boundary condition would therefore represent a considerable operational advantage, allowing significantly reduced grid generation time and computational expense by exploiting the advantages of high-Re boundary conditions. The formulation, validation and demonstration of such a solution is presented in the following sections.

2 Formulation and Implementation

The formulation of the hybrid adaptive boundary condition is strongly based on the methodology developed by Rung et al. [8] with some modifications made

[1] This is indeed the "natural" boundary condition type for the Spalart–Allmaras and k-ω families of models implemented in FLOWer.

to eliminate the need for an infrastructure modification in the FLOWer solver. Consequently, the specification of the velocity at the wall-neighbouring point is achieved using the turbulent viscosity rather than the momentum equation coefficients. As will be seen, the hybrid formulation is based on an extension of the high-Re boundary condition into the low-Re zone. Hence, the implementation of a high-Re boundary condition in FLOWer was a sensible precursor step, and will be presented first.

In the following, local wall coordinates will be used, with the x direction tangential and y normal to the wall. The corresponding mean velocities U and V are relative to the wall velocity. Values at the neighbouring grid point to the wall are denoted (P), while (B) refers to values at the solid boundary. The implementation has been carried out for both the cell-centered and cell-vertex discretisation and is valid for all implemented k-ω models.

2.1 High-Reynolds Boundary Condition

The derivation of conventional high-Re boundary conditions is based on simplified two-dimensional planar shear flow. In the absence of strong external forces, e.g. strong pressure gradients, self-similar profiles are present in the wall-tangential directions, as a consequence of which the number of independent variables reduces to one [10]. The flow is characterised by a simple turbulent field dominated by an approximate equilibrium of production and dissipation. Due to these simplifications, the high-Re boundary condition is of questionable validity when these physical conditions are violated, however the reduction to a minimal number of variables is practically desirable.

As outlined in Sect. 1, the high-Re boundary condition bridges the semi-viscous near wall region through a parametric description of the flow, and the velocity profile is logarithmic:

$$U = \frac{U_\tau}{\kappa}\left[\ln Y^+ + \kappa B\right] = \frac{U_\tau}{\kappa}\left[\ln(EY^+)\right] ,$$
$$\kappa = 0.41 , \qquad B = 5.2 , \qquad E = \mathrm{e}^{\kappa B} . \tag{1}$$

As mentioned above, the velocity at the wall-neighbouring grid point is fixed by specifying the eddy viscosity μ_t, which is given by

$$\mu_{t(P)} = \frac{U_\tau \kappa \varrho n}{\ln\left(EY^+\right)} - \mu . \tag{2}$$

To resolve the dependency of (2) on U_τ, a value of U_τ is obtained through the parameterisation of the turbulent field. Assuming that the effective shear stresses between the turbulent and semi-viscous regions are constant gives

$$-\varrho\overline{uv}_{(P)} = \tau_w \quad = \varrho_{(P)} U_\tau^2 \qquad \Rightarrow \qquad {U_\tau}^2 = \tau_w/\varrho = |\overline{uv}|_{(P)} . \tag{3}$$

Applying the common empirical observation that $|\overline{uv}|/k \approx 0.3$ [7] gives the following expression for U_τ:

$$k_{(P)} = \frac{|\overline{uv}|_{(P)}}{\sqrt{c_\mu}} \qquad \Rightarrow \qquad U_\tau = c_\mu^{1/4} \sqrt{k_{(P)}} \,, \tag{4}$$

where the parameter $c_\mu = 0.09$ is used.[2]

For the turbulent kinetic energy k, the value $k_{(P)}$ is taken directly from the model transport equation and a Neumann boundary condition is applied such that $k_{(B)} = k_{(P)}$. In addition, the production term P_k of the k-equation is modified according to

$$P_{k(P)} = -\overline{uv}\frac{\partial U}{\partial n} = \frac{\tau_w}{\varrho}\frac{U_\tau}{\kappa n} \,. \tag{5}$$

The definition of eddy viscosity for k-ω models together with a mixing length scale for the log-law region are combined to give an expression for the turbulent frequency ω in the wall-neighbouring cell:

$$\omega_{(P)} = \left(\frac{k^{1/2}}{c_\mu^{1/4} \kappa n} \right)_{(P)} \,. \tag{6}$$

The value of ω on the wall is of no significance to the solution, and is singular by definition. For cosmetic purposes $\omega_{(B)} = \omega_{(P)}$ is used.

2.2 Hybrid Adaptive Boundary Condition

The high-Re formulation must be extended such that it blends seamlessly into a low-Re formulation inside the viscous sublayer. The dimensionless wall-normal distance in the log-law region is given by the following rearrangement of (1):

$$Y^+ = \frac{e^{\kappa U^+}}{E} \,. \tag{7}$$

A Taylor series expansion of the numerator of (7) for small values of kU^+ about the origin combined with the low-Re value of Y^+ (i.e. $Y^+ = U^+$) delivers the hybrid expression for the dimensionless wall distance

$$Y^+_{\text{hyb}} = U^+ + \frac{1}{E}\left[e^{(\kappa U^+)} - \left(1 + \kappa U^+ + \frac{(\kappa U^+)^2}{2!} + \frac{(\kappa U^+)^3}{3!} + \cdots \right) \right] . \tag{8}$$

Armed with this relatively simple expression for Y^+_{hyb}, the hybrid adaptive boundary condition can be formulated. This is activated in place of the high-Re boundary condition when the criterion

$$Y^+_{(P)} = \left(\frac{n\, c_\mu^{1/4} \sqrt{k}\, \varrho}{\mu} \right)_{(P)} \leq 15 \tag{9}$$

is met. The mechanism of the boundary condition is given by the following:

[2] Although many of the k-ω models employ a variable c_μ parameter, the boundary value must be fixed in the high-Re formulation.

$$
\begin{aligned}
Y^+_{\text{hyb}} = U^+ + \frac{1}{E}\left(e^{\kappa U^+} - 1 - \kappa U^+ - \dots - \frac{(\kappa U^+)^9}{9!}\right) \\
U^+ = \left(1 - e^{-0.14 Y^+_{\text{hyb}}}\right)\frac{\ln(E Y^+_{\text{hyb}})}{\kappa} \\
\varphi = \left(1 - e^{-0.09 Y^+_{\text{hyb}}}\right)^2 \\
\theta = (1-\varphi)\frac{\mu}{n} + \varphi \frac{\varrho \kappa U_\tau}{\ln(E Y^+_{\text{hyb}})} \\
\tau_w = \theta(U_{(P)} - U_{(B)})
\end{aligned} \tag{10}
$$

The blending function φ varies smoothly between near-zero in the viscous sublayer to unity deep inside the log law region, at approximately $Y^+ > 100$. This serves to define the blending between laminar and turbulent viscosity in the low-Re and high-Re regions respectively, which is effective in the parameter θ. This in turn gives the effective viscosity divided by the wall normal distance for determination of the shear stress τ_w. Due to the implicit $U^+ = U^+(Y^+)$ formulation of the system of expressions, these must be solved by iteration at every wall point and time step in an explicit solver implementation such as FLOWer. For practical robustness reasons, the value of the dimensionless wall-normal distance is limited to $0.3558 \leq Y^+ \leq 400$, and an initialisation value of $\tau_w = 0.0015 \varrho_\infty U_\infty^2$ has proven effective.

The treatment of the turbulent kinetic energy and its production term is unchanged with respect to the high-Re boundary condition, i.e. (5). The boundary condition for ω is defined by

$$
\omega_{(P)} = \left(\frac{k^{1/2}}{c_\mu \, L_\omega}\right)_{(P)} , \tag{11}
$$

where

$$
L_\omega = \kappa c_\mu^{-3/4} n \left(1 - e^{-\gamma_\omega \text{Re}_k}\right) , \quad \text{Re}_k = \frac{\sqrt{k}\, n}{\nu} , \quad \gamma_\omega = 0.057 . \tag{12}
$$

3 Validation

3.1 Transonic Airfoil Flow: RAE 2822 Case 9

For the initial validation of the newly implemented boundary conditions, the well known two-dimensional RAE 2822 test case [3] was selected. This is characterised by transonic flow conditions at Ma $= 0.73$ and Re $= 6.5 \times 10^6$, with a shock located at the 55% chord position on the upper surface. The interaction of the shock with the boundary layer causes a weak or non-existent flow separation at the angle of attack of $\alpha = 2.8°$ studied. The transition position was user-specified at the 3% chord location.

The original grid provided was suitable for low-Re calculations and consisted of 353×65 vertices in a C-topology. In order to generate a high-Re type grid with

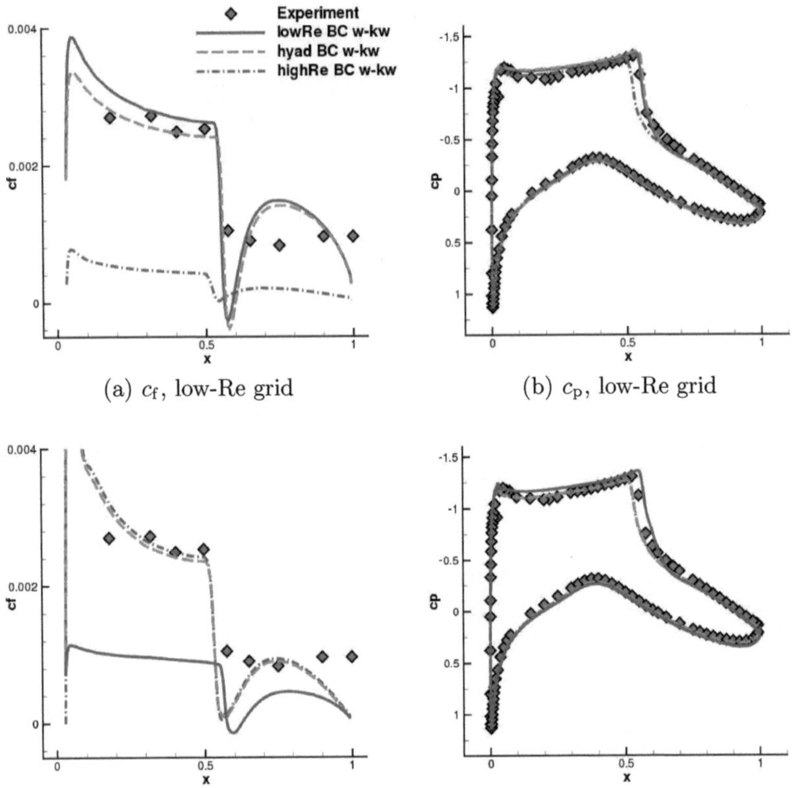

(a) c_f, low-Re grid

(b) c_p, low-Re grid

(c) c_f, high-Re grid

(d) c_p, high-Re grid

Fig. 2. RAE 2822 case 9, friction coefficient c_f *(left)* and pressure coefficient c_p *(right)* profiles computed on the low-Re *(above)* and high-Re *(below)* grids using the Wilcox k-ω model

otherwise minimal difference to the low-Re grid, the innermost eight cells nearest to the wall were removed, resulting in 353×57 points for the high-Re grid.

In Fig. 2 the chordwise distributions of pressure and friction coefficient computed using the Wilcox k-ω model [13] are compared to the experimental data for both grids and all three boundary condition formulations. A dramatic demonstration is given for the deterioration of the results when the traditional low-Re and high-Re formulations are applied to the inappropriate grid type; the skin friction coefficient is massively underpredicted by a factor of approximately five in both cases and a significant shift in the shock position is also seen in the plots of c_p. In contrast, the hybrid-adaptive boundary condition gives reliable results irrespective of the grid used, which are comparable in quality to the application of either low-Re or high-Re conditions on their native grids. This characteristic was furthermore observed for a multitude of different k-ω type models tested, which is demonstrated in Sect. 3.2.

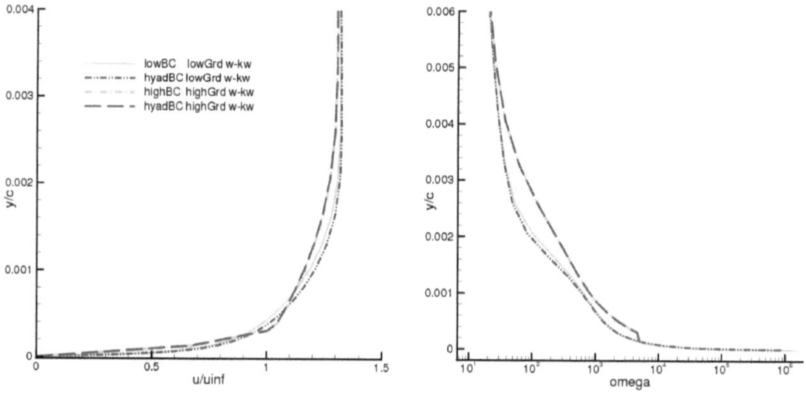

Fig. 3. RAE 2822 case 9: normalised velocity U/U_∞ *(left)* and turbulent frequency ω *(right)* at about 10% chord length (on the upper airfoil surface) along a wall-normal grid line

Boundary layer profiles of U/U_∞ and ω are compared between the different boundary conditions along a wall-normal grid line in Fig. 3. The results from the adaptive formulation are almost identical to those from the corresponding low-Re or high-Re condition, and hence the solution can be said to be dominated by the grid rather than the boundary condition formulation. The value of ω specified in the first cell of the high-Re grid is seen to agree very well with the equivalent value on the low-Re grid, however some deviation is seen in the gradient towards the neighbouring cells.[3] It is suspected that this may be caused by the uneven grid spacing; whereas simply deleting the inner grid cells to create the high-Re grid was done to ensure that both grids are otherwise comparable, this gives rise to a significant jump in the cell spacing. In the natural application of high-Re boundary conditions, the wall-normal distance of the first cell would be specified together with a constant stretching ratio away from this.

3.2 Transonic Wing Flow: ONERA M6

The flow around the three-dimensional ONERA M6 swept finite wing geometry [11] has been computed as a further validation test case. At the computed conditions of Ma $= 0.84$, Re $= 11 \times 10^6$ and $\alpha = 3.06°$, the flow exhibits a weak separation due to shock – boundary layer interaction and a lambda double-shock structure inboard that unifies at around the $\eta = 0.85$ span position (Fig. 4). The calculation was conducted fully turbulent, and the high-Re grid was once again generated via the removal of the eight inner cells from the low-Re grid, giving rise to respectively $225 \times 49 \times 33$ and $225 \times 41 \times 33$ vertices for the low-Re and high-Re grids.

[3] This behaviour has been demonstrated to be entirely independent of the ω value at the wall, which is set equal to the first cell value purely for "cosmetic" reasons, as stated in Sect. 2.

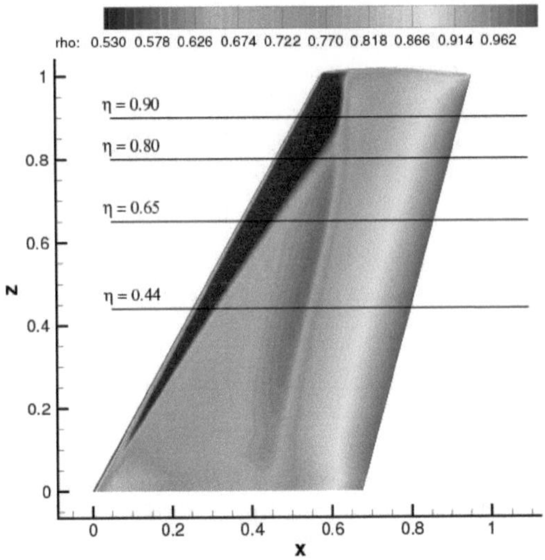

Fig. 4. ONERA M6: Surface contours of density ϱ computed with the Wilcox k-ω model and hybrid adaptive boundary condition on the high-Re grid

The pressure coefficient distributions at various spanwise locations are plotted in comparison with the experimental data in Fig. 5. As seen for the two-dimensional RAE 2822 flow case, the agreement between the low-Re boundary condition on the low-Re grid and the hybrid boundary condition on the high-Re grid is excellent (shown for the Wilcox k-ω model). The functionality of the hybrid-adaptive boundary condition for other k-ω models is also shown, with the profiles from the linearised explicit algebraic (LEA) model [9] and explicit algebraic Reynolds stress (EARSM) model [12] given for the high-Re grid.

To demonstrate the reduction in computation time achievable by the application of high-Re boundary conditions, the convergence of computations using the low-Re and high-Re grids are compared in Fig. 6 (low-Re and hybrid adaptive boundary conditions, respectively). The convergence of the solution is much more rapid on the high-Re grid, achieving the same level of convergence with much less iterations. Due to the 16% saving in grid points, each of these iterations furthermore requires less CPU time. The number of iterations and CPU time required to achieve the same level of convergence in the k-equation are listed in Tab. 1: the necessary CPU time is reduced by 90% by the introduction of a high-Re grid and suitable boundary condition. This is naturally of significant benefit to industrial application.

3.3 Industrial Conditions

To demonstrate some of the advantages that can arise from application of the adaptive wall function in an industrially-relevant environment, some example results from the European REMFI project [1] will be shown. The focus of the

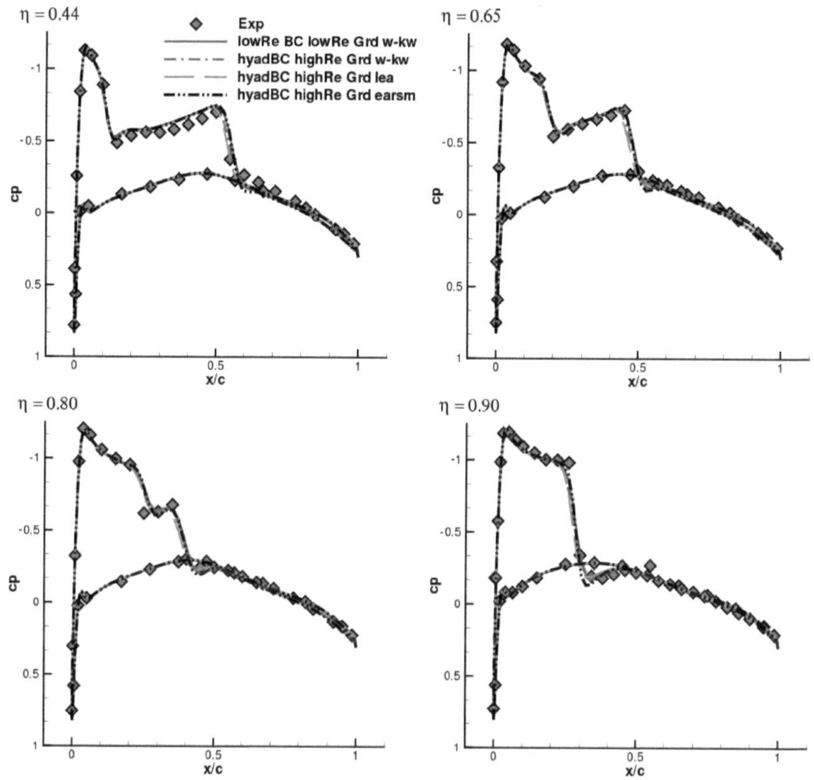

Fig. 5. ONERA M6: Comparison of chordwise c_p distributions at the $\eta = 0.44, 0.65, 0.80$ and 0.90 spanwise positions. Results from the high-Re and low-Re grid are shown for the Wilcox k-ω model and those from the LEA and EARSM models are shown for the high-Re grid only.

Table 1. ONERA M6: Number of iterations and computation time required to achieve the same level of convergence in the k-equation

	Iterations	CPU sec
Low-Re BC on low-Re grid	10000	54000
Hybrid BC on high-Re grid	1500	5000

project was the investigation of improved numerical and experimental methods dedicated to the prediction of empennage performance for transport aircraft. A significant part of this involved the application of twin-sting rigs to support wind tunnel models with minimal disturbance of the tailplane flow. The emphasis of the relevant work package was to describe and quantify the aerodynamic interference mechanisms of such wind tunnel supports using computational methods

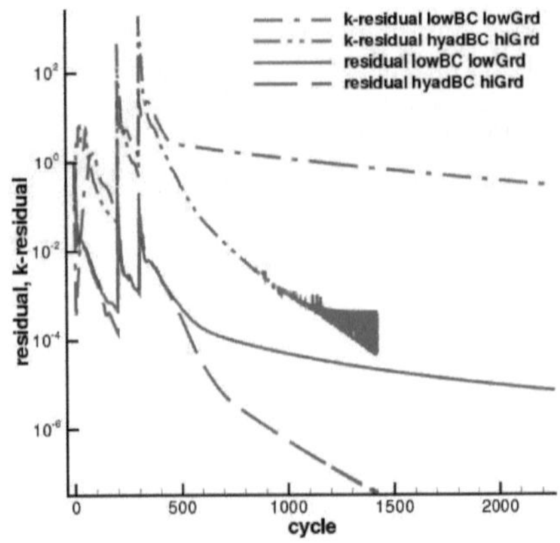

Fig. 6. ONERA M6: Residuals of density ϱ and turbulent kinetic energy k from computations with the Wilcox k-ω model. Comparison of the hybrid boundary condition on a high-Re grid with the low-Re boundary condition on a low-Re grid.

in the framework of a joint numerical and experimental study [6]. As such, the turbulent flow around a wing-body-tail configuration of a transport aircraft was computed at a range of transonic flow conditions. To provide information on the disturbance effect, simulations of the clean aircraft were compared to those with the twin sting support geometry included, and an important part of the investigation was the capturing of Reynolds number effects. A range of Reynolds numbers $3 \times 10^6 \leq \text{Re} \leq 38 \times 10^6$ were therefore to be computed, which for the standard low-Re boundary condition would have meant a succession of increasingly refined wall-normal grid spacings to maintain $Y^+ \leq 1$ as Re increases. In addition to the increase in computational expense, the grid generation effort tends to increase as the near-wall cell aspect ratio increases; the thinner the cells become, the greater the frequency of "negative volume" generation in the vicinity of uneven surface contours.

Considerable project delays could hence be recovered through the introduction of the hybrid-adaptive wall boundary condition, by enabling multiple Reynolds number values to be computed on the same grid. The distribution of the Y^+ value of the wall-adjacent cell is shown in the contour plots of Fig. 7 for the $\text{Re} = 3 \times 10^6$ computation (for which the grid was designed) and the highest $\text{Re} = 38 \times 10^6$ value. Although the level of empricism increases as more "wall-function activity" is introduced, the results at higher Re are nonetheless valid. Attempting this with conventional boundary conditions would not be possible as the majority of the flow exhibits Y^+ values that fall in the "gap" between low and high-Re formulations.

A further feature of the adaptive boundary condition that was exploited in REMFI is the ability to blend high and low-Re boundary condition types seamlessly within a single simulation. The higher physical validity (and higher computational expense) of the low-Re formulation can be limited to more important

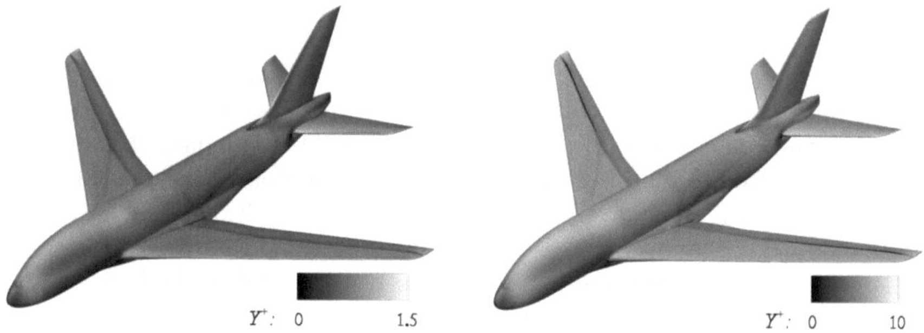

(a) Re = 3×10^6 (b) Re = 38×10^6

Fig. 7. Transport aircraft wing-body-tail configuration: Surface contours of Y^+ values obtained from valid computations of varying Reynolds numbers using the same grid using the hybrid-adaptive boundary condition and the FLOWer solver (Wilcox k-ω model)

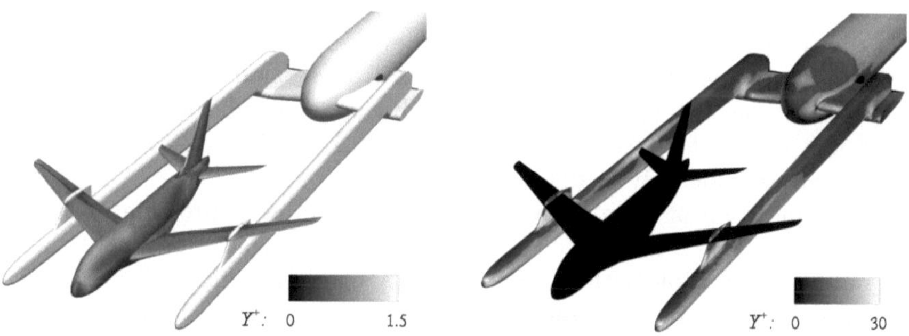

(a) Contours $0 \leq Y^+ \leq 1.5$ (b) Contours $0 \leq Y^+ \leq 30$

Fig. 8. Transport aircraft wing-body-tail model mounted on twin-sting wind tunnel support rig: Surface contours of Y^+ values obtained from a computation at Re $= 3 \times 10^6$ using the hybrid-adaptive boundary condition and the FLOWer solver (Wilcox k-ω model)

or physically complex flow regions, with coarser wall-normal distributions targeting high-Re boundary conditions elsewhere. Precisely this was done for the computations that include the twin sting support rig: the same fine grid resolution was applied to the aircraft geometry as used for the free aircraft, and a much coarser grid was applied to the twin-sting booms and the yoke plate and boss to which they are attached downstream. Two visualisations of the surface Y^+ distribution (scaled to show the two target regions) are shown in Fig. 8 for one of the twin-sting rig configurations at Re $= 3 \times 10^6$.

4 Conclusion

A hybrid-adaptive wall boundary condition has been succesfully implemented into the FLOWer flow solver, which is applicable to all k-ω models and both the cell-vertex and cell-centered discretisation schemes. The boundary condition has been demonstrated to be robust and to successfully address all of the shortcomings identified with the conventional two boundary condition types in Sect. 1. The new boundary condition gives valid solutions for any value of Y^+ presented by the grid, tending to the low-Re solution as $Y^+ \to 1$ and to the high-Re solution as $Y^+ > 15$. The problem of the gap between low-Re and high-Re boundary conditions has been solved through the seamless formulation. Compared to the previous state of affairs with the FLOWer solver, in which only low-Re formulations were available, the new boundary condition offers tremendous advantages in an industrial environment:

- Structured grid design is often referred to as the bottle-neck in the industrial CFD design process. The new boundary condition alleviates this situation by allowing coarser near-wall cells in the vicinity of highly uneven surface contours.
- The grid generation burden is further alleviated, through the removal of the necessity to remesh a new configuration if it is found that the Y^+ values exceed the acceptable limit for low-Re boundary conditions. A similar argument applies for unsteady flows where the Y^+ values can be expected to fluctuate considerably at different time steps.
- The application of large Y^+ values allows very rapid approximate solutions to be obtained (a 90% reduction in CPU time is typical, Sect. 3.2), which would be impossible with low-Re boundary conditions alone.
- Approximate coarse grid computations would be of particular use for "quick look" initial computations of a number of different prototype configurations, as well as for coupled fluid-structure-interaction, multi-disciplinary optimisation and unsteady manoeuvre calculations.
- Reynolds number effects can be captured on a single grid; previously, a new grid would be required for increased Reynolds numbers.
- Mixed high/low-Re grids can be generated to focus the numerical expenditure on the most important flow regions.

It must be emphasised that a move from low-Re to high-Re boundary conditions brings an accompanying loss of generality: the simple high-Re boundary conditions introduced here are based upon such physical assumptions as low surface curvature, local-equilibrium turbulence and zero tangential pressure gradient. However, a range of enhanced wall functions have been proposed in the literature that attempt to re-introduce these physical effects. Such enhancements could in principle be incorporated into the hybrid adaptive boundary condition formulation with ease.

The practical advantages of the hybrid adaptive wall boundary condition have been clearly demonstrated in the REMFI project, where its application made a successful completion of the computational objectives possible. The project progress would have suffered tremendously from an excessive grid generation work load if only the original low-Re boundary condition were available.

Acknowledgements

The application of the adaptive wall function to the industrial test cases of Sect. 3.3 was carried out within the European REMFI project (contract AST3-CT-2004-502895), 6th EU framework programme. Many of the computations presented in Sect. 3.3 were carried out on the IBM pSeries 690 supercomputer of the North German Supercomputing Complex – HLRN (*Norddeutscher Verbund für Hoch- und Höchstleistungsrechnen*). The assistance of both of these is acknowledged with thanks.

References

1. Abbas, A., Dias, J., Cabello, J.: REMFI: Rear-fuselage and empennage flow investigation. In: Proc. Council of European Aerospace Societies / Thematic Network on Key Aerodynamic Technologis (CEAS-KATnet) Conference, Bremen (2005)
2. Casey, M., Wintergerste, T.: Best Practice Guidelines, ERCOFTAC Special Interest Group on Quality and Trust in Industrial CFD (2000)
3. Cook, P., McDonald, M., Firmin, M.: Aerofoil RAE 2822 – Pressure Distributions and Boundary Layer Wake Measurements. In: AGARD AR-138 (1979)
4. Kim, J., Moin, P., Moser, R.: Turbulence statistics of a fully developed channel flow at low Reynolds number. J. Fluid Mech. 177, 133–166 (1987)
5. Launder, B., Spalding, D.: Lectures in mathematical models of turbulence. Academic Press, London (1972)
6. Mockett, C., Frederich, O., Schmidt, T., Thiele, F.: Numerical prediction of the aerodynamic interference of twin-sting model supports on empennage measurements. Submitted to 7th International ERCOFTAC Symposium on Engineering Turbulence Modelling and Measurements, Cyprus, Greece (2008)
7. Pope, S.: Turbulent Flows. Cambridge University Press, Cambridge (2000)
8. Rung, T., Luebcke, H., Thiele, F.: Universal wall-boundary conditions for turbulence-transport models. Zeitschrift für angewandte Mathematik und Mechanik 81(1), 1756–1758 (2000)
9. Rung, T.: Entwicklung anisotroper Wirbelzähigkeitsbeziehungen mit Hilfe von Projektionstechniken. Dissertation, Technische Universität Berlin (2000)
10. Schlichting, H.: Grenzschichttheorie. Braun-Verlag, Karlsruhe (1982)
11. Schmitt, V., Charpin, F.: Pressure Distributions on the Onera M6 Wing at Transonic Mach Numbers. In: AGARD AR-138 (1979)
12. Wallin, S., Johannson, A.: An Explicit Algebraic Reynolds Stress Model for Incompressible and Compressible Turbulent Flows. J. Fluid Mech. 403, 89–132 (2000)
13. Wilcox, D.: Turbulence modeling for CFD. DCW Industries Inc., La Cañada (1993)

Acceleration of CFD Processes for Transport Aircraft

Eberhard Elsholz

Airbus, Airbusallee 1, 28199 Bremen, Germany

Summary. A number of possible techniques to accelerate common CFD processes based on the DLR TAU code [1] is addressed, the outcome is briefly discussed and illustrated. As a result of these investigations, considerable increase of computational efficiency has been identified and demonstrated.

1 Introduction

Since the use of CFD within the aerodynamic design process increased more and more over the last years, permanent effort is dedicated to improve the CFD tools in terms of accuracy and efficiency. So an important part of work within the MEGADESIGN project was dedicated to the acceleration of the industrial CFD processes in general, especially with respect to the CFD tools used in transport aircraft design process.

2 Overview

Concentrating on improvements of the efficiency of CFD tools applied within the complete process chain, there were considered three main aspects:

- Computational acceleration was investigated by utilizing improved features of the TAU code such as implicit solution (LUSGS: Lower/Upper Symmetric Gauss–Seidel, see [2] rather than sticking to the (well validated) explicit (Runge–Kutta time stepping) method. This attempt aimed to directly decrease the necessary computing time.
- A serious attempt was made to introduce further automatization into the process chain wherever possible. This idea was linked to decreasing the input queueing time and by this the overall turn-around time of a complete computation.
- Simultaneous application like the prediction of several polar points at the same time was demonstrated in order to reduce the overall turn-around time drastically. However, this approach requires a considerable amount of hardware available, that must be reserved for a number of specific batch jobs throughout the overall turn-around time.

N. Kroll et al. (Eds.): MEGADESIGN and MegaOpt, NNFM 107, pp. 35–40.
springerlink.com

It has been shown that by all of these techniques, significant acceleration had been achieved – and clearly the ultimate combination of the different approaches would promise dramatic speed increase for the near-future industrial design process.

3 Utilization of Improved Code Features

Up to now, most industrial applications of the TAU code relied on the well validated and stable explicit Runge–Kutta time stepping method. Although this technique is limited to CFL (Courant–Friedrich–Lewy) numbers in the order of 2, stable and safe day-by-day computations were done at Airbus within this CFL range, starting up the full multi-grid process even with CFL < 1.

However, considerable advantages can be drawn from the utilization of the implicit LUSGS Iteration, which should allow much higher CFL numbers and by this much faster convergence of the solution.

Two different investigations have been performed on this specific topic comparing also the solutions obtained from Spalart–Almaras 1-equation turbulence model including Edwards modification (SAE) versus the Menter SST 2-equation "Shear Stress Transport" turbulence model.

The investigations were performed on a variety of relevant test cases, ranging from 2D multi-element profile up to 3D High-lift configurations (Fig. 1)

Summarizing across two aforementioned investigations, these test cases were chosen as follows:

1. 2D High-lift	$Ma = 0.185$	$Re = 2.5 \times 10^6$
2. W/b coarse mesh (0.2 Mpoints)	$Ma = 0.8$	$Re = 5.5 \times 10^6$
3. W/b fine mesh (5.2 Mpoints)	$Ma = 0.78$	$Re = 4 \times 10^6$
4. W/b-engine (10.3 Mpoints)	$Ma = 0.8$	$Re = 28 \times 10^6$
5. W/b-flaps-tfn (11.7 Mpoints)	$Ma = 0.2$	$Re = 1.5 \times 10^6$
6. F4 w/b (≈ 20 Mpoints)	$Ma = 0.2$	$Re = 3 \times 10^6$

For two different turbulence models, i.e. the 1-equation SAE model and the 2-equation Menter–SST model, a typical comparison of performance of the different solution procedures is given in Fig. 2. For each of the test cases, both methods were computed up to 10000 iterations for comparison, where at 2000 cycles a switch towards target lift was set. Using different CFL numbers, i.e. CFL $= 0.5 \ldots 2.0$ in case of Runge–Kutta time stepping and up to CFL $= 50$ for the LUSGS method, demonstrates a considerable speed-up potential of the implicit LUSGS method, reaching Ru–Ku's accuracy of $10e^{-6}$ after 10000 cycles already within less than 3000 cycles. With respect to the turbulence models tested, the SAE model allowed higher CFL numbers for the LUSGS method than the 2-equation SST model does.

In general, the resulting differences from SAE- and SST-turbulence models are negligible, they mainly occur at the leading-edge suction peak and are smaller than the differences obtained from different windtunnel tests (ONERA and NLR windtunnels) in this region.

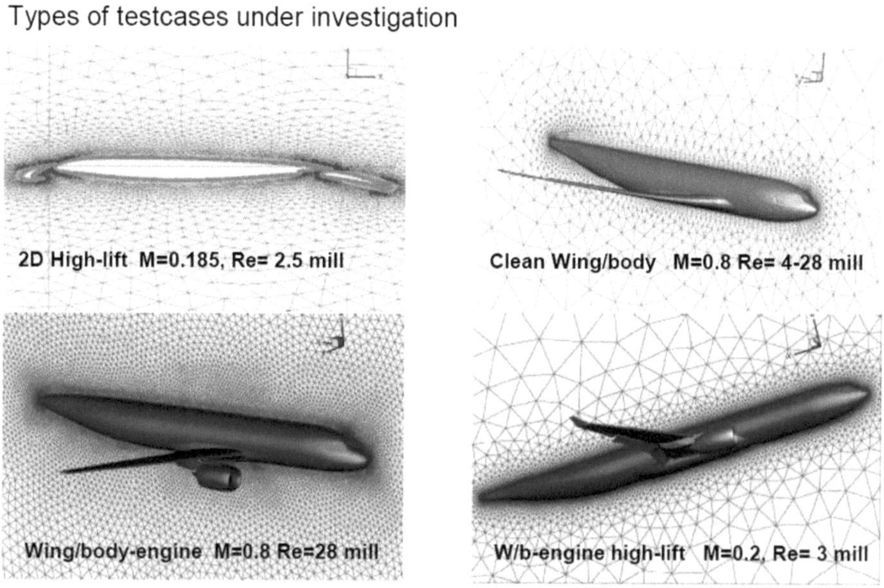

Fig. 1. Different test cases used for code feature comparisons

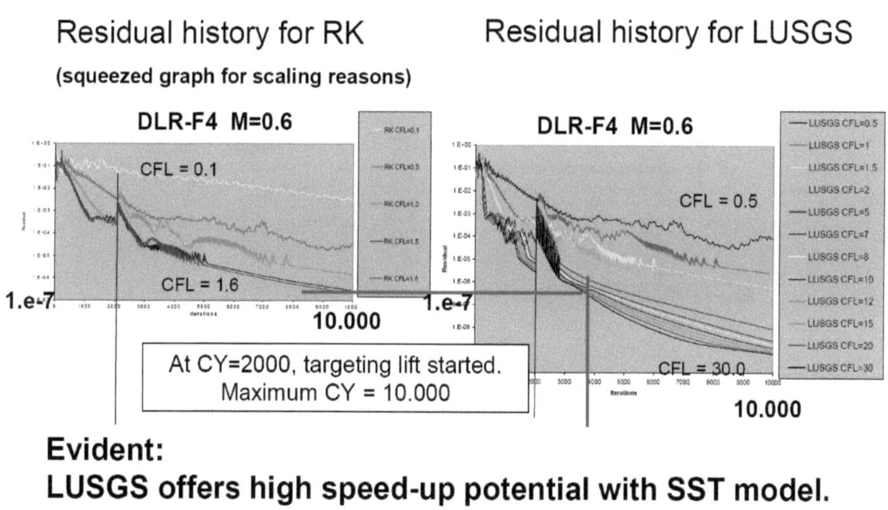

Fig. 2. Residual history of explicit and implicit methods for 2-equation SST turbulence model (Testcase DLR-F4)

4 Automatization of Numerical Process Chain

Considerable progress in automatization of the process chain has been achieved by introducing the Python script "Setup_Taujob.py". This script addresses different pre- and post-processing tools, such as mesh-deformation, drag decomposition and boundary layer analysis in an easy way to use. The script then generates standardized data structures and input files for the flow solver from the user's requirements and finally submits the complete job to the parallel-processing computing system.

This technique allows highly efficient computations of

- complete parameter series like polars
- the full process chain including pre- and post-processing
- coupled CFD-CSM loops via a simplified beam bending approach

The script was developed as part of MEGADESIGN work to reach the complex goals of the project.

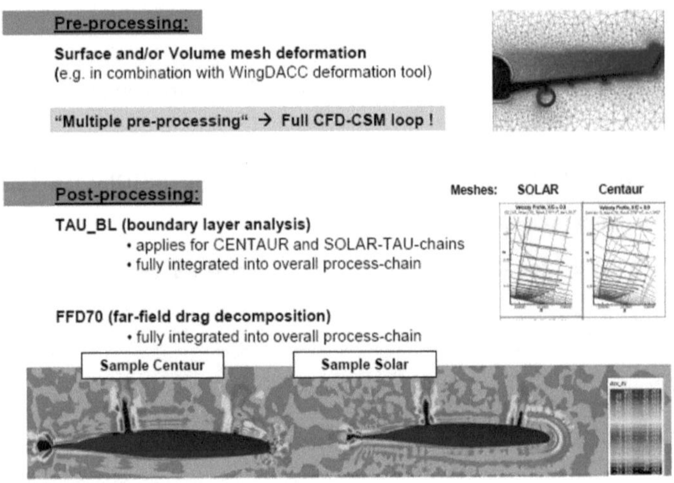

Fig. 3. Collection of typical pre- and post-processing results obtained via Setup_Taujob script

Figure 3 presents an overview of typical results obtained from mesh deformation pre-processing as well as from boundary layer analysis and drag decomposition post-processing. A short summary of flexible wing computations is given in Fig. 4, showing the possible improvements of solutions along a complete polar in comparison to the experiments (project milestone M 6.1).

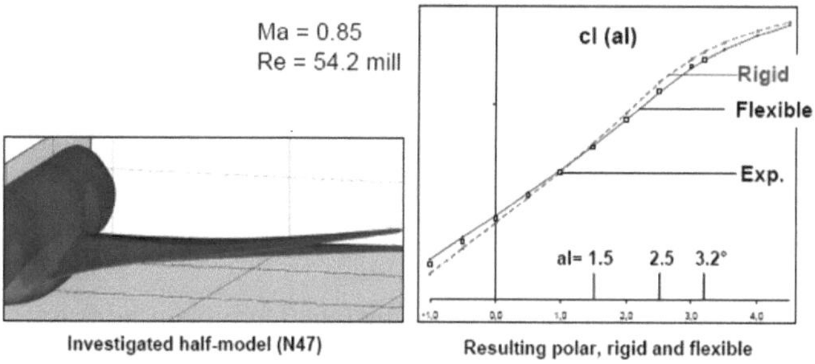

Fig. 4. Coupled CFD-CSM within Setup_Taujob script: Typical results

5 Simultaneous Approach

As described above, productive computations can be prepared/submitted/computed very efficiently by utilizing the Setup_Taujob.py script. For demonstration purposes, this technique had been used in a simultaneous effort to obtain a complete polar computation as fast as possible, i.e. nearly over night. For this demonstration a complex aircraft in landing configuration was used, i.e. including slats and flaps as well as engine/pylon installation, see Fig. 5. In this case, the computational mesh consists of approximately 12 million nodes. Simultaneous computations were performed for 6 different incidences along the polar, each of them starting from scratch, i.e. free-stream conditions. Stable results were obtained after 4000–6000 iterations, depending on the specific incidences. All this was done within 12.9 hours of computation, utilizing up to 288 processors (48 per computed incidence) at a time.

Fig. 5 also summarizes the typical convergence behaviour, skin-friction pattern at $\alpha = 15.18°$ and the resulting polars in terms of lift and drag versus the

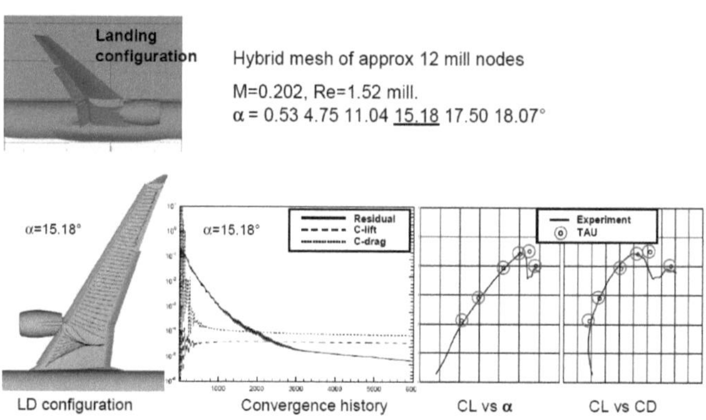

Fig. 5. Test case for simultaneous approach: Residual history, skin-friction at $\alpha = 15.18°$, lift and drag polars

incidence. Comparison with the experimental data shows reasonable quality of the achieved results.

This demonstration was performed in order to reach the project milestone M 4, which was fully achieved.

6 Summary/Conclusions

Three possible ways to accelerate CFD solutions have been addressed and their outcome has been illustrated. From these investigations, the main conclusions are as follows:

- Implicit time iteration schemes within the TAU code can speed-up computations by a factor up to 3 compared to the commonly used explicit Runge–Kutta time-stepping method.
- Increasing the automatization including the existing pre-/post-processing tools can save considerable turn-around time of computations.
- Simultaneous computations can produce a complete polar "over night", depending on the overall availability of involved hardware (processors) in the parallel computing environment.

The investigations showed a great potential in accelerating CFD in general, however, the specific techniques used in future applications may depend on the user's requirements.

Acknowledgement

The results and techniques discussed here were provided and supported by S. Albensoeder, H. Jakob, D. John, O. Knobloch, S. Noske, G. Rodriguez, G. Vidjaja.

References

1. Gerhold, T.: Overview of the Hybrid RANS Code TAU. In: Kroll, N., Fassbender, J.K. (eds.) Notes on Numerical Fluid Mechanics and Multi-disciplinary Design, vol. 89, pp. 81–92. Springer, Heidelberg (2005)
2. Yoon, S., Kwak, D., Chang, L.: LUSGS implicit algorithm for implicit three dimensional Navier–Stokes equations with source term. AIAA Paper 89-1964-CP (1989)

Efficient Combat Aircraft Simulations with the TAU RANS Code

H. Rieger and K. Sørensen

Aerodynamics and Methods, EADS Military Air Systems, Germany

Summary. Advanced aircraft design is driven by accurate analysis tools which are combined to an optimization process. Modern aerodynamic analysis for complex aircraft configurations are mostly dependent on flowfield simulation results with small uncertainty levels. A software package which more and more fulfills those requirements is the TAU Code developed at the DLR over many years. The aim of this contribution is to demonstrate the ability of the TAU code to compute physically complex flow problems which occur in flow field simulation tasks for combat aircraft at high angle of attack. It is shown that the improvements developed during the Megadesign project have led to efficiency gains beyond a factor of 2 with respect to run time. This progress has contributed to two main milestones of the project and fulfilled the requirements posed.

1 Background

The design challenges and problems which have to be tackled in aircraft design for EADS Military Air Systems (EADS-MAS) are illustrated in Fig. 1. It is shown that the designs of interest span from small aspect ratio wing combat aircraft to high altitude long endurance aircraft systems including highly integrated configurations with extreme small radar cross section properties.

The complexity of aircraft configurations can differ strongly dependent on the development status of the project. An example configuration resulting in geometrically complex aerodynamic flowfield analysis tasks is shown in Fig. 2. To make flowfield simulations around such configuration successful in an industrial sense the corresponding methods and tools have to be efficient and robust.

To achieve those objectives EADS Military Air Systems has developed over the years a hybrid unstructured mesh generation system called MESHER [1] which is designed and developed such that the user interaction becomes minimal. The geometric configuration in Fig. 2 consisted of a CatiaV4 description file of 140 MB in size, including the definition of 2749 topological faces and 6788 topological edges. The relevant surface mesh generation user input was

N. Kroll et al. (Eds.): MEGADESIGN and MegaOpt, NNFM 107, pp. 41–52.
springerlink.com

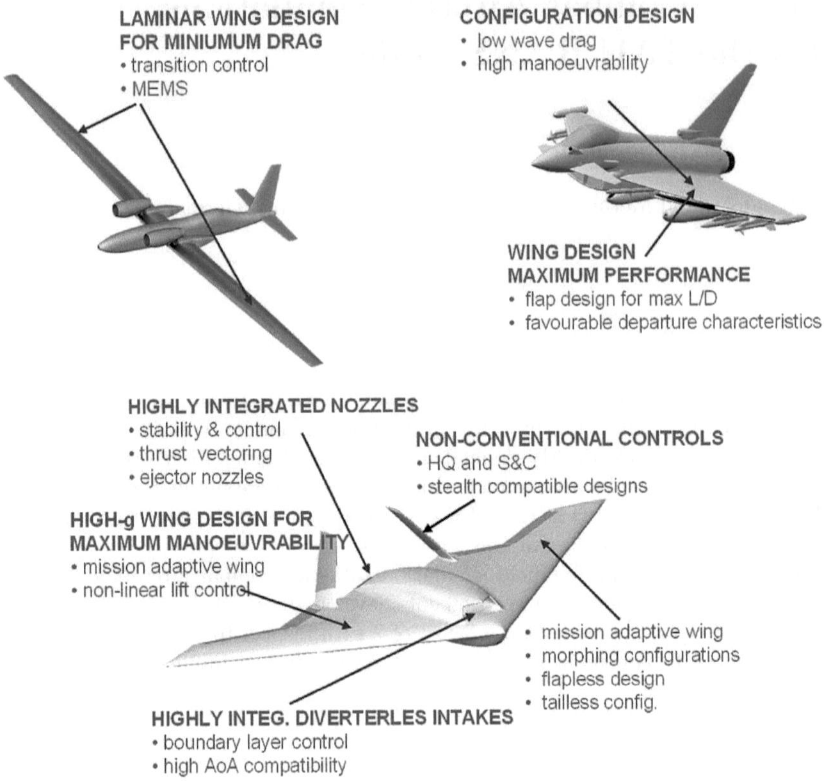

Fig. 1. Design challenges for EADS-MAS

the parameter for the minmal surface edge length for the whole aircraft and a minimal edge length for the four air data transducers of 0.5 mm which can be seen as small devices under the forebody. For the configuration in Fig. 2 the surface mesh had 2.3 million triangles. It can be observed that the edge length distribution is dependent on the local curvature (Fig. 2 left) which is a key prerequisite for achieving almost automatic surface mesh generation. The righthand figure shows the flat shaded surface mesh illustrating the smooth and accurate geometric resolution at surface locations of high local geometric curvature.

Based on the computational meshes generated by the MESHER system Reynolds-Averaged-Navier-Stokes (RANS) flowfield simulations were run successfully with the TAU code [2] developed by the DLR. An example for the complexity of simulation runs as they are performed at EADS-MAS frequently for feasibility studies is depicted in Fig. 3.

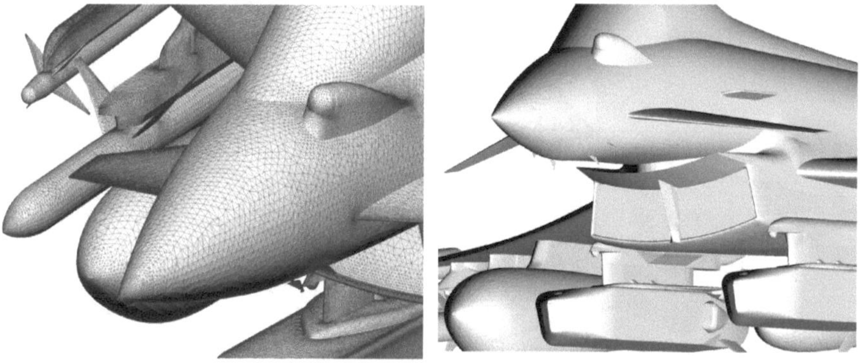

Fig. 2. Configuration complexity representative for aerodynamic studies at EADS-MAS

Fig. 3. Combat aircraft in potential store configuration

2 Objectives

One major objective of the Megadesign project was to improve the simulation efficiency in terms of runtime savings of the TAU code. The DLR as the developer of the TAU code worked on several aspects of the software implementation. Corresponding efforts were directed on possible improvements in the data structures for cache-based processors and on the improvements of the solution algorithms of the resulting algebraic non-linear equation problem. Two milestones were dedicated to the proof of the related efforts from the industrial end-users of the TAU code.

3 TAU Code Efficiency Improvements

The first major milestone M1 "Reduction of simulation time for a full aircraft configuration at least by a factor of 2" should confirm that for an industrial representative configuration a runtime efficiency gain by a factor of two could

Fig. 4. Combat aircraft configuration with (left) and without controls (right) deployed

be realized. For that exercise EADS-MAS chose a representative aircraft configurations being used for internal studies. Test case 1 consisted of a baseline zero-control configuration whereas test case 2 represented a configuration with partially deployed controls. Both test case configurations are depicted in Fig. 4 with their respective surface meshes.

In specific test case 1 simulated the flow at Mach 0.65, $\alpha = 12°$ and $14°$ while Case 2 solved the flow for Mach 0.85, $\alpha = 22°$ and $24°$ for a half model configuration with both leading-edge slats and canard deployed. For both test cases a fully turbulent flow was assumed and corresponding RANS simulations were performed at $\frac{\mathrm{Re}}{m} = 8 \cdot 10^6$. For all simulations the Spalart-Almaras turbulence model after Edwards was used. The hybrid meshes shown in Fig. 4 used approximately $12 \cdot 10^6$ mesh-points for case 1 and almost $13 \cdot 10^6$ for case 2.

Fig. 5 illustrates the resulting flow simulation solutions in form of the surface pressure distributions. A detailed analysis of the solutions revealed that the leeside flow is well ordered and characterized by a stable leading edge vortex system. This physical behaviour is not observed for test case 2 where the flow at higher angles of attack leads to a vortex breakdown which is shifted towards the

Fig. 5. Surface pressure distribution for test case 1 (left) and test case 2 (right)

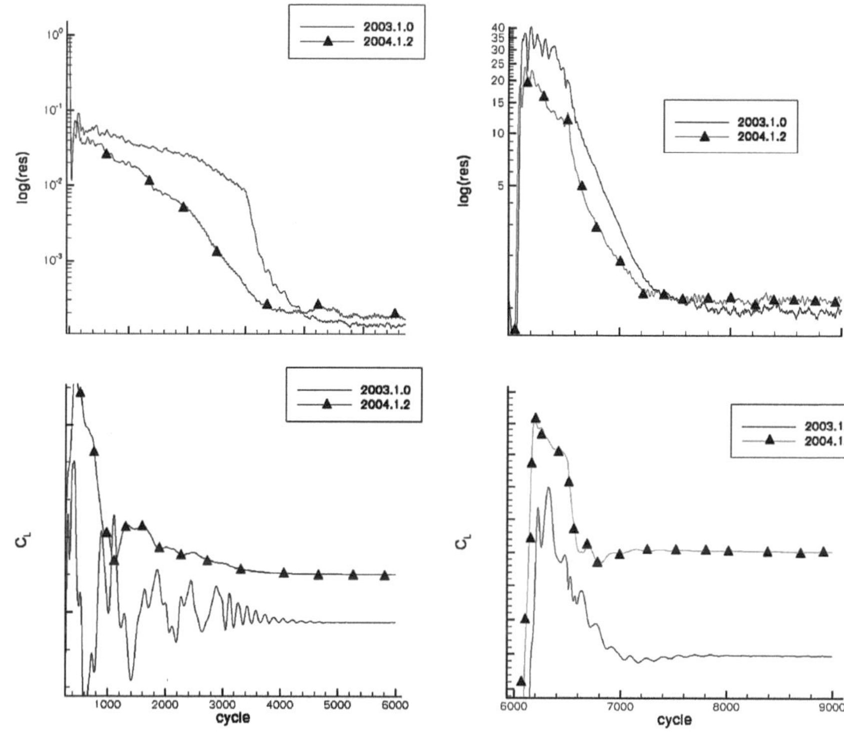

Fig. 6. Comparison of convergence characteristics for test case 1, Start-Up Phase (left column) and Restart-Phase (right column)

apex of the wing. Although the configuration has the slat system fully deployed and the foreplane turned into a nominally neutral position, the vortex burst phenomenon cannot be avoided. So unsteady modes are introduced into the RANS solution which produce oscillations around the average values for forces and moments.

Those flow conditions are reflected in the convergence characteristics of the two test cases. In Fig. 6 the convergence results for test case 1 are depicted using two versions of the TAU code. As the reference the TAU code version 2003.1.0 was used as this was the actual version at the beginning of the project. For performance comparisons the TAU version 2004.1.2 was chosen according to the milestone time schedule. The major difference in between both versions are the implementation of a relaxation method (LU-SGS) and improved local data structures for cache-based processing for the 2004.1.2 version.

On the left hand side of Fig. 6 the convergence behaviour for the density residuals and total lift is displayed for the start-up phase whereas the right hand side shows the corresponding curves for the restart of the computation with different parameters settings. The staged computational procedure is derived form the

Table 1. Case 1

(a)

TAU - version	time per sg-cycle [*sec*]	time per 2v-cycle [*sec*]	time per 3w-cycle [*sec*]
2003.1.0	21.5	33.5	37.0
2004.1.2	6.3	12.2	13.7

(b)

Phase	TAU-version	angle of attack	CFL fine mesh	CFL coarse meshes	cycles	equivalent 3w cycles
Start	2003.1.0	12°	1.0 / 1.0 / 1.0	- / 0.5 / 1.0	2000 sg /1000 2v / 3000 3w	5067
	2004.1.2	12°	1.5 / 1.5 / 1.5	- / 1.0 / 1.5	2000 sg /1000 2v / 2500 3w	4311
Restart	2003.1.0	14°	1.0 / - / 1.0	- / - / 1.0	500 sg / - / 1500 3w	1791
	2004.1.2	14°	1.5 / - / 1.5	- / - / 1.5	500 sg / - / 1000 3w	1230

experience that for complex configurations at higher angles of attack efficient parameter settings are rarely stable. To circumvent unsuccessful computations first a reference solution (e.g. at a smaller angle of attack) is produced using conservative numerical parameters. Any variations of the flow state nearby the reference solution is tried with more efficient parameters such like activation of a multigrid solution process or the use of higher CFL numbers, etc..

All computations were performed on a Linux-cluster at EADS MAS. The computing node consisted of two Intel Xeon processors being operated at 2.66 Ghz. The various nodes are linked together over a high performance switch from Quadrics delivering a latency time for small data packages in the order of 6μsec and a bi-directional bandwidth of almost 400 MB/sec. The code was compiled with the standard Intel compiler. For all computations 32 processors or 16 computing nodes were used.

In the following the obtained performance data are arranged in tables 1 and 2 in appropriate form. Here sg identifies a computation on the finest grid level alone (single grid), 2v stands for a simple 2-level multigrid cycle and 3w means a 3-level multigrid cycle. Experience shows that computational efficiency is improved by using the multigrid technique on coarser levels.

The tables 1 and 2 for case 1 respective case 2 show that the relation of the 2003.1.0 / 2004.1.2 Tau versions indeed produce a CPU time reduction by a factor of 3.2 for the start case for case 1 and 2.6 for case 2, while the restart case 1 achieves a factor of 3.9 and the one of case 2 a factor of 3.3. In tables 1(b) and 2(b) the corresponding numerical parameters and procedures used are given in full detail. The figures presented allow to draw the conclusion that the run time efficiency improvements are linked evenly to the developments of improved data structures for cache-based processor architectures and to the introduction of implicit relaxation methods (LU-SGS). However the results also show that the simulation results do not coincide for same parameter choices. This issue has to be clarified by the code developers.

Table 2. Case 2

(a)

TAU - version	time per sg-cycle [*sec*]	time per 2v-cycle [*sec*]	time per 3w-cycle [*sec*]
2003.1.0	21.5	33.5	37.0
2004.1.2	6.9	13.4	14.6

(b)

Phase	Tau-version	angle-of-attack	CFL fine mesh	CFL coarse meshes	cycles	equivalent 3w cycles
Start	2003.1.0	22°	1.0 / 1.0 / 1.0	- / 0.5 / 1.0	2000 sg /1000 2v / 5000 3w	7067
	2004.1.2	22°	1.5 / 1.5 / 1.5	- / 1.0 / 1.0	2000 sg/1000 2v / 5000 3w	6862
Restart	2003.1.0	24°	1.0 / - / 1.0	- / - / 1.0	500 sg / - / 2000 3w	2291
	2004.1.2	24°	1.0 / - / 1.0	- / - / 1.0	500 sg / - / 1500 3w	1736

Nevertheless, these figures demonstrate the substantial progress being achieved on the TAU code algorithmic efficiency within the Megadesign project.

4 Polar Simulations for High Angle-of-Attack Cases

Another demonstration of efficiency gains being realized by TAU code improvements is the computation of an angle-of-attack polar over night for a partially trimmed aircraft. This demonstration was linked to the major milestone M4. The following parameters were chosen allowing a comparison with experimental values. Some details of the configuration are illustrated in Fig. 7.

Angle-of-attack	α	-4°, 0°, 4°, 8°, 12°, 16°
Mach	Ma	0.85
Reynoldsnumber	$Re_{L_{ref}}$	$3.8 \cdot 10^6$
Referencelength (1:15 Model)	L_{ref}	386 mm
Reference-temperature	T_{ref}	300 K
Transition		fully turbulent
Slats	ε	-19.5°
Canard		-20°

The computational mesh for the configuration was also generated with the EADS Mesher tool [1]. According to experience the surface triangular mesh was refined in critical regions of the configuration. This is especially true at the stagnation points, wing-flaps and wing-fuselage intersections and the leeside of

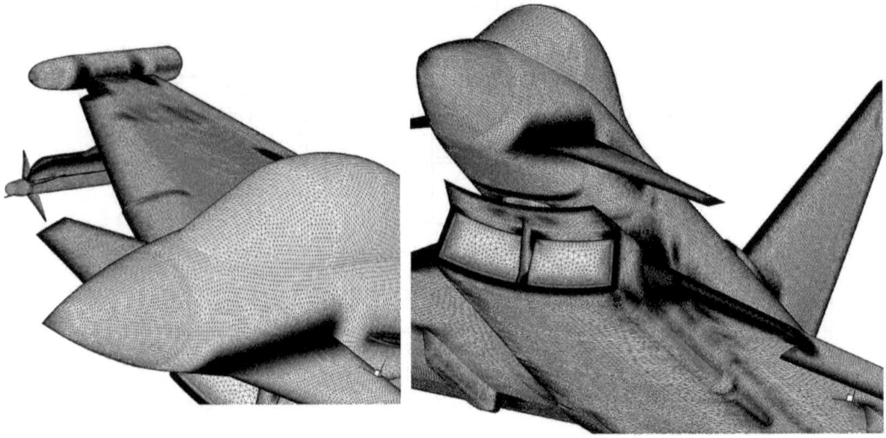

Fig. 7. Details of the surface mesh for a partially trimmed configuration specified in Table 4

the wings. Almost 650000 surface grid points were used to describe the half model configuration accordingly. In total the volume mesh consisted of 16 million mesh points.

For the exercise the TAU code version 2006.1.0 was applied. Experience has shown that for simulations in the transonic regions the k, ω-LEA turbulence model reveal better agreements with experimental results as compared with the standard k, ω-model. Therefore all polar computations were based on the linearized, explicit and algebraic variant of the k, ω-model. All computations were run on 48 partitions on the cluster system described before.

To achieve the aims efficiently a baseline simulation was performed at an angle of attack of 8° which was used as the necessary solution for the intended restart procedure of all other polar points being defined in Table 4. This computational technique is necessary to keep numerical stability for reasonable code parameter values as the CFL number or convergence acceleration technique like multigrid. It turned out that code instabilities for complex configuration like a combat aircraft are linked to the retarded establishment of the turbulence solution field. If an adequate turbulent solution field is available computations could be started with more efficient parameter values.

Fig. 8 displays the convergence characteristics for the baseline computation. The first 1000 iterations were performed without using the multigrid technique, during the following 9000 iterations the 3w-cycle was applied to enhance numerical convergence. It is observed that after 5000 cycles the density residual has reached an asymptotic level whereas the force and moment coefficients tend to reach their final values after 6000 cycles. The computational time to achieve such a result after 6500 iteration cycles was almost 43 hours on the hardware

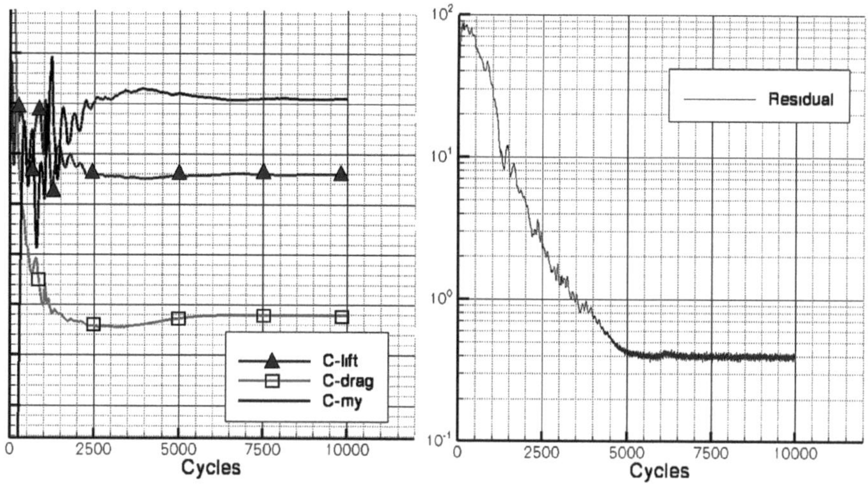

Fig. 8. Convergence characteristics for force and moments (left) and density residual (right) at $\alpha = 8°$

resources available. Fig. 9 shows the resulting distribution of the y^+-values on the upper and lower side of the configuration. For two-equation turbulence models those values should not exceed a value of 1 which is achieved reasonably well in the computation.

Fig. 10 shows the lift and pitching moment of the simulations in comparison with wind-tunnel results. Given the fact, that the flow field in many cases is dominated by separated vortex flows with embedded shock systems the agreement of in both cases is quite satisfactory. It is observed that between $20° < \alpha < 30°$ a stabilization of the pitching moment occurs. The capability to re-predict such effects would certainly increase the confidence into the simulation approach. This was demonstrated successfully in in-house studies. However, those computations last far longer than the time given for the Megadesign exercise and therefore were not taken into account for the polar demonstration.

Fig. 11 shows the pressure and skin friction distribution for a medium $\alpha = 8°$ and higher $\alpha = 16°$ angle-of-attack case. For the $\alpha = 16°$ case it is obvious that leading edge vortex systems develop on the leeside wing. The trace of these vortices can be seen very clearly on the skin friction maps shown in Fig. 11 (right hand side, bottom view). From Fig. 12 the effect of the restart solution procedure is illustrated in form of the convergence characteristics for forces and moments. Whereas almost 8000 iterations were needed to achieve at a converged state the restart immediately allows to choose more efficient numerical parameters. This is reflected by a quasi-steady state being reached in less than 3000 cycles.

In total it can be stated that all "restart" computations could be run in less than 20 hours wall clock time. This allows to run 5 parallel jobs on the available cluster hardware with 256 processors. Such an approach requires a

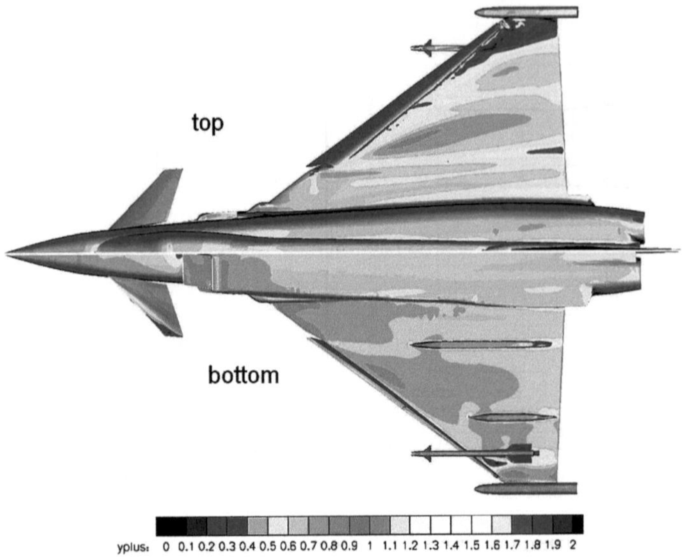

Fig. 9. Y^+-Distribution on the combat aircraft configuration at Mach = 0.85, $\alpha = 8°$

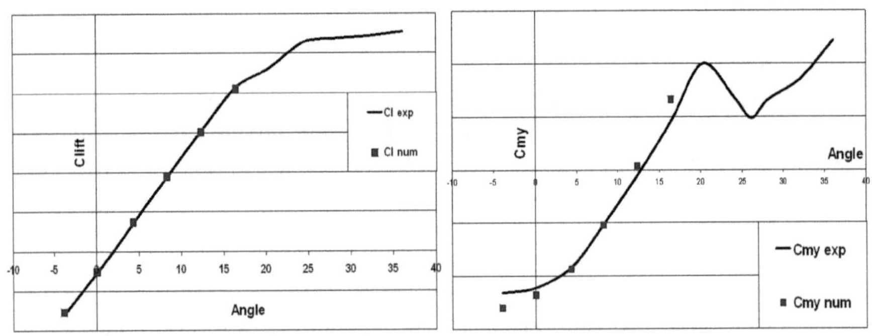

Fig. 10. Lift and pitching moment of a combat aircraft at Mach 0.85

baseline solution nearby the targeted flight conditions. As explained above such a baseline simulation requires much longer time for convergence. Therefore the aim to perform a 6-point polar computation over night (12 hours wall clock time) could not be reached. It was estimated that with the hardware resources installed approximately 430 processors would be needed to exactly fulfill the requested milestone. However the runtime figures obtained show clearly that with reasonable hardware resources given such requirements could be easily achieved today.

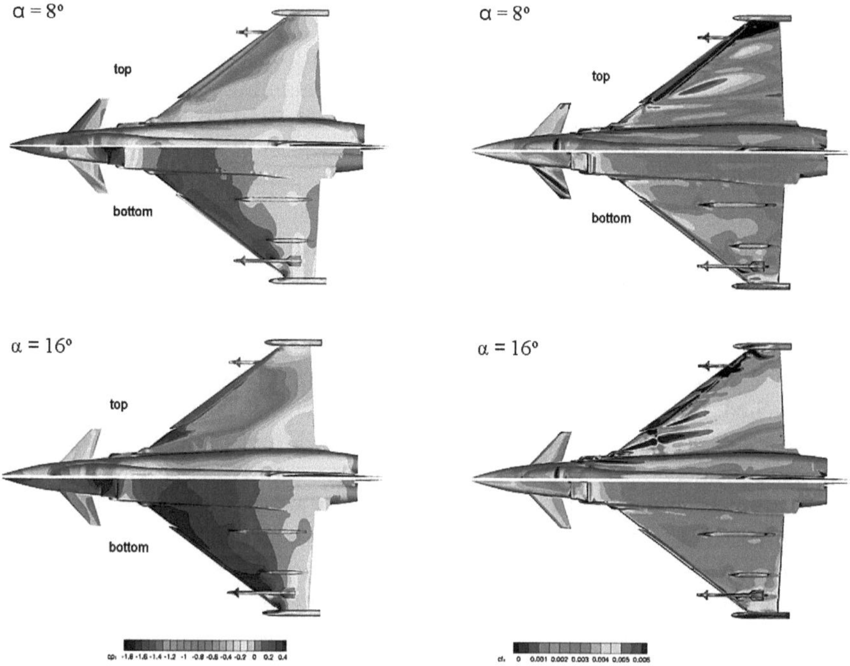

Fig. 11. Surface pressure (left) and skin-friction magnitude (right) on the combat aircraft configuration at Mach = 0.85 and $\alpha = 8°$ (top) and 16° (bottom)

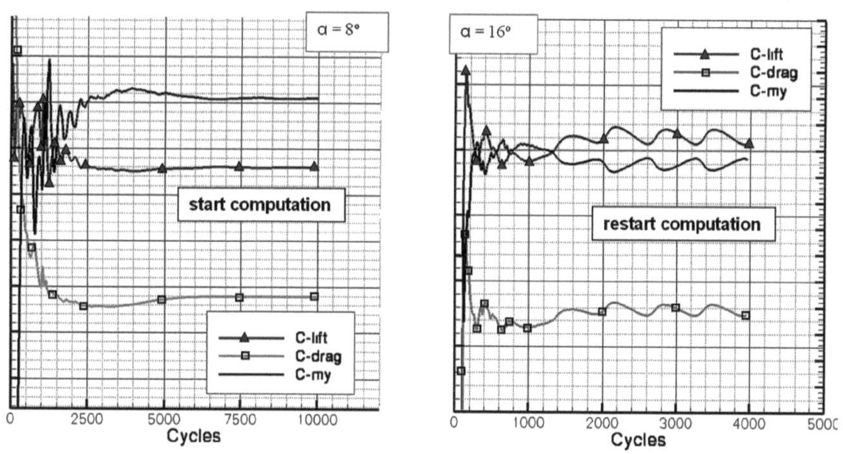

Fig. 12. Convergence characteristics for force and moments at $\alpha = 8°$ (left) and for $\alpha = 16°$ (right) restarting the computation from the $\alpha = 8°$ solution

5 Conclusions

One of the relevant objectives of the Megadesign project was to demonstrate that the TAU code is capable to cope with the demanding industrial simualtion tasks. Those demonstrations were given by industry and DLR frequently in recent years. However, in the industrial context a CFD code should exhibit features like robustness, accuracy and efficiency. Efficiency improvements were in the focus of the Megedesign project. Several measures have contributed to convincing runtime improvements for the TAU code. EADS Military Air Systems has used a complex combat aircraft configuration for corresponding computational exercises. The baseline runtime improvements of the TAU code for cache–based processors could be demonstrated in a convincing manner. In all computations a factor of 2-3 could be verified and therefore all development aims were successfully reached. Such a progress of the TAU code made it possible to successfully tackle the milestone linked to another industrial relevant requirement. This is to compute routinely aerodynamic polars to support aerodynamic data set generation by numerical means. In the course of the project it was demonstrated that the computation of an aerodynamic polar for a complex combat aircraft could be realized with success. Certainly those requests depend on the availability of corresponding computer hardware. However the required hardware performance gains are achievable in industry by modest investments. Overall the work being done in the Megadesign project to improve the TAU performance is an important step forward for industrial use in an operational project context.

References

1. Tremel, U., Deister, F., Hassan, O., Weatherill, N.P.: Automatic Unstructured Surface Mesh Generation for Complex Configurations. Intern. Journal for Numerical Methods in Fluids 44 (2004)
2. Gerhold, T., Friedrich, O., Evans, J., Galle, M.: Calculation of Complex Three-Dimensional Configurations Employing the DLR TAU-Code. AIAA 97-0167 (1997)

Part II

Improvement of Simulation Quality

Universal Wall Functions for Aerodynamic Flows: Turbulence Model Consistent Design, Potential and Limitations

Tobias Knopp

Institute of Aerodynamics and Flow Technology, DLR Göttingen,
Bunsenstrasse 10, 37073 Göttingen, Germany
Tobias.Knopp@dlr.de

Summary. A universal wall-function method for RANS turbulence modelling is presented which allows for a considerable solver accelaration and reduction of memory consumptions at only a small loss in accuracy even in flows with separation and reattachment. The range of validity of the approach in investigated by considering the near-wall RANS solutions in regions of strong adverse pressure gradient leading to separation and near stagnation points. The method is then applied successfully to aerodynamic flows with separation including transonic flows with shock induced separation and a subsonic highlift airfoil close to stall. From this some first best practice guidelines are suggested.

1 Introduction

This contribution is dedicated to turbulence-model consistent universal wall-functions which allow for a considerable solver accelaration and reduction of memory consumptions at only a small loss in accuracy even for flows with separation and reattachment. The huge computing costs for 3D flow simulations in complex geometries are still a major limiting factor in the "appropriate" usage (in terms of the numerical discretization error) of CFD tools in industry. Additional needs for faster CFD solutions arise as CFD-solvers are used more and more as part of optimization processes and from engineering studies which have to cover a wide range of flight conditions e.g., Mach number, Reynolds number, angle of attack, and possibly flap deflection angle.

Denote $y^+(1) = y(1)u_\tau/\nu$ the distance of the first node above the wall in viscous length-scales where $y(1)$ denotes the wall distance, u_τ is the friction velocity and ν is the viscosity. The aim of hybrid wall functions is to provide a boundary condition for solid walls that enables flow solutions independent of the location of the first grid node above the wall, allowing large values of $y^+(1)$ in regions of attached flow. Classical low-Re boundary conditions require a so-called low-Reynolds grid with $y^+(1) \approx 1$. Then up to 50% of the boundary-layer nodes reside in the near-wall region. Moreover, this increases the numerical stiffness due to the small cell height in conjunction with the steep gradients. As a remedy, wall-functions are considered. The major short-coming of standard

N. Kroll et al. (Eds.): MEGADESIGN and MegaOpt, NNFM 107, pp. 55–71.
springerlink.com

wall-functions based on the log-law is that the underlying high-Re grid requirement ceases to be valid inevitably in flows with separation. Historically, the first improvement over standard wall-functions was to use a so-called hybrid formulation, see e.g. [4, 15, 16] which removes the high-Re grid constraint in a natural way. Another approach by [7, 22] is discussed in [10]. Hybrid wall-functions are approximative velocity profiles for the entire near-wall region down to the wall. In turbulent boundary layer flows close to equilibrium, on high-Re grids they give grid-independent solutions (similar to standard wall-functions) but a significant grid dependent spreading can be observed if the first off-wall node is located in the buffer layer. Hence the second crucial modification was to devise universal wall-functions which are consistent with the turbulence model of the global (outer) flow, as revealed first by [8]. Such wall-functions are based on the universality of the near-wall RANS solutions for each given one- respectively two-equation model in zero-pressure gradient (ZPG) turbulent boundary layer flows.

Application to aerodynamic flows requires further investigation. Such flows include stagnation points and subsequent not fully developed turbulent flow, regions of strong pressure gradient with a large pressure gradient parameter and regions of separated flow. A detailed investigation of the range of validity of model-consistent wall-functions in such flow situations and a discussion of the treatment of the turbulence variables in wall-function methods is given in [10, 11].

This contribution is organized as follows. Section 2 gives the governing equations for compressible fluid flow and RANS turbulence modelling. In Sect. 3 the proposed wall-function method is described and validated for a flat plate turbulent boundary layer flow at zero pressure gradient. The role of the pressure gradient parameter is discussed in Sect. 4. The numerical method is described in Sect. 5. In Sect. 6 the method is applied to aerodynamic flows with separation. Section 7 considers the combination of wall-functions and adaptation of the prismatic boundary layer mesh in wall-normal direction. Some first best practice guidelines are given in Sect. 8 and conclusions are drawn in Sect. 9.

2 Compressible RANS Equations

We consider the steady-state Favre-averaged compressible Navier–Stokes equations in a bounded Lipschitz domain $\Omega \subset \mathbb{R}^d$ $(d = 2, 3)$. We use the eddy-viscosity assumption for the Reynolds-stress tensor and the gradient-diffusion approximation for the turbulent heat-flux vector. We seek velocity $\mathbf{u}$, density ϱ, pressure p, and temperature ϑ s.t.

$$\nabla \cdot (\varrho \mathbf{u}) = 0 \text{ in } \Omega \,, \tag{1}$$

$$\nabla \cdot (\varrho \mathbf{u} \otimes \mathbf{u}) - \nabla \cdot [2\mu_{\mathrm{e}} \mathsf{T}(\mathbf{u})] + \nabla p = 0 \text{ in } \Omega \,, \tag{2}$$

$$\nabla \cdot (\varrho \mathbf{u} H) - \nabla \cdot [\mathbf{u}\,(2\mu_{\mathrm{e}} \mathsf{T}(\mathbf{u}))] - \nabla \cdot (\kappa_{\mathrm{e}} \nabla \vartheta) = 0 \text{ in } \Omega \,. \tag{3}$$

We use the Sutherland law for molecular viscosity μ and the equations of state $p = \varrho R \vartheta$, $e = c_{\mathrm{v}} \vartheta$ for specific internal energy, and $h = e + p/\varrho = c_{\mathrm{p}} \vartheta$ for specific enthalpy, with gas constant R, specific heat at constant volume c_v, specific heat at constant pressure c_p. Denote $H = h + \frac{1}{2}\mathbf{u}\cdot\mathbf{u}$ the total enthalpy and moreover we define the strain rate tensor $\mathsf{T}(\mathbf{u})$ by

$$\mathsf{T}(\mathbf{u}) \equiv \mathsf{S}(\mathbf{u}) - \frac{1}{3}\nabla\cdot\mathbf{u}\,\mathsf{I}\,, \qquad \text{with } \mathsf{S}(\mathbf{u}) = \frac{1}{2}\left(\nabla\mathbf{u} + (\nabla\mathbf{u})^T\right)\,.$$

Moreover we introduce effective viscosity $\mu_\mathrm{e} = \mu + \mu_\mathrm{t}$ and effective thermal conductivity $\kappa_\mathrm{e} = \kappa + \kappa_\mathrm{t}$ where $\kappa = c_\mathrm{p}\mu/\mathrm{Pr}$, $\kappa_\mathrm{t} = c_\mathrm{p}\mu_\mathrm{t}/\mathrm{Pr}_\mathrm{t}$ with laminar and turbulent Prandtl numbers $\mathrm{Pr} = 0.72$ and $\mathrm{Pr}_\mathrm{t} = 0.85$, respectively.

The Spalart-Allmaras type one-equation turbulence models [3, 19] compute the eddy viscosity μ_t from the relation $\mu_\mathrm{t} = \varrho\nu_\mathrm{t}$ with $\nu_\mathrm{t} = f_{\mathrm{v}1}\max(\tilde{\nu};0)$ where $\tilde{\nu}$ is the solution of the transport equation

$$\nabla\cdot(\varrho\mathbf{u}\tilde{\nu}) - \nabla\cdot\left(\frac{\mu+\varrho\tilde{\nu}}{\sigma}\nabla\tilde{\nu}\right) - \varrho\frac{c_{\mathrm{b}2}}{\sigma}(\nabla\tilde{\nu})\cdot(\nabla\tilde{\nu}) = c_{\mathrm{b}1}\varrho\tilde{S}\tilde{\nu} - c_{\mathrm{w}1}\varrho f_\mathrm{w}\left(\frac{\tilde{\nu}}{d}\right)^2$$

with d being the distance to the closest wall and near-wall damping function $f_{\mathrm{v}1} = \chi^3/(\chi^3 + c_{\mathrm{v}1}^3)$ with $\chi = \tilde{\nu}/\nu$. On walls $\tilde{\nu} = 0$ is prescribed.

Several k-ω model versions (e.g. [13, 14, 17, 23]) have evolved from the original proposal [25] where $\mu_\mathrm{t} = \varrho k/\omega$ and k, ω are the solution of

$$\begin{aligned}\nabla\cdot(\varrho\mathbf{u}k) - \nabla\cdot((\mu+\sigma_k\mu_\mathrm{t})\nabla k) &= 2\mu_\mathrm{t}\mathsf{T}(\mathbf{u}):\nabla\mathbf{u} - \beta_k\varrho k\omega\\ \nabla\cdot(\varrho\mathbf{u}\omega) - \nabla\cdot((\mu+\sigma_\omega\mu_\mathrm{t})\nabla\omega) &= 2\gamma\varrho\,\mathsf{T}(\mathbf{u}):\nabla\mathbf{u} - \beta_\omega\varrho\omega^2\end{aligned}$$

with constants β_k, β_ω, γ, σ_k and σ_ω. We impose $k = 0$ on solid walls.

Regarding ω, asymptotic theory (cf. [25]) gives the near-wall behaviour $\omega = 6\nu/(\beta_\omega y^2)$ as $y \to 0$ which becomes singular at the wall. In industrial RANS solvers, the boundary condition equation (26) in [14] (abbreviated Menter b.c.) is very popular. Alternatively, Wilcox [25] suggests to prescribe ω (least least) at the first grid point above the wall Γ_δ at wall-distance y_δ located in the viscous sublayer (Wilcox b.c.):

$$\text{Menter b.c.: } \omega = C_\mathrm{w}\omega_\delta \text{ on } \Gamma_\mathrm{w}, \text{ with } \omega_\delta = \frac{6\nu}{\beta_\omega y_\delta^2}\,, \quad C_\mathrm{w} = 10\,, \tag{4}$$

$$\text{Wilcox b.c.: } \omega = \omega_\delta \quad \text{on } \Gamma_\delta\,, \text{ with } \omega_\delta = \frac{6\nu}{\beta_\omega y_\delta^2}\,. \tag{5}$$

3 Turbulence Model Consistent Universal Wall Functions

The aim of wall-functions is to remedy the no-slip boundary condition for the momentum equation (2) and the required near-wall resolution. In order to simplify the presentation, an adiabatic wall boundary condition is assumed for the energy equation (3).

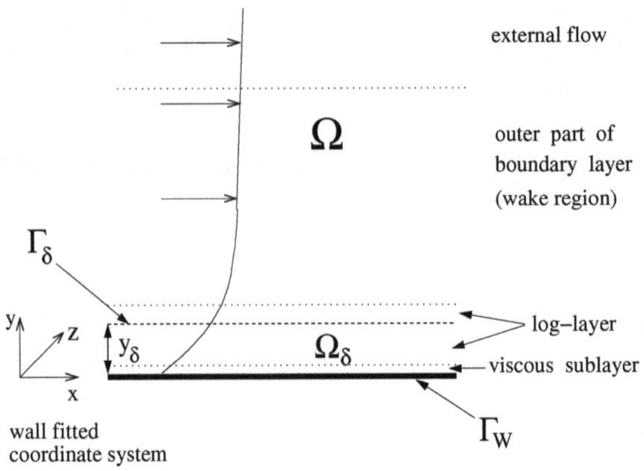

Fig. 1. Domain decomposition with full overlap in the near-wall region

3.1 Wall Function Formulation

For this purpose, we couple the global RANS problem solved in the entire computational domain Ω with a problem to be solved in the near-wall region Ω_δ via a domain-decomposition with full overlap as sketched in Fig. 1, see [12, 21]. For this purpose let Γ_δ denote the artificial inner boundary located within or below the logarithmic part of the boundary layer.

1. *Global RANS problem.* Solve the full compressible RANS plus turbulence model equations in the whole domain Ω with modified wall boundary condition for the momentum equation, where the wall-shear stress is prescribed instead of imposing no-slip:

$$\mathbf{u}\cdot\mathbf{n}=0, \quad (\mathsf{I}-\mathbf{n}\otimes\mathbf{n})2\mu\mathsf{T}(\mathbf{u})\mathbf{n}=-\tau_{\mathrm{w}}^{\mathrm{bl}}\mathbf{u}_{\mathrm{t}} \text{ on } \Gamma_{\mathrm{w}}\,. \tag{6}$$

   Therein, $\mathsf{I}-\mathbf{n}\otimes\mathbf{n}$ is the projection operator onto the tangential space of Γ_{w} and $\mathbf{u}_{\mathrm{t}}$ denotes the unit velocity vector in streamwise direction

$$\mathbf{u}_{\mathrm{t}}=\frac{\mathbf{v}_{\mathrm{t}}}{|\mathbf{v}_{\mathrm{t}}|}, \quad \mathbf{v}_{\mathrm{t}}=(\mathsf{I}-\mathbf{n}\otimes\mathbf{n})\mathbf{u}|_{\Gamma_\delta} \text{ with } (\mathsf{I}-\mathbf{n}\otimes\mathbf{n})_{ij} = \delta_{ij}-n_in_j \tag{7}$$

   with $\delta_{ij}=1$, if $i=j$ and zero otherwise ($1\le i,j\le d$).
2. *Near-wall RANS problem.* Solve the full compressible RANS plus turbulence model equations in the near-wall region Ω_δ with no-slip condition at the wall and matching of global flow and near-wall problem at the inner matching boundary Γ_δ

$$\mathbf{u}^{\mathrm{bl}}=\mathbf{0} \text{ on } \Gamma_{\mathrm{w}}\,, \qquad \mathbf{u}^{\mathrm{bl}}=\mathbf{u} \text{ on } \Gamma_\delta\,.$$

3. *Compute wall shear stress.* The wall shear stress $\tau_{\mathrm{w}}^{\mathrm{bl}}$ is determined from the near-wall solution $\mathbf{u}^{\mathrm{bl}}$ via

$$(\mathsf{I}-\mathbf{n}\otimes\mathbf{n})2\mu\mathsf{T}(\mathbf{u}^{\mathrm{bl}})\mathbf{n}=-\tau_{\mathrm{w}}^{\mathrm{bl}}\mathbf{u}_{\mathrm{t}} \text{ on } \Gamma_{\mathrm{w}}\,.$$

3.2 Boundary-Layer Approximation for Universal Wall Functions

The idea of the wall-function method is to compute τ_{w} from the solution of a simplified set of equations called boundary layer equations. Numerical tests show that effects of compressibility in the near-wall region of equilibrium flows are negligible for Mach numbers smaller 1.4. Moreover, in regions of attached equilibrium flow the near-wall solution is already surprisingly well described by the Prandtl boundary-layer equations.

Then, for each $\mathbf{x}_{\mathrm{w}} \in \Gamma_{\mathrm{w}}$ and given $u_\delta = ||\mathbf{v}_{\mathrm{t}}||$ from the global RANS solution in (7), seek the wall-parallel component of velocity $u^{\mathrm{bl}}(y)$ such that

$$\frac{\partial}{\partial y}\left((\nu + \nu_{\mathrm{t}}^{\mathrm{bl}})\frac{\partial u^{\mathrm{bl}}}{\partial y}\right) = f \text{ in } \{\mathbf{x}_{\mathrm{w}} - y\mathbf{n} \mid y \in (0, y_\delta)\}\,, \tag{8}$$

$$u^{\mathrm{bl}}(0) = 0\,, \quad u^{\mathrm{bl}}(y_\delta) = u_\delta\,, \tag{9}$$

where the right hand side $f = 0$ or $f = 1/\varrho\, \mathrm{d}p/\mathrm{d}x$ is assumed to be independent of y and given from the global RANS solution at Γ_δ. Therein, $\nu_{\mathrm{t}}^{\mathrm{bl}}$ is the near-wall turbulence model to be studied later and $\mathbf{n}$ is the outer surface normal vector. Moreover, denote $y_\delta = \mathrm{dist}(\mathbf{x}_{\mathrm{w}}, \Gamma_\delta)$. The variant $f = 0$ is called equilibrium stress balance model leading to universal near-wall solutions.

It is well-known that for equilibrium boundary layers, e.g., the fully developed turbulent flow over a flat plate at zero pressure gradient, in the region between the wall and the outer edge of the logarithmic layer, the profiles for mean flow u and turbulence quantities k, ω, $\tilde{\nu}$ and hence ν_{t} are universal, i.e., they collapse when scaled with friction velocity u_τ and viscosity $\nu = \mu/\varrho$

$$u^+ = \frac{u}{u_\tau}\,, \quad y^+ = \frac{y u_\tau}{\nu}\,, \quad \nu_{\mathrm{t}}^+ = \frac{\nu_{\mathrm{t}}}{\nu}\,, \quad k^+ = \frac{k}{u_\tau^2}\,, \quad \omega^+ = \frac{\omega\nu}{u_\tau^2}\,.$$

These universal near-wall profiles may be obtained by integration of (8) with $f = 0$ and the corresponding one-dimensional boundary-layer equations for k and ω, respectively $\tilde{\nu}$. The universal profiles for ν_{t}^+ are turbulence model specific in detail, but in equilibrium flow situations close to the mixing-length relation $\nu_{\mathrm{t}}^+ = \kappa y^+$ with near-wall damping.

Moreover it can be seen from (8) in plus-units which reads

$$(1 + \nu_{\mathrm{t}}^{\mathrm{bl},+})\frac{\mathrm{d}u^{\mathrm{bl},+}}{\mathrm{d}y^+} = 1 + p^+ y^+ \text{ in } (0, y_\delta^+), \quad p^+ = \frac{\nu}{\varrho u_\tau^3}\frac{\mathrm{d}p}{\mathrm{d}x}\,, \tag{10}$$

that it is the pressure gradient parameter p^+ which controls the validity of the equilibrium stress balance assumption.

3.3 Model-Consistency of Universal Wall Functions and Grid-Independent Predictions

Universal wall functions are the solution of (8) with $f = 0$ which reads in non-dimensional form

$$(1+\nu_{\mathrm{t}}^{\mathrm{bl},+})\frac{du^{\mathrm{bl},+}}{dy^+} = 1 \text{ in } (0, y_\delta^+)\,. \tag{11}$$

A universal wall-function method is called *consistent* with respect to the turbulence model used, if the low-Re RANS solution for a flat-plate turbulent boundary layer flow at zero-pressure gradient also solves (11). This implies $\nu_{\mathrm{t}}^{\mathrm{bl}} = \nu_{\mathrm{t}}$, i.e., wall functions have to be turbulence-model specific, as demonstrated first by [8]. Consistency ensures that $u = u^{\mathrm{bl}}$ in the entire near-wall region Ω_δ, except a deviation due to numerical errors. Then predictions for surface transfer coefficients like c_{p} and c_{f} are also independent of the location of the matching boundary Γ_δ which consists of the first grid nodes above the wall. Thus turbulence-model consistent universal wall functions should give (almost) grid-independent predictions for flows close to equilibrium.

Except the noval approach [8], existing wall-function methods are not model-consistent. Instead, they are based e.g. on the algebraic model for $\nu_{\mathrm{t}}^{\mathrm{bl}}$ by Spalding [20], see [4], or they use an approximative hybrid law of the wall for u^{bl}, see e.g. [15].

As the near-wall profiles of different versions of the Spalart-Allmaras model respectively the k-ω model almost collaps (see [9, 11]) it is sufficient to determine one model-consistent universal wall-function for the Spalart-Allmaras model and one for the k-ω model

$$\mathrm{F}_{\mathrm{SA,a}} = (1-\varphi_{\mathrm{SA}})\mathrm{F}_{\mathrm{Sp,5}} + \varphi_{\mathrm{SA}}\mathrm{F}_{\mathrm{Rei,m}},\ \varphi_{\mathrm{SA}} = \tanh\left(\left(\frac{y^+}{24}\right)^3\right) \tag{12}$$

$$\mathrm{F}_{k\omega,\mathrm{a}} = (1-\varphi_{k\omega})\mathrm{F}_{\mathrm{Sp,3}} + \varphi_{k\omega}\mathrm{F}_{\mathrm{Rei,m}},\ \varphi_{k\omega} = \tanh\left(\left(\frac{y^+}{50}\right)^2\right) \tag{13}$$

Therein, we use Reichardt's law of the wall $u^+ = \mathrm{F}_{\mathrm{Rei}}(y^+)$ with

$$\mathrm{F}_{\mathrm{Rei}}(y^+) \equiv \frac{\ln(1+0.4y^+)}{\kappa} + 7.8\left(1 - \mathrm{e}^{-\frac{y^+}{11.0}} - \frac{y^+}{11.0}\mathrm{e}^{-\frac{y^+}{3.0}}\right)\,. \tag{14}$$

and use the fact that Reichardt's law blended with the classical log-law $\mathrm{F}_{\log} = \ln(y^+)/\kappa + 5.1$ gives an excellent agreement in the log-layer when using the formula

$$\mathrm{F}_{\mathrm{Rei,m}} = (1-\varphi_{\mathrm{b1}})\mathrm{F}_{\mathrm{Rei}} + \varphi_{\mathrm{b1}}\mathrm{F}_{\log}\ ,\quad \varphi_{\mathrm{b1}} = \tanh\left(\left(\frac{y^+}{27}\right)^4\right) \tag{15}$$

Spaldings law [20] with parameter $N \in \{3,4,5\}$ is given by the inverse formula $y^+ = F_{\mathrm{Sp,N}}^{-1}(u^+)$ with

$$F_{\mathrm{Sp,N}}^{-1}(u^+) \equiv u^+ + \mathrm{e}^{-5.2\kappa}\left(\mathrm{e}^{\kappa u^+} - \sum_{n=0}^{N}\frac{(\kappa u^+)^n}{n!}\right)\,. \tag{16}$$

For a general approach for arbitrary RANS turbulence models we refer to [8].

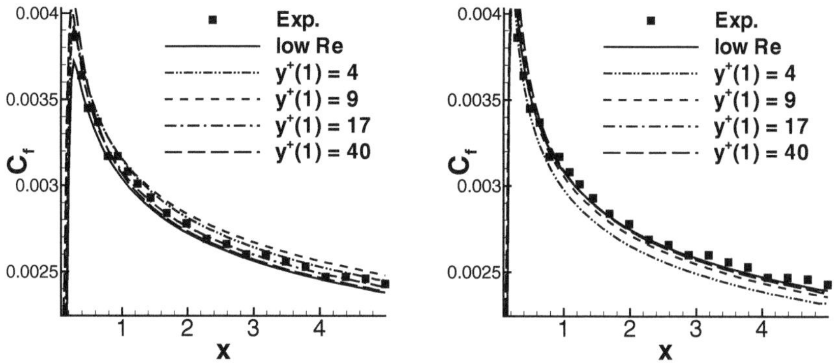

Fig. 2. Analytical model-consistent wall-function (12) for SA-E model (left) and (13) for baseline k-ω model (right)

3.4 Flat Plate Turbulent Boundary Layer at Zero Pressure Gradient

The ability of the present wall function proposal to give solutions independent of the wall-normal grid spacing of the first mesh node is studied for the boundary layer flow at zero pressure gradient by Wieghardt and Tillmann, recorded in [24], pp. 98–123, as Flow 1400.

The model-consistent formulas (12) and (13) give almost grid independent predictions for c_f for the SA-Edwards (SA-E) model and the baseline k-ω model, respectively (see Fig. 2). We mention that the remaining slight grid-dependence is due to numerical errors, see [8] for a detailed discussion. In order to obtain grid independent results for k-ω type models, it is crucial to replace the standard (sqrt) blending [22] by a new proposal, as the standard blending deviates discernibly from the low-Re RANS solution for ω in the buffer layer. The following formulas are considered, where latter is preferred:

$$\text{Standard (sqrt) blending:} \quad \omega = \sqrt{\omega_{\text{vis}}^2 + \omega_{\text{log}}^2}\,,$$

$$\text{New proposal:} \quad \omega = \varphi\omega_{\text{b1}} + (1-\varphi)\omega_{\text{b2}}, \quad \varphi = \tanh\left(\left(\frac{y^+}{10}\right)^4\right)$$

with the blending formula and the asymptotic relations

$$\omega_{\text{b1}} = \omega_{\text{vis}} + \omega_{\text{log}}\,, \quad \omega_{\text{b2}} = \left(\omega_{\text{vis}}^{1.2} + \omega_{\text{log}}^{1.2}\right)^{\frac{1}{1.2}}\,,$$

$$\omega_{\text{vis}} = \frac{6\nu}{\beta_\omega y^2}\,, \quad \omega_{\text{log}} = \frac{u_\tau}{\sqrt{\beta_k}\kappa y}\,.$$

The standard blending causes a significant grid-dependent spreading of the results. Moreover, we recommend to use Wilcox b.c. (5) instead of Menter b.c. (4) (recall that the latter was designed for low-Re grids). We finally mention that formula (13) gives almost grid-independent results also for other variants of the k-ω model. For more details see [10, 11].

4 Near-Wall Behaviour of RANS Models in Situations of Non-equilibrium Flow

Aerodynamic flows around airfoils and full aircraft configurations are characterised by non-equilibrium flow situations which cannot be described by the stress equilibrium assumption (11), and the purpose of this section is to investigate its range of validity. Of major interest are stagnation points and subsequent not yet fully developed turbulent flow, regions of adverse pressure gradient with relatively large pressure gradient parameter, and regions of separation and reattachment. For the near-wall behaviour of RANS turbulence models in a recirculation region see [8].

In regions of adverse pressure gradient (APG), the important parameter is the pressure gradient parameter p^+, see (10). Due to its definition, p^+ goes to infinity as the stagnation and separation points are approached. Figure 3 (left) shows the Reynolds number dependence of the p^+-distribution on the upper side of the A-airfoil at $\alpha = 13.3°$. As can be seen from (10), in attached fully developed turbulent boundary layer flow, the range of validity of zero-pressure gradient wall-functions is determined by the size of p^+. Note that p^+ is positive for adverse pressure gradients and negative for favourable pressure gradients.

The near-wall solutions for both the SA-E model and the SST k-ω model show an increasing deviation from the universal solution with increasing p^+, see Fig. 3 (right) and Fig. 4 (left) respectively. For the turbulence quantities $\tilde{\nu}$ and k, the deviation from their universal solution is much larger than for the corresponding velocity profiles. However, in the viscous sublayer ($y^+ < 5$) the agreement with the universal ZPG solution is still very close, except at very large p^+-values very close before separation. For a detailed discussion we refer to [10, 11]. The

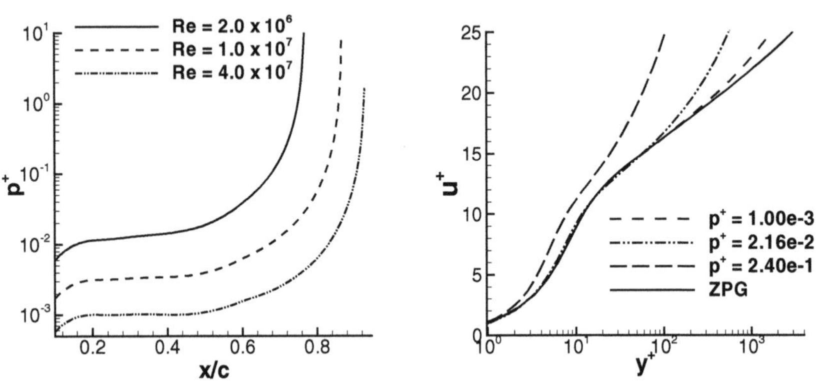

Fig. 3. Reynolds number dependence of p^+ (left) and p^+-dependent near-wall behaviour (right) for SA-E model for adverse pressure gradient (APG) flow [8]

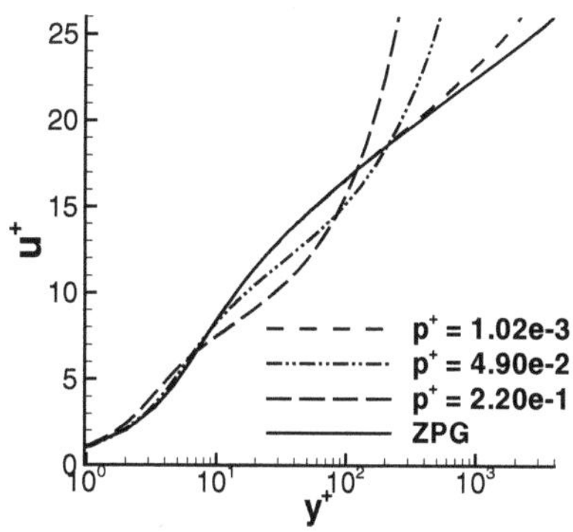

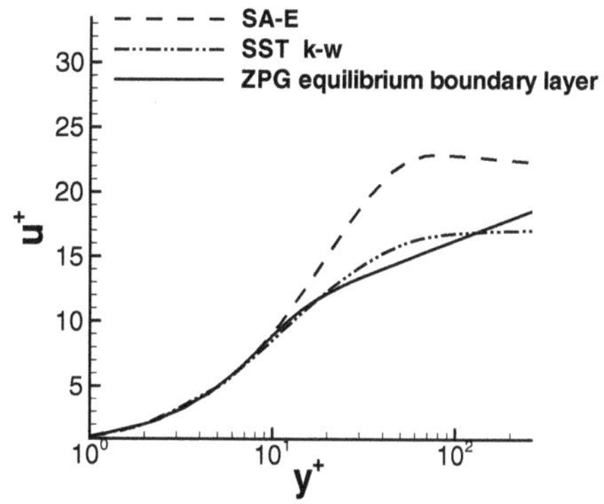

Fig. 4. Left: Near-wall solutions for SST k-ω model at different p^+-stations for APG flow [8]. Near-wall solutions in leading edge region of ONERA A-airfoil at incidence of $13.3°$.

near-wall behaviour of the SA-E and the SST k-ω model at a certain position in the leading edge region of the ONERA A-airfoil at an incidence of $13.3°$ and $\text{Re} = 2 \times 10^6$ is shown in Fig. 4. For $y^+ \lessapprox 10$ the universal velocity profile is a good approximation to the RANS solution if the airfoil is treated fully turbulent. For larger y^+-values the agreement is poor, in particular for the SA model. Moreover it has to be pointed out that the boundary layer thickness in viscous units is extremely thin in the leading edge region, i.e., $y_{99}^+ \lessapprox 100$.

From these observations it may be concluded that very good agreement between the low-Re solution and the solution with wall-functions can be expected only if the grid constraint $y^+(1) \lessapprox 10$ is satisfied. This constraint has to be satisfied in particular for SA-type models at small Reynolds numbers, say $\text{Re} \lessapprox 3 \times 10^6$. Otherwise the solution can exhibit spurious oscillations near the leading edge. On the other hand, for transonic flows at large Reynolds numbers, very good agreement between the low-Re solution and the solution with wall-functions can be obtained even on meshes with $y^+(1) \lessapprox 70$.

We remark that off-wall boundary conditions for k and $\tilde{\nu}$ using the universal ZPG solution for k and $\tilde{\nu}$ are additional sources of grid-dependent results, see also [11]. Instead, we prescribe $k = 0$ and $\tilde{\nu} = 0$ on Γ_w.

5 Numerical Method

The numerical results are obtained using the DLR TAU-code. The wall-shear stress prescribed in (6) is given by $\tau_\text{w}^\text{bl} = \varrho u_\tau^2$, where u_τ is computed as follows. Suppose a solution of (8), (9) with $f = 0$ is known in either of the two closed forms

$$\frac{u}{u_\tau} = \text{F}\left(\frac{y u_\tau}{\nu}\right) \text{ or } \frac{y u_\tau}{\nu} = \text{F}^{-1}\left(\frac{u}{u_\tau}\right),$$

then the matching condition $u^{\rm bl} = u_\delta$ on Γ_δ and the relation $u^{\rm bl} = u_\tau {\rm F}(yu_\tau/\nu)$ imply

$$\mathrm{F}\left(\frac{y_\delta u_\tau}{\nu}\right) = \frac{u_\delta}{u_\tau} \text{ resp. } \mathrm{F}^{-1}\left(\frac{u_\delta}{u_\tau}\right) = \frac{y_\delta u_\tau}{\nu} ,$$

which can be solved for u_τ using Newton's method. Denote $\mathrm{TM} \in \{\mathrm{SA}, \mathrm{k}\omega\}$ and $\mathrm{N} \in \{3,5\}$. For the numerical solution of

$$\frac{u_\delta}{u_\tau} = \mathrm{F_{TM}}\left(\frac{y_\delta u_\tau}{\nu}\right) , \quad \mathrm{F_{TM}} = (1-\varphi_{\rm TM})\mathrm{F_{Sp,N}} + \varphi_{\rm TM}\mathrm{F_{Rei,m}}$$

we proceed as follows:

1. From the initial guess $u_\tau^0 = u_\delta/y_\delta$, seek $u_{\tau,\rm Rei}$ as solution of $u_\delta/u_\tau = \mathrm{F_{Rei,m}}(y_\delta u_\tau/\nu)$.
2. Using the new initial guess $u_\tau^0 = u_{\tau,\rm Rei}$, seek $u_{\tau,\rm Sp}$ as solution of $y_\delta u_\tau/\nu = \mathrm{F}^{-1}_{\rm Sp,N}(u_\delta/u_\tau)$.
3. Compute $\varphi_{\rm TM}$ from (12) respectively from (13) and set $u_\tau = (1-\varphi_{\rm TM})u_{\tau,\rm Sp} + \varphi_{\rm TM}u_{\tau,\rm Rei}$.

Convergence in steps (1) and (2) is achieved within three to four iteration steps.

6 Validation for Aerodynamic Flows

In this section the method is validated for aerodynamic flows. For all cases we use a series of hybrid-Re grids of O-type with varying $y^+(1)$, generated using the commercial grid generation tool CentaurSoft (*www.centaursoft.com*). The grids are built such that the thickness of the prismatic layer has an almost constant value and fully contains the boundary layer.

6.1 Transonic Airfoil Flows RAE-2822 Cases 9 and 10

We apply the method to the transonic airfoil flows RAE-2822 case 9 (no/small separation region at Ma = 0.73, Re = 6.5×10^6 and angle of attack $\alpha = 2.8°$) and case 10 (shock induced separation at Ma = 0.75, Re = 6.2×10^6 and $\alpha = 2.8°$) studied experimentally in [2].

The predictions for the pressure coefficient $c_{\rm p}$ and the local skin friction coefficient $c_{\rm f}^{(\rm loc)}$ (based on the local dynamic pressure at the boundary layer edge $q_{\rm P}$, [2]) are given in Figs. 5(a)–5(f).

The predictions of $c_{\rm p}$ are remarkably grid-independent, in particular regarding the shock position. The predictions for $c_{\rm f}$ with wall-functions are in fair agreement with the low-Re results provided $y^+(1) \lessapprox 10$. In the separation region, the agreement for $y^+(1) \lessapprox 10$ is surprisingly good, whereas on the coarser grids with $y^+(1) \gtrapprox 20$ the differences become more pronounced.

The aerospace engineer is more interested in the integral lift and drag coefficients, which are given in Table 1. Moreover the results on a low-Re mesh of H-type using both the low-Re condition and the wall-function (also called hybrid Reynolds number) boundary condition (denoted by hybRe) are given. It can be

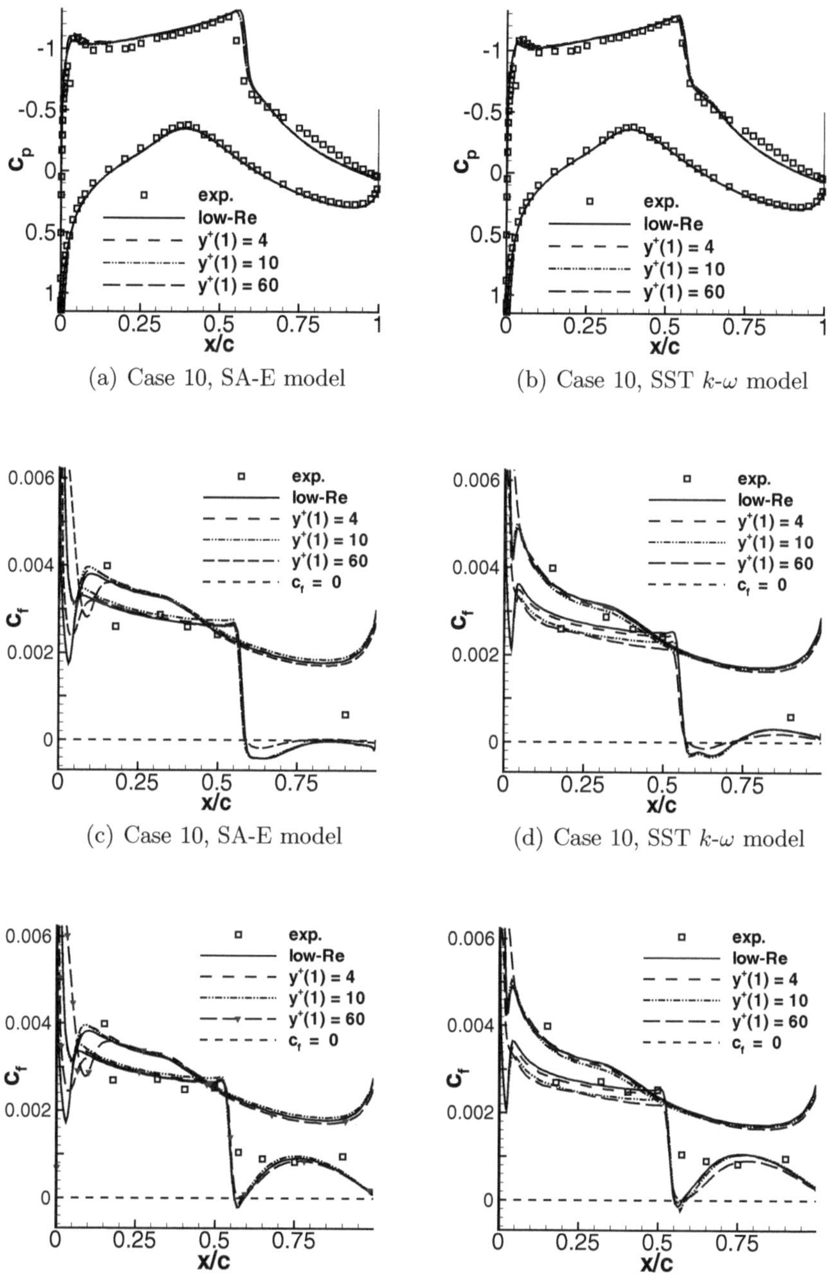

(a) Case 10, SA-E model

(b) Case 10, SST k-ω model

(c) Case 10, SA-E model

(d) Case 10, SST k-ω model

(e) Case 9, SA-E model

(f) Case 9, SST k-ω model

Fig. 5. RAE-2822 case 9 & 10: Distribution of c_p and c_f

Table 1. Aerodynamic coefficients for RAE-2822 case 10. Left: Predictions using wall functions on series of O-type meshes with varying $y^+(1)$. Right: Spreading depending on different turbulence models and on the choice of the momentum boundary condition on an H-type mesh.

$y^+(1)$	c_l(SA-E)	c_l(SST)	c_d(SA-E)	c_d(SST)
low-Re	0.79461	0.76483	0.02708	0.02575
1	0.79458	0.76507	0.02707	0.02574
4	0.79135	0.76912	0.02696	0.02577
7	0.78389	0.77166	0.02685	0.02582
10	0.78377	0.77459	0.02681	0.02586
20	0.79098	0.77531	0.02691	0.02580
40	0.78993	0.76363	0.02682	0.02534
60	0.79205	0.76763	0.02662	0.02567

Turb. model	C_l	C_d
Exp.	0.743	0.0242
SA-E	0.761	0.0254
SST low-Re	0.737	0.0247
SST hyb-Re	0.725	0.0238
TNT low-Re	0.769	0.0260
TNT hyb-Re	0.756	0.0251

seen that the grid-dependent spreading of the aerodynamic coefficients using wall-functions is not larger than other sources of uncertainties, e.g. the turbulence model, the grid topology or the implementation of the boundary condition.

6.2 Subsonic A-Airfoil in Highlift Configuration

In this section we consider the subsonic flow around the "A-airfoil" in highlift configuration at $\mathrm{Ma} = 0.15$, $\mathrm{Re} = 2.0 \times 10^6$, and $\alpha = 13.3°$, investigated experimentally in [5, 6]. Due to the large angle of attack, a strong adverse pressure gradient arises on the upper side of the airfoil, which causes the turbulent boundary layer to separate close to the trailing edge. In the experiment, transition was prescribed at $x/c = 0.3$ on the lower side and free transition was observed at $x/c = 0.12$ on the upper side, but in the present computations the airfoil surface is treated fully turbulent, as the wall-function approach relies on the assumption of a fully developed turbulent flow. It is noteworthy that neglecting transition increases the deviation of the low-Re results from the experimental data significantly.

The results are shown in Figs. 6(a)–6(d). Therein the value for $y^+(1)$ is taken on the lower side, where $y^+(1)$ is almost constant. Around the suction peak on the upper side, $y^+(1)$ is almost three times larger. The SA solutions with wall functions suffer from spurious oscillations in c_p for $y^+(1) \gtrsim 70$ in the leading edge region. This can be explained by the large deviation of the SA solution from the universal profile, see Fig. 4. On the meshes with smaller $y^+(1)$, the agreement in c_p and c_f with the low-Re solution is good. The solution of the SST k-ω model does not exhibit oscillations, but the agreement in the leading edge region and also in the prediction of the separation point is better for smaller $y^+(1)$-values. For a detailed discussion see [10, 11]. It is worthwhile mentioning that for the SST k-ω model, the spreading in the separation point is 1% on the meshes with nominal $y^+(1) \leq 20$ and 4% on the coarsest mesh with (nominal)

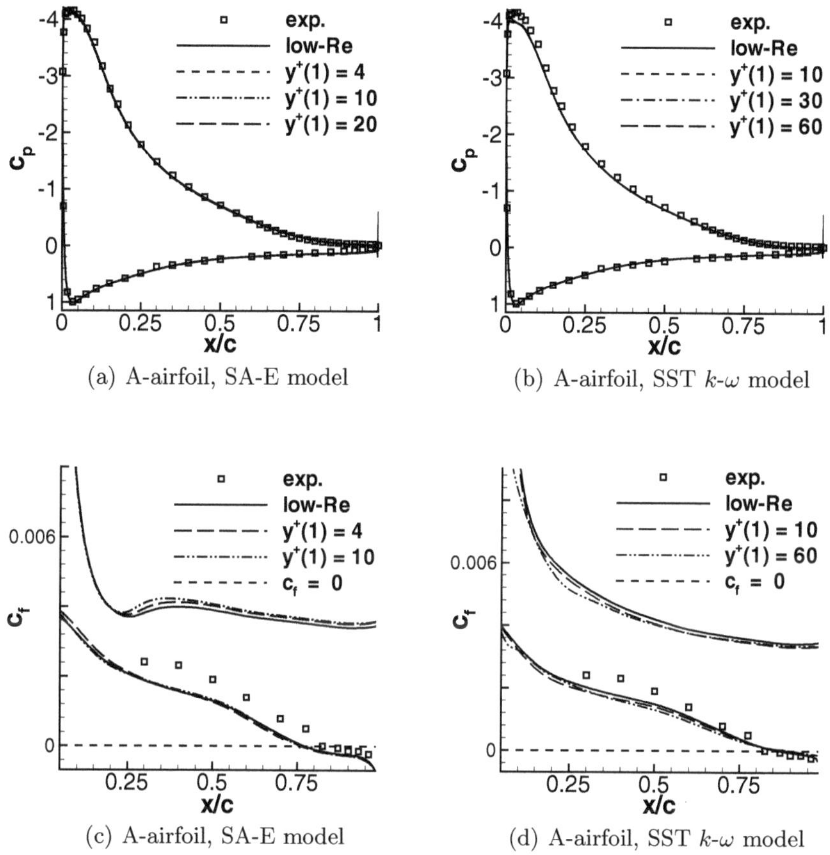

Fig. 6. ONERA A-airfoil at $\alpha = 13.3°$ incidence: Distribution of c_p and c_f

$y^+(1) = 60$. In constrast, the predictions of the separation point between SST k-ω and SA-E model differ by 11%.

6.3 Application to 3D Testcases

The method has also been applied successfully to 3D testcases, viz., the ONERA M6-wing and the F6 (clean) wing-body configuration, but the results will be presented elsewhere. The ONERA M6 wing is a low aspect ratio swept, tapered semi-span wing with no twist generated from the symmetric ONERA D-airfoil, see [18]. The flow conditions considered are Ma $= 0.8395$, Re $= 11.72 \times 10^6$, and incidence $\alpha = 3.06°$. The results fully support the observations made for the RAE-2822 airfoil. The shock position is predicted almost independent of the $y^+(1)$ spacing for meshes with $y^+(1) \lessapprox 80$ in the leading edge region. On meshes with $y^+(1) \gtrapprox 120$, the solution starts to show oscillations in the region of the suction peak.

7 Combination of Wall-Functions and y^+-Adaptation

From the present investigation we may conclude that the most critical region for the application of universal wall-functions is the leading edge region. Due to the stagnation point, the flow is not fully developed turbulent. Moreover, for large incidences, the favourable pressure gradient may be very large. In contrast, regions of moderate adverse pressure gradient, shocks, separation and reattachment can be treated surprisingly well using such simple boundary layer approximation.

Additional testcases suggest that the problems in the leading edge region become much more severe if the Reynolds number is relatively small, say $\mathrm{Re} \lesssim 2.0 \times 10^6$. Revisiting the ONERA A-airfoil, Fig. 7(a) shows the y^+-distribution after an adaptation of the nodes inside the prismatic boundary layer grid by changing their distribution along wall-normal lines. By shifting the first off-wall

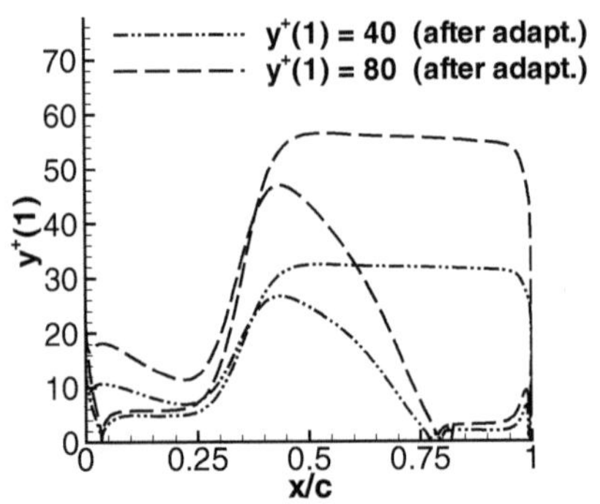

(a) ONERA A-airfoil, y^+ after adaptation (for SA-E model)

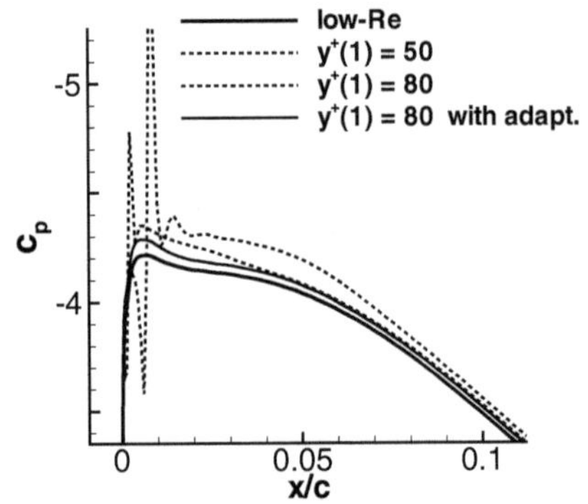

(b) ONERA A-airfoil, c_p near leading edge for SA-E model

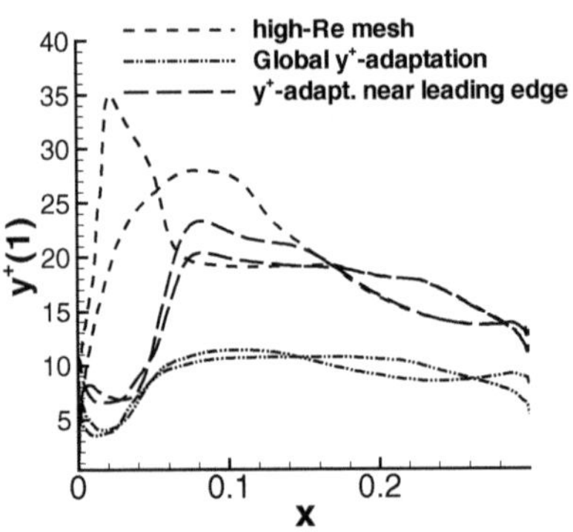

(c) VC-Opt, SA-E model

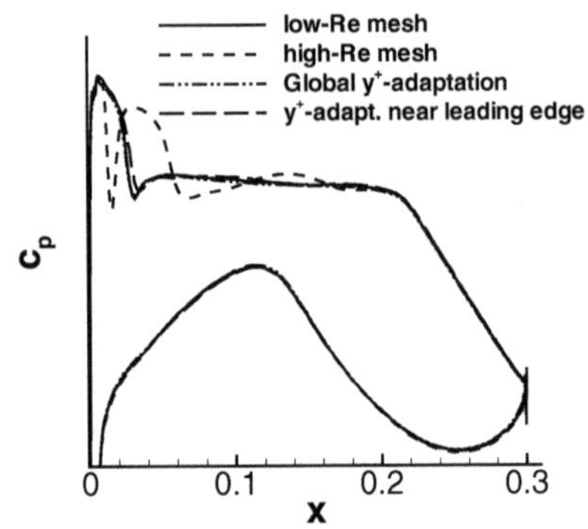

(d) VC-Opt, SA-E model

Fig. 7. Combination of wall-functions and y^+-adaptation for ONERA A-airfoil and VC-Opt profile nodes towards the viscous sublayer, the spurious oscillations in the solution can be remedied, see Fig. 7(b).

Secondly, this method is applied to the VC-Opt profile, which is similar to an AIRBUS profile. The flow conditions are $\mathrm{Re} = 2 \times 10^6$, $\mathrm{Ma} = 0.757$ and $T_\infty = 217\,\mathrm{K}$. As industrial data are confidential, the scaling in Fig. 7(d) for c_p is not shown. Due to the relatively low Reynods number (stemming from wind-tunnel conditions) the standard wall-function method fails unless the y^+-distribution is adapted. In Fig. 7(c) two techniques are shown. The first strategy is to limit the maximum $y^+(1)$ in the leading edge region and leaving the prismatic mesh unchanged otherwise (This method is denoted by "y^+-adapt. near leading edge" in Fig. 7(c) (left)). The second approach is to set $y^+_\mathrm{target} = 4$ near the leading edge and $y^+_\mathrm{target} = 10$ otherwise (this strategy is denoted by "Global y^+ adaptation" in Fig. 7(c) (left)). The second strategy gives best agreement with the low-Re solution, see Fig. 7(d).

More generally speaking, for flows with separation it was observed that best agreement between the wall-function results and the low-Re solution is obtained if $y^+(1) \in [5, 10]$ near the leading edge, and $y^+(1) \in [1, 5]$ close to the separation point and in regions of separated flow, see [11] for details.

The increasing effect of non-equilibrium terms with decreasing Reynolds number may be explained by Fig. 3 (left), which shows that the significance of the pressure gradient parameter is increasing largely, if the Reynolds number is decreased. For more details we refer to [1].

8 Best Practice Guidelines

In this section some first guidelines for best usage of universal wall functions are summarized. So far the method has been applied only to problems with steady state solution. At the present stage of knowledge, the critical flow parameter concerning the applicability of wall-functions is the Reynolds number. The critical parameter to be controlled by the user is the spacing $y^+(1)$ of the first node above the wall. For $\mathrm{Re} \lessapprox 2.0 \times 10^6$, the wall-function results are close to the low-Re results if the mesh satisfies the requirement $y^+(1) \lessapprox 10$. This observation is valid for both subsonic highlift and transonic cruise flight conditions. For transonic flows at high Reynolds numbers, say $\mathrm{Re} \gtrapprox 6.0 \times 10^6$, the predictions for pressure coefficient are close to the low-Re solution even on meshes with $y^+(1) \approx 50$ in the region before the shock. But close agreement in skin friction requires that $y^+(1) \lessapprox 10$ if flow separation occurs.

9 Conclusions

A turbulence-model specific wall function method was demonstrated to give successful predictions for aerodynamic flows with separation in the subsonic and transonic regime. Some first best practice guidelines are given, which are based on an investigation of the near-wall behaviour of the RANS turbulence models in non-equilibrium flow situations and on the application of the wall-function method to several test cases. Wall-functions only allow for a reduction of nodes in the near-wall region in wall-normal direction. Reduction of the number of nodes in streamwise and spanwise direction and in the region of the outer flow (e.g., for shocks) requires to employ additional mesh adaptation techniques using e.g. feature- or adjoint-based sensors.

References

1. Alrutz, T., Knopp, T.: Near-wall grid adaptation for turbulent flows. International Journal of Computing Science and Mathematics xxx:xx–xx, 200x
2. Cook, P.H., McDonald, M.A., Firmin, M.C.P.: Aerofoil RAE 2822 – Pressure distributions and boundary layer and wake measurements. Technical report, AGARD Advisory Report AR-138, pp. A6.1–A6.77 (1979)
3. Edwards, J.R., Chandra, S.: Comparison of eddy viscosity-transport turbulence models for three-dimensional, shock separated flowfields. AIAA Journal 34, 756–763 (1996)
4. Frink, N.T.: Tetrahedral unstructured Navier-Stokes method for turbulent flows. AIAA Journal 36, 1975–1982 (1998)
5. Gleyzes, C.: Opération décrochage – Résultats des essais à la soufflerie F2. Technical report, RT-DERAT 55/4004 DN, ONERA (1988)
6. Gleyzes, C.: Opération décrochage – Résultats de la 2ème campagne d'essais à F2 - Mesures de pression et vélocimétrie laser. Technical report, RT-DERAT 55/5004 DN, ONERA (1989)
7. Grotjans, H., Menter, F.: Wall functions for general application cfd codes. In: ECCOMAS 1998, vol. 1, pp. 1112–1117 (1998)
8. Kalitzin, G., Medic, G., Iaccarino, G., Durbin, P.: Near-wall behaviour of RANS turbulence models and implications for wall functions. Journal of Computational Physics 204, 265–291 (2005)
9. Knopp, T.: A new adaptive wall-function method for subsonic and transonic turbulent flows. Technical report, DLR IB 224-2005 A 14 (2005)
10. Knopp, T.: On grid-independence of RANS predictions for aerodynamic flows using model-consistent universal wall-functions. In: Proceedings of the European Conference on Computational Fluid Dynamics, ECCOMAS CFD 2006 (2006)
11. Knopp, T., Alrutz, T., Schwamborn, D.: A grid and flow adaptive wall-function method for RANS turbulence modelling. Journal of Computational Physics 220, 19–40 (2006)
12. Knopp, T., Lube, G., Gritzki, R., Rösler, M.: A near-wall strategy for buoyancy-affected turbulent flows using stabilized FEM with applications to indoor air flow simulation. Computer Methods in Applied Mechanics and Engineering 194, 3797–3816 (2005)
13. Kok, J.C.: Resolving the dependence on freestream values for the k/ω turbulence model. AIAA Journal 38, 1292–1295 (2000)
14. Menter, F.R.: Zonal two-equation k/ω turbulence models for aerodynamic flows. AIAA Paper 1993-2906 (1993)
15. Mohammadi, B., Medic, G.: A critical evaluation of the classical k/ϵ model and wall-laws for unsteady flows over bluff bodies. Technical report, INRIA, Rapport de recherche no. 3056 (1996)

16. Rung, T.: Formulation of universal wall boundary conditions for transport equation turbulence models. Technical report, Hermann-Föttinger-Institut, Technische Universität Berlin (1999)
17. Rung, T., Lübcke, H., Franke, M., Xue, L., Thiele, F., Fu, S.: Assessment of explicit algebraic stress models in transonic flows. In: Engineering Turbulence Modelling and Experiments 4, Proc. 4th Int. Symposium on engineering Turbulence Modelling and Measurements, Corsica, France, pp. 659–668 (1999)
18. Schmitt, V., Charpin, F.: Pressure distributions on the ONERA-M6-Wing at transonic Mach numbers. In: Barche, J. (ed.) Experimental Data Base for Computer Program Assessment. Report of the Fluid Dynamics Panel Working Group 04, pp. B1-1–B1-44. AGARD-AR-138 (1979)
19. Spalart, P.R., Allmaras, S.R.: A one-equation turbulence model for aerodynamics flows. AIAA Paper 1992-0439 (1992)
20. Spalding, D.B.: A single formula for the law of the wall. Journal of Applied Mechanics 28, 455–457 (1961)
21. Tidriri, M.D.: Domain decomposition for compressible Navier–Stokes equations with different discretizations and formulations. Journal of Computational Physics 119, 271–282 (1995)
22. Vieser, W., Esch, T., Menter, F.R.: Heat transfer predictions using advanced two-equation turbulence models. Technical report, CFX Technical Memorandum CFX-VAL10/0602 (2002)
23. Wallin, S., Johansson, A.V.: An explicit algebraic Reynolds stress model for incompressible and compressible turbulent flows. Journal of Fluid Mechanics 403, 89–132 (2000)
24. Wieghardt, K., Tillman, W.: On the turbulent friction layer for rising pressure. In: Coles, D.E., Hirst, E.A. (eds.) Computation of Turbulent Boundary Layers - 1968 AFOSR-IFP-Stanford Conference, vol. II, pp. 98–123. Thermosciences Division, Department of Mechanical Engineering, Stanford University (1969)
25. Wilcox, D.C.: Turbulence modeling for CFD. DWC Industries, La Canada (1998)

Computational Modelling of Transonic Aerodynamic Flows Using Near-Wall, Reynolds Stress Transport Models

S. Jakirlić[1], B. Eisfeld[2], R. Jester-Zürker[1], C. Tropea[1], and N. Kroll[2]

[1] Fachgebiet Strömungslehre und Aerodynamik,
Technische Universität Darmstadt
Petersenstr. 30, 64287 Darmstadt, Germany
[2] Institute of Aerodynamics and Flow Technology,
DLR – German Aerospace Center
Lilienthalplatz 7, 38108 Braunschweig, Germany

Summary. The present work reports on the further development of the Hanjalić–Jakirlić (1998) near-wall, second-moment closure (SMC) model in the RANS (Reynolds-Averaged Navier–Stokes) framework, updated to account for a wall-normal free, non-linear version of the pressure-strain term model, its implementation into the DLR-FLOWer code and its validation in computing some (compressible) transonic flow configurations. Furthermore, the wall boundary condition is based on the asymptotic behaviour of the Taylor microscale λ and its exact relationship to the dissipation rate ε in the immediate wall vicinity. In addition, the calculations were performed using the DLR-FLOWer's default Reynolds stress transport model (Eisfeld, 2006), representing a numerically robust combination of the Launder–Reece–Rodi (1975) model resolving the near-wall layer with the Speziale–Sarkar–Gatski (1991) model being employed in the outer region. The flow geometries considered in this work include the transonic RAE 2822 profiles (cases 9 and 10), the ONERA M6 wing and the DLR-ALVAST wing-body configuration. The model results are analysed and discussed in conjunction with available experimental databases and the results of two widely used eddy-viscosity-based models, the one-equation Spalart–Allmaras model (1994) and the two-equation k-ω model of Wilcox (1988). The SMC predictions show encouraging results with respect to the shock position, shock-affected flow structure and the strength of the wing-tip vortex.

1 Introduction

Computational Fluid Dynamics (CFD) is nowadays regarded as an indispensable tool in aerodynamic design and optimization. The application of efficient CFD methods in aircraft aerodynamics and related disciplines enables reduction of costs and shortens substantially the time for development in aircraft industry. Furthermore, the employment of CFD methods supports to a large extent a detailed evaluation of new technologies and concepts as well as alternative configurations with respect to security, environmental (e.g., noise pollution) and economic aspects of transport and passenger aircrafts.

N. Kroll et al. (Eds.): MEGADESIGN and MegaOpt, NNFM 107, pp. 73–92.
springerlink.com

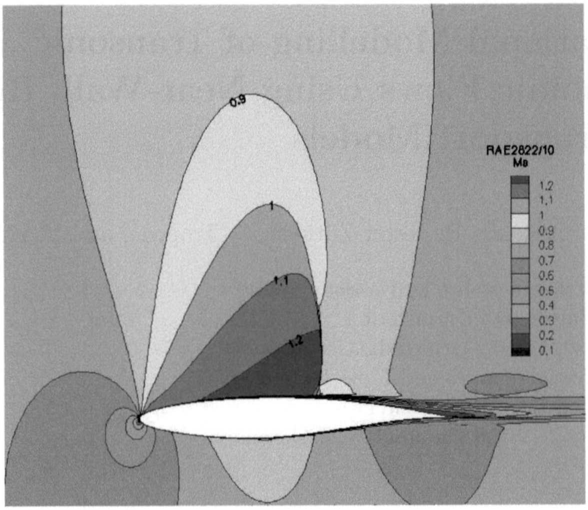

Fig. 1. Mach-number contours in the transonic flow around RAE2822 profile

Together with the high-lift configurations corresponding to take-off and landing, the high-speed, cruise-flight configurations play an important role in the entire operational range of an aircraft. Transport and passenger aircrafts operate at cruise flight velocities corresponding to a Mach number interval Ma $\in [0.75, 0.85]$. The associated flow regime past an aircraft wing is of transonic nature. It is characterized by the development of a closed, wall-bounded supersonic flow region at the upper wing surface, Fig. 1 (its appearance on the lower wing surface is also possible depending on its curvature and the angle of attack). The flow is strongly accelerated, reaching sonic velocities at the leading edge of the wing. The supersonic region spreads behind the so-called sound line, denoting the isoline with Ma $= 1.0$. The pressure increase behind the maximum profile thickness causes the formation of a shock wave which closes the supersonic region. The shock front proceeds almost orthogonally to the wing profile contour. Transonic flows also exhibit boundary-layer/shock interaction, i.e. a boundary layer thickening due to the adverse pressure gradient, and in some cases shock-induced separation occurs[1]. The change of the flow regime – from subsonic regime to supersonic and back – implies the solving of a combined elliptic/hyperbolic flow problem. These features as well as the compressible properties of the flow in general pose a special challenge not only for the numerical treatment but also for turbulence models. The present work focuses on the application of two near-wall second-moment closure (SMC) models, one accounting separately for both viscous effects and kinematic wall blockage with

[1] The transonic flow past aircraft wings can also exhibit unsteady shock/boundary-layer interaction caused by the shock position change due to self-excited oscillations. Readers interested in these, so-called transonic buffet flows are referred to the work of Barakos and Drikakis (2000).

respect to the anisotropic nature of the Reynolds-stress and stress dissipation fields (Jakirlić and Hanjalić, 1995; Hanjalić and Jakirlić, 1998) and a second model, being a default Reynolds-stress transport model in the compressible flow solver DLR-FLOWer (Eisfeld, 2006). Very extensive work on the computation of different shock-affected flow configurations, including also the RAE2822 airfoil, has been performed by Leschziner and co-workers (Leschziner et al., 2000 and Batten et al., 1999). They applied several linear and non-linear, near-wall eddy-viscosity models and second-moment closure models, among others also the Jakirlić and Hanjalić (1995) model, whose updated version (see Sect. 2 for more details) is applied in the present work. The performance of the original linear HJ model was evaluated by computing the transonic flows over a plane channel bump (internal flow) and over an axisymmetric bump (both configurations are characterized by a strong pressure increase in the supersonic portion of the flow, which is additionally enhanced by a much stronger cross-section constriction compared to the cases considered in the present work). These results were inferior to those obtained using the non-linear MCL (Modified Craft–Launder) model, especially with respect to the low intensity of the shock-induced flow reversal and pressure recovery. Batten et al. (1999) blamed the linear model of the redistribution term and conventional wall-reflection term due to Gibson and Launder (1978) used in the HJ model, the latter returning an excessive turbulence intensity in flows affected by strong deceleration (e.g. stagnation flow regions).

2 Computational Method

The continuity, momentum and energy equations governing the compressible flow are given in differential form in the Reynolds-Averaged Navier–Stokes framework as follows:

$$\frac{\partial \overline{\varrho}}{\partial t} + \frac{\partial}{\partial x_j}\left(\overline{\varrho}\widetilde{U}_j\right) = 0 \tag{1}$$

$$\frac{\partial\left(\overline{\varrho}\widetilde{U}_i\right)}{\partial t} + \frac{\partial}{\partial x_k}\left(\overline{\varrho}\widetilde{U}_i\widetilde{U}_k + \overline{\varrho}\widetilde{u_i''u_k''}\right) = -\frac{\partial \overline{P}}{\partial x_i} + \frac{\partial \overline{\tau}_{ik}}{\partial x_k} \tag{2}$$

$$\frac{\partial\left(\overline{\varrho}\widetilde{E}\right)}{\partial t} + \frac{\partial}{\partial x_k}\left(\overline{\varrho}\widetilde{H}\widetilde{U}_k + \overline{\varrho}\widetilde{u_i''u_k''}\widetilde{U}_i\right) = \frac{\partial}{\partial x_k}\left(\overline{\tau}_{ik}\widetilde{U}_i\right) - \frac{\partial}{\partial x_k}\left(\overline{q}_k + \overline{q}_k^{(\mathrm{t})}\right) + \overline{\varrho}D_{kk} \tag{3}$$

with total energy $\widetilde{E} = \widetilde{e} + \widetilde{U}_k\widetilde{U}_k/2 + \widetilde{k}$ ($\widetilde{e} = C_\mathrm{v}\widetilde{T}$) and total enthalpy $\widetilde{H} = \widetilde{h} + \widetilde{U}_k\widetilde{U}_k/2 + \widetilde{k}$ ($\widetilde{h} = C_\mathrm{p}\widetilde{T}$). In these equations the overbar ($\overline{\phi}$) and the tilde ($\widetilde{\phi}$) denote simple and mass weighted averages, respectively. Here, a Newtonian fluid with the stress tensor and a Fourier type heat flux

$$\overline{\tau}_{ij} = 2\overline{\mu}\widetilde{S}_{ij}\,;\ \widetilde{S}_{ij} = \frac{1}{2}\left(\frac{\partial \widetilde{U}_i}{\partial x_j} + \frac{\partial \widetilde{U}_j}{\partial x_i} - \frac{2}{3}\frac{\partial \widetilde{U}_k}{\partial x_k}\delta_{ij}\right) \tag{4}$$

$$\overline{q}_i = -\overline{\lambda}\frac{\partial \widetilde{T}}{\partial x_i} \tag{5}$$

is assumed. In these material laws the dynamic viscosity follows from Sutherland's formula and the heat conductivity is related to the dynamic viscosity via a constant Prandtl number. Thermodynamic closure is achieved by the assumption of an ideal ($\overline{P} = \overline{\varrho}\mathcal{R}\widetilde{T}$), calorically perfect gas. Note, that the contribution of turbulent diffusion $\overline{\varrho}D_{kk}$ ($= \partial(\overline{\varrho u_i'' u_i'' u_k''/2} - \overline{u_i''\tau_{ik}})/\partial x_k$) to the total energy equation (3) and the contributions of the kinetic energy of turbulence to the total energy $\widetilde{E}$ and total enthalpy $\widetilde{H}$ are usually neglected. This simplification was followed also in the present work.

The turbulent heat flux is modelled in analogy to Fourier type heat conduction

$$\overline{q}_i^{(\mathrm{t})} \left(= \overline{\varrho h'' u_i''}\right) = -\overline{\lambda}^{(\mathrm{t})}\frac{\partial \widetilde{T}}{\partial x_i} \tag{6}$$

where the eddy conductivity is computed from an equivalent eddy viscosity via the definition of the turbulent Prandtl number, i.e.

$$\overline{\lambda}^{(\mathrm{t})} = \frac{C_{\mathrm{p}}\overline{\mu}^{(\mathrm{t})}}{Pr_{\mathrm{t}}} \tag{7}$$

with $\overline{\mu}^{(\mathrm{t})}$ determined by an eddy viscosity model scheme. The following sections outline the description of the turbulence models used and numerical method applied.

2.1 Turbulence Modelling

The computations were performed with a wall-normal free version of the low-Re number Second-Moment (Reynolds stress) closure model (RSM) due to Hanjalić and Jakirlić (HJ, 1998) which is based on the model equations governing the Reynolds stress tensor $\widetilde{u_i'' u_j''}$ and the dissipation rate of the kinetic energy of turbulence ε. The model satisfies all important requirements, with a specific emphasis on limiting states of turbulence (vanishing and very high turbulence Re numbers, two-component limit, etc.), reproducing the laminar-to-turbulent and reverse transition (by-pass and shear-generated transition with minimum background turbulence), appropriate reproduction of effects of extra strain rates (transverse shear, skew-induced three-dimensionality), high acceleration (including laminarization), high deceleration (approaching separation), swirl effects, mean compression, flow separation, recirculation and reattachment, see e.g. Hanjalić and Jakirlić (2002). The complete specification of its updated version is given in the work of Jakirlić et al. (2007). Here, only some specific features will be highlighted.

The stress dissipation tensor was modelled by using the following anisotropic formulation

$$\varepsilon_{ij} = f_{\mathrm{s}}\widetilde{u_i'' u_j''}\frac{\widetilde{k}}{\varepsilon} + (1 + f_{\mathrm{s}})\,\frac{2}{3}\varepsilon\delta_{ij} \tag{8}$$

with $f_{\mathrm{s}} = 1 - \sqrt{A}E^2$ (one should note that this formulation doesn't result in the correct asymptotic behaviour of the ε_{22}, ε_{12} and ε_{23} components, but this fact caused only a slight near-wall imbalance in the equations governing the stress components comprising the normal-to-the-wall fluctuation and has marginal effect on other components) and the pressure-rate-of-strain term model, extended to account for the non-linearity in line with the proposal of Speziale et al. (1991) reads as follows[2]

$$\begin{aligned}
\Phi_{ij} =&\Phi_{ij,1} + \Phi_{ij,2} + \Phi_{ij,w} \\
\Phi_{ij,1} =& -\varepsilon\left[C_1 a_{ij} - C_1'\left(a_{ik}a_{jk} - \frac{1}{3}\delta_{ij}A_2\right)\right] \\
\Phi_{ij,2} =& -C_2' a_{ij}\mathcal{P}_k + C_3\widetilde{k}\widetilde{S}_{ij} + C_4\widetilde{k}\left(a_{ip}\widetilde{S}_{pj} + a_{jp}\widetilde{S}_{pi} - \frac{2}{3}a_{pq}\widetilde{S}_{pq}\delta_{ij}\right) \\
&- C_5\widetilde{k}\left(a_{ip}\widetilde{W}_{pj} + a_{jp}\widetilde{W}_{pi}\right) \\
\Phi_{ij,w} =&C_1^w f_w\frac{\varepsilon}{\widetilde{k}}\left(\widetilde{u_k''u_m''}n_k n_m\delta_{ij} - \frac{3}{2}\widetilde{u_i''u_k''}n_k n_j - \frac{3}{2}\widetilde{u_k''u_j''}n_k n_i\right) \\
&+ C_2^w f_w\frac{\varepsilon}{\widetilde{k}}\left(\Phi_{km,2}^{\mathrm{IP}}n_k n_m\delta_{ij} - \frac{3}{2}\Phi_{ik,2}^{\mathrm{IP}}n_k n_j - \frac{3}{2}\Phi_{kj,2}^{\mathrm{IP}}n_k n_i\right)
\end{aligned} \tag{9}$$

The model coefficients and functions are summarized as follows:

$$\begin{aligned}
&C_1 = C + \sqrt{A}E^2\ ,\ C = 2.5AF^{1/4}f\ ,\ C_1' = \max\left[0.8A_2; 0.5\right]C_1 \\
&C_2 = 0.8A^{1/2}\ , C_2' = 0\ , C_3 = 1.067A^{1/2}\ , C_4 = C_5 = 0.8A^{1/2} \\
&C_1^w = \max\left[0.9 - 0.7C; 0.3\right]\ ,\ C_2^w = \min\left[A; 0.3\right]
\end{aligned} \tag{10}$$

$$\begin{aligned}
F &= \min\{0.6; A_2\}\ ,\ f = \min\left\{\left(\frac{\mathrm{Re_t}}{150}\right)^{3/2}; 1\right\} \\
\mathrm{Re_t} &= \frac{\widetilde{k}^2}{\nu\varepsilon}\ ,\ f_w = \min\left[\frac{\widetilde{k}^{3/2}}{2.5\varepsilon y_n}; 1.4\right]
\end{aligned} \tag{11}$$

Prior to the computations, the performance of the updated HJ model for the stress dissipation were checked *a priori* using the DNS database of some related flows being affected by separation (flow over a backward-facing step flow; DNS by Le et al., 1997), Fig. 2-left. The model captured the dissipation anisotropy in good agreement with the DNS data. The results obtained with the Rotta's isotropic model ($\varepsilon_{11} = \varepsilon_{22} = \varepsilon_{33} = 2\varepsilon/3$; $\varepsilon_{12} = \varepsilon_{13} = \varepsilon_{23} = 0$) are also shown for

[2] with mean vorticity tensor $\widetilde{W}_{ij} = 0.5(\partial\widetilde{U}_i/\partial x_j - \partial\widetilde{U}_j/\partial x_i)$, Reynolds stress and stress dissipation anisotropy tensors $a_{ij} = \widetilde{u_i''u_j''}/\widetilde{k} - 2\delta_{ij}/3$ and $e_{ij} = \varepsilon_{ij}/\varepsilon - 2\delta_{ij}/3$ and their two-componentality parameters A and E ($A = 1 - 9(A_2 - A_3)/8$, $A_2 = a_{ij}a_{ji}$, $A_3 = a_{ij}a_{jk}a_{ki}$; $E = 1 - 9(E_2 - E_3)/8$, $E_2 = e_{ij}e_{ji}$, $E_3 = e_{ij}e_{jk}e_{ki}$). The linear, *isotropization-of-production* (IP, Launder et al., 1975) form of the rapid pressure-strain model $\Phi_{ij,2}^{\mathrm{IP}} = -C_2(P_{ij} - 2P_k\delta_{ij}/3)$ (with C_2 defined in (10)) was retained in the term denoting the wall influence on the rapid pressure scrambling process.

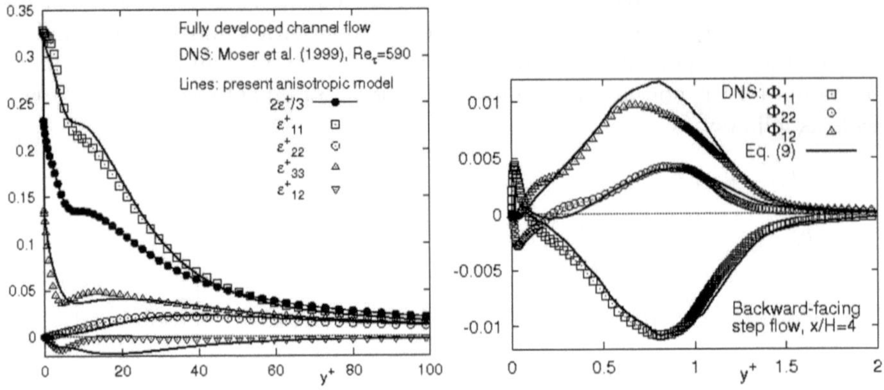

Fig. 2. Stress dissipation components obtained by the present anisotropic model (8) and pressure-strain-term components obtained by (9)–(11)

comparison. It is clearly seen, that the anisotropies of the dissipation correlation are not only pronounced in the immediate wall vicinity, but they also affect the core flow. This feature is beyond the reach of the Rotta's proposal. The effects of the non-linear additions on the redistribution among the normal stress components are unambiguous, Fig. 2-right. The wall reflexion term model was made wall-normal free by introducing a unit vector pointing into the direction of the non-homogeneity of the turbulence field, in line with the proposal of Gerolymos and Vallet (2002): $n_i = \nabla l / |\nabla l|$. Hereby, a modified length scale $l = \widetilde{Ak^{3/2}}/\varepsilon$ was adopted as a parameter reflecting the turbulence non-homogeneity. As illustration, the contours of the n_i-components obtained using the DNS database of the backward-facing step flow (Le et al., 1997) are displayed in Figs. 3. The functional dependency of the coefficient C_1' in the nonlinear part of the slow term is obtained by a method for an *a priori* determination of the model coefficients, Jakirlić (2004), Fig. 4-left. The C_2' profiles obtained from the DNS database of the plane channel flow for $\mathrm{Re}_\tau = 180$ (Moser et al., 1999) using the same method[3] exhibit values between -0.2 and 0.2, missing by far the positive value in, e.g., the Speziale, Sarkar and Gatski (1991) – SSG model ($C_2' = 0.9$), whereas the evaluated profiles of the coefficients C_4 and C_5 exhibit the values agreeing well with the SSG proposal, Fig. 4-right. The appropriate value and even the sign of the coefficient C_2' remain to be clarified. Because of this uncertainty its value was set to zero in the present work, i.e. the linear model for the

[3] For each combination of three different components (e.g., $ij = 11$, $ij = 22$ and $ij = 12$; other combinations can also be used) of the rapid term, the model formulation for $\Phi_{ij,2}$ (9) can be written as a system of three equations with three unknowns C_2', C_4 and C_5 (C_3 coefficient has been taken as known, e.g., $C_3 = 0.8 - 0.65\sqrt{A_2}$ in the SSG model, Fig. 4-right). The input data for all variables (also for $\Phi_{ij,2}$) were taken from the DNS database of a fully-developed channel flow.

rapid part was applied. The wall boundary condition is based on the asymptotic behaviour of the Taylor microscale λ ($\lambda = \sqrt{5}y + \cdots$, Fig. 5-left) and its exact relationship to the dissipation rate in the immediate wall vicinity: $\varepsilon = 10\nu\widetilde{k}/\lambda^2$ (see Fig. 5-right for comparison of the dissipation rate obtained by this formulation and the DNS database in all characteristics region of a backward-facing step flow). Its linear dependency on the wall distance through the entire viscous sublayer (even up to $y/H \approx 0.04$ – H being the step height – corresponding to $y^+ = 10$ at $x/H = 19$, Fig. 5-right) enables the wall-closest grid node to be positioned immediately below the edge of the viscous sublayer, leading to a substantial coarsening of the near-wall grid resolution. The profiles of all quantities in the remainder of the cross-section follow closely the result obtained on the finest grid, as it is demonstrated on the semi-log profile of the mean axial velocity, Fig. 6-left. This approach provides bridging of a portion of the viscous sublayer, higher grid flexibility with respect to flow regions featured by different phenomena and weaker sensitivity against the grid non-uniformities in the near-wall regions. The best illustration for the model's ability to account properly for a large deviation from the equilibrium conditions is given in Fig. 6-right, where semi-log plots of the mean velocity are presented for several boundary layer flows subjected to different pressure gradients (favourable and adverse) featuring the flow phenomena pertinent to aircraft aerodynamics. Another interesting model feature accounts separately for the wall effects on anisotropy of stress bearing and dissipative scales by introducing both the turbulent stress and dissipation rate anisotropy, in addition to viscosity effects taken into account via $\mathrm{Re_t} = \widetilde{k}^2/(\nu\varepsilon)$. The strong difference in the anisotropy rates of both tensors (a_{ij} and e_{ij}) expressed in terms of their two-componentality factors A and E and the turbulence model capability to capture them correctly is shown in Fig. 7-left. Fig. 7-right displays also the normal Reynolds stress intensities across the zero-pressure gradient boundary layer at the location corresponding

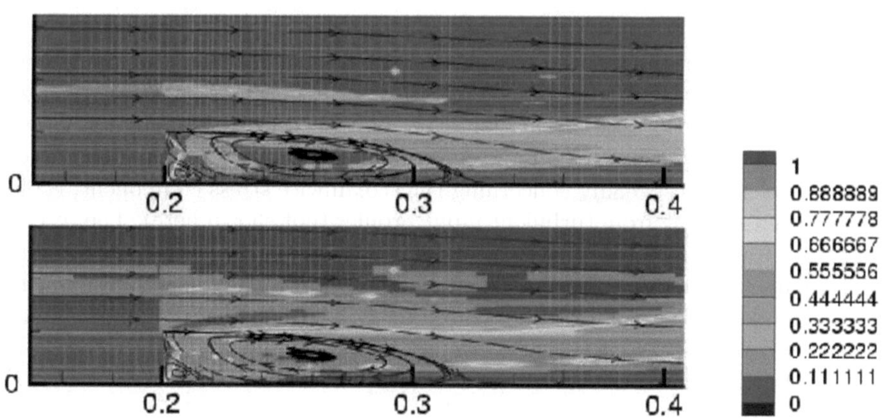

Fig. 3. Contours of the n_1 (upper) and n_2 (lower) components of the unit vector n_i in the flow over a backward-facing step

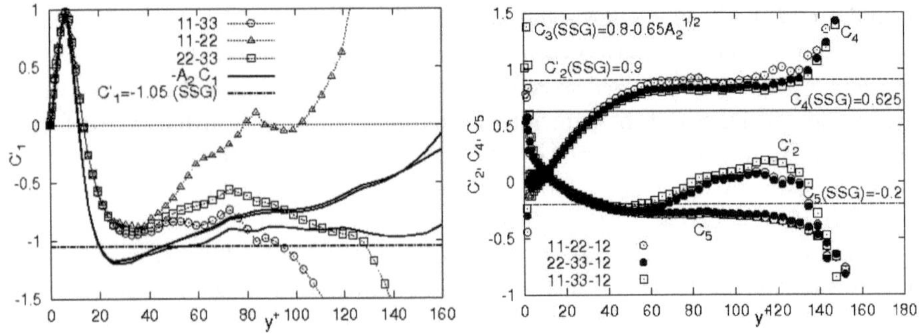

Fig. 4. Profiles of C_1', C_2', C_4 and C_5 in the fully developed channel flow obtained by application of an *a priori* method on (9) (Jakirlić, 2004)

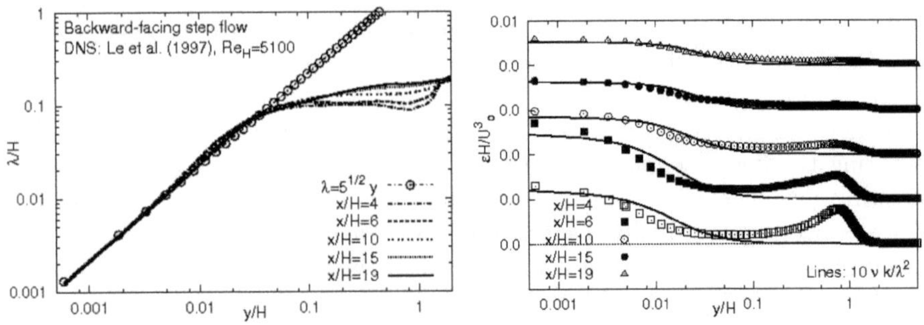

Fig. 5. Near-wall behaviour of Taylor microscale λ (left) and dissipation rate profiles at selected locations in the backward-facing step flow (right) ($x/H \in \{4, 6\}$ – recirculation zone; $x/H \in \{10, 15, 19\}$ – recovery region)

to $\mathrm{Re}_\theta = 1410$. It should be noted that this result is obtained by starting from the laminar flow in front of the flat plate and prescribing the location where an abrupt transition occurs (the source terms in the Reynolds stress equation – production, redistribution and dissipation – are multiplied by a step-function taking zero value in the laminar region and unity value in the turbulent region). Even better agreement, especially concerning the streamwise stress component, is obtained when starting from turbulent input profiles (not shown here). The second near-wall, Reynolds stress transport model used in the present work (denoted as SSG/LRR-ω throughout the manuscript) represents a default second-moment closure model in the compressible computer code DLR-FLOWer. This model scheme, proposed by Eisfeld (2006), represents a numerically robust combination of the Launder–Reece–Rodi (LRR, 1975) model (with linear formulation of the slow part) resolving the near-wall layer with the Speziale–Sarkar–Gatski (SSG, 1991) model being employed in the outer region. The coefficients in this hybrid model are weighted between the values used in the basic models by applying following blending relationship $C_i = F_1 C_i^{\mathrm{LRR}} + (1 - F_1) C_i^{\mathrm{SSG}}$, with the

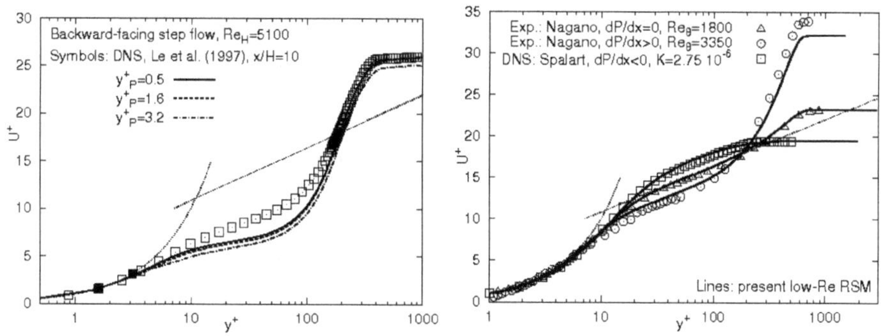

Fig. 6. Semi-log profile of the mean axial velocity at $x/H = 10$ in the backward-facing step flow (left) and in the boundary layers subjected to different pressure gradients (right)

blending function F_1 defined after Menter (1994). Menter's baseline ω-equation ($\omega = \varepsilon/(C_\mu \widetilde{k})$, $C_\mu = 0.09$) is exploited for supplying the length scale. It should be noted, that the coefficients in the LRR model arose from its ω-based, near-wall adaptation proposed by Wilcox (1998). Hereby, the wall reflection term was omitted. The Rotta's isotropic model was adopted for the dissipation correlation: $\varepsilon_{ij} = 2\varepsilon\delta_{ij}/3$ (i.e. $\varepsilon_{ij} = 2C_\mu\widetilde{k}\omega\delta_{ij}/3$). The model equation governing the specific time scale ω was formulated in line with the proposition due to Menter (1994). Because of the constant values of the model coefficients in the pressure redistribution model (C_i^{LRR} and C_i^{SSG}), its slow part doesn't vanish at the wall for all components. It means that the two-componentality parameter A doesn't approach the zero value at the wall, reflecting the model's non-capability to fully resolve the stress anisotropy in the thin, immediate wall vicinity. Beyond this region, characterized by a strong peak of the streamwise stress component, the Reynolds stress intensities display good agreement with the DNS results, Fig. 7-right. Despite this deviation the model is capable of correctly capturing the pressure redistribution in flows affected by shock/boundary-layer interaction and related phenomena, as it will be illustrated in Sect. 3. The readers are referred to the works of Eisfeld (2006) and Jakirlić et al. (2007) for detailed specification of the model.

In addition, all cases considered were computed by two popular eddy-viscosity-based models being extensively used for the aerodynamics applications: the Spalart–Allmaras eddy-viscosity transport model (1994) and the Wilcox's (1988) k-ω model.

2.2 Numerical Method

All computations were performed by using the DLR FLOWer code, which is well established in the academic research and aeronautical industry. The DLR FLOWer code is density-based, solving the coupled system of conservation equations for

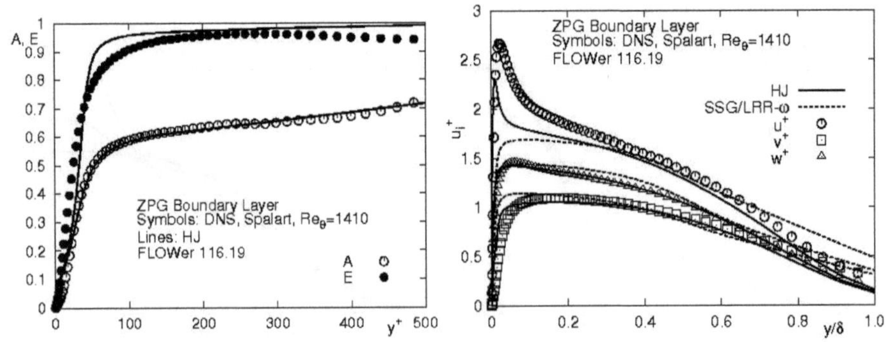

Fig. 7. Two-componentality parameters A and E (left) and Reynolds stress components (right) in a zero-pressure gradient boundary layer

mass, momentum, total energy and turbulent quantities in their time-dependent, integral formulation. It employs a variety of turbulence models, ranging from eddy viscosity models (Baldwin–Lomax, Spalart–Allmaras type, k-ω type) to full Reynolds stress models (Wilcox $\widetilde{u_i''u_j''} - \omega$, SSG/LRR-$\omega$), which have been validated in a large number of test cases, ranging from baseline to complete aircraft flow configurations, Kroll et al. (2002), Eisfeld (2006), Eisfeld and Brodersen (2005). The FLOWer code uses a cell-centered, finite-volume discretization method for block-structured meshes (Kroll et al., 1995). For the RANS equations different spatial discretization schemes are available, where the most common one is a second-order central scheme with artificial dissipation. The mean flow equations are integrated in time by an explicit five-stage hybrid Runge–Kutta scheme, which is accelerated by local time stepping, an implicit smoothing of the residuals and a multi-grid algorithm (Jameson et al., 1981). According to a detailed analysis of Fassbender (2004), the equations for turbulence quantities are integrated by an implicit scheme on the finest grid level only, where the source terms are linearized in time, which has been shown to be a highly efficient and robust approach. The readers are referred to the works of Kroll et al. (1995) and Fassbender (2004) for more specific details about the numerical method used.

3 Results and Discussion

The performance of the turbulence models presented are illustrated by computing three transonic flow cases: two-dimensional flow over the RAE2822 airfoil (Cook et al., 1979), three-dimensional flow past the ONERA M6 wing (Schmitt and Charpin, 1979) and the flow over the DLR-ALVAST wing-body configuration representing a generic transport aircraft model, Burgsmüller and Hoheisel (2000). Selected results concerning the wall pressure distribution and mean velocity and turbulence fields are displayed and discussed in the following subsections.

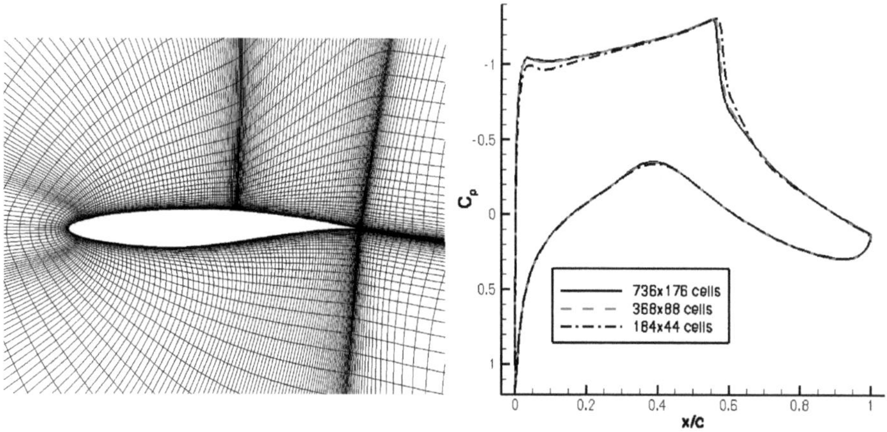

Fig. 8. Blow-up of the C-type grid (736×176 cells) used for the computation of the flow past RAE2822 airfoils (every second grid line is shown) and effect of grid refinement on pressure distribution (the SSG/LRR-ω model was used)

3.1 RAE2822

The first example is the transonic flow around the RAE2822 airfoil investigated experimentally by Cook et al. (1979). A fairly strong shock is created near the mid chord ($x/c = 0.55$). Two cases, denoted by Case 9 and 10, characterized by $\text{Ma}_\infty = 0.73$, $\text{Re}_\infty = 6.5 \times 10^6$ and $\text{Ma}_\infty = 0.75$, $\text{Re}_\infty = 6.2 \times 10^6$ respectively, were considered. Both airfoil configurations are at incidence angle of $\alpha = 2.8°$. Important difference between the two flows is a thin, post-shock separation occurring in the case 10. Two-dimensional calculations were performed on a C-type grid, Fig. 8-left. Although there is some doubt regarding the two-dimensionality of the experiments (see e.g. Haase et al., 2006), turbulence models that are capable to predict both cases reasonably well usually perform also well for complex industrial applications in the transonic flow regime. Three grid resolutions were used in the framework of the grid dependence study: coarse (consisting of 184×44 grid cells), medium (368×88 grid cells) and fine (736×176 grid cells). It was checked using the SSG/LRR-ω model that the pressure distributions at the finest and second finest grid level are almost identical (see Fig. 8-right). Finally, the finest grid, being regarded to be fine enough to ensure grid converged solutions, was adopted for the computations. The solution domain is extended to 20 chord lengths in all directions. At the solution domain boundaries the far-field boundary conditions based on the theory of characteristics were applied. Transition to turbulence in the experiment was induced by tripping the flow near the leading edge at $x/c = 0.03$ on both upper and lower surfaces of the airfoil. The computational treatment of this transition location was accounted for by multiplying the entire source term (production, redistribution and dissipation) in the equations governing the Reynolds stresses and dissipation rate by an appropriate step function, providing its zero value in the laminar flow part and a

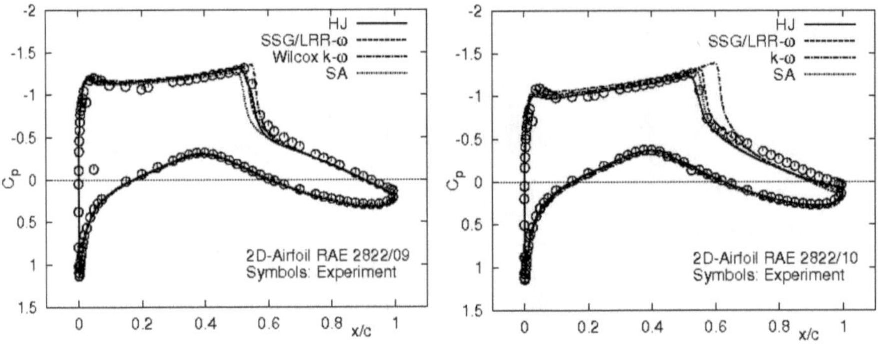

Fig. 9. Pressure coefficient distribution for RAE2822 airfoil: case 9 (left) and case 10 (right)

unit value in the fully turbulent flow region. Furthermore, it should be noted that the corrections of the computational results obtained under free-flight conditions with respect to the wind tunnel confinement (experimental conditions) were introduced according to the proposal of Haase et al. (1993).

Figure 9 shows the comparison of the computed pressure coefficient distribution with available experimental data for both flow configurations considered. Continuous flow acceleration causes considerable pressure rise in the supersonic flow part which ends up in a shock appearance after reaching an appropriate level. The overall agreement between Reynolds-stress model results and experiment is satisfactory for both configurations in contrast to the results obtained by SA and k-ω models. Whereas the SA model application resulted in a premature shock in the case 9 and correct shock position in the case 10, the k-ω model results indicate a shock positioned too far downstream in the case 10, while returning a correct shock location in the case 9. A slight deviation with respect to the maximum pressure level on the suction side immediately after the leading edge of airfoil (Fig. 9-right) in the case 10 was obtained by both second-moment closures. The position of the pressure peak coincides with the onset of the turbulent region immediately after transition ($x/c = 0.03$). A fairly crude transition treatment can certainly be the reason for this departure, since the process of the Reynolds stress component generation pertinent to second-moment closure models is especially sensitive to. A slight discrepancy with respect to the premature shock location obtained by the HJ model in the case 9 is noticeable in the mean velocity field, Fig. 10-left. The velocity profile at the location E corresponding to the shock position indicates a somewhat stronger momentum loss. Interestingly, these circumstances don't influence the correct capturing of the flow recovery in the post-shock region. The evolution of the mean velocity field indicates clearly the boundary layer thickening starting at the shock foot, being further enhanced by the adverse pressure gradient effects corresponding to the destabilized wall curvature, i.e. to the continuous cross-section expansion towards the trailing edge of the airfoil. This predicted behaviour is in good agreement with

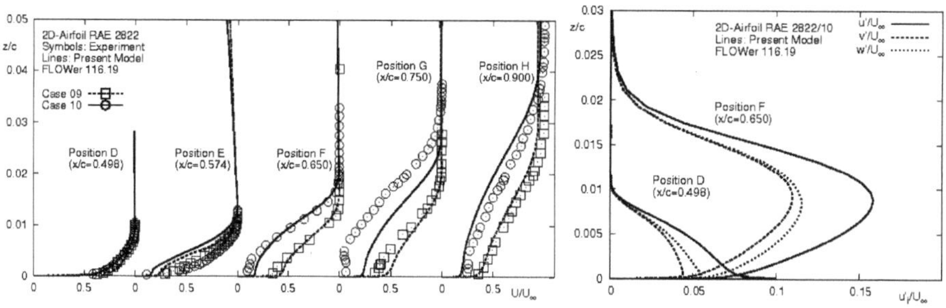

Fig. 10. Mean axial velocity profile evolution in the transonic flow over the upper RAE2822-profile surface for the cases 9 and 10 (left) and RMS of the normal components of the Reynolds stress tensor at two selected locations at the RAE2822/10-profile before and after the shock (right)

experiment. Both the pre-shock suction pressure and the velocity profile immediately after the shock wave (location F) were captured well in the case 10. However, important discrepancies concerning the post-shock region with respect to the local, low-intensity flow reversal situated around the location G were revealed. It should be recalled here that the experimentally obtained velocity field doesn't indicate the appearance of flow separation. In some previous computational studies, e.g. Bardina et al. (1997), a slightly higher angle of attack (up to 3.10) was imposed in order to accommodate the experimental conditions more appropriately. It should also be noted that the present near-wall, Reynolds-stress model due to Hanjalić and Jakirlić has never been tested before in the flows influenced by compressibility. Furthermore, the model version used here doesn't comprise the two additional terms in the dissipation equation which improve its sensitivity against adverse pressure gradient effects with respect to the intensified normal straining and excessive length-scale increase (see e.g., the works of Jakirlić et al., 2002 and Apsley and Leschziner, 1999). The analysis of the model with respect to the compressibility effects is in progress. The boundary layer structure is strongly influenced by the sudden pressure increase. Figure 10-right shows the profiles of the normal Reynolds stress components at two streamwise locations corresponding to the positions D and F, with the shock wave occurring in between. The effects of the increasingly varying adverse pressure gradient in the region of shock are demonstrated through the intensive turbulence production (all stress components experienced strong increases) due to strongly enhanced irrotational straining with respect to the rapid flow deceleration.

3.2 ONERA M6 Wing

Flow over the ONERA M6 semi-span wing (Schmitt and Charpin, 1979) is calculated at $Ma_\infty = 0.84$, $Re_\infty = 11.72 \times 10^6$ and $\alpha = 3.06°$ angle of incidence, Fig. 11-left. The fully-turbulent calculations were performed at the C-O-type grid comprising $240 \times 64 \times 52$ cells, Fig. 11-right. This grid was used as the

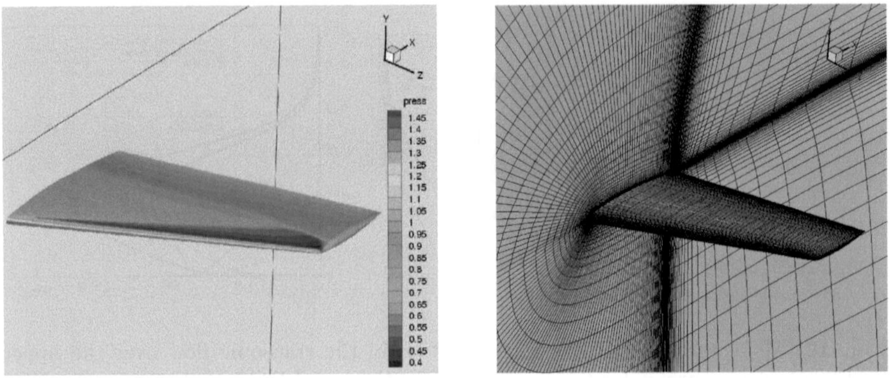

Fig. 11. ONERA M6 wing: wall pressure contours and computational grid used (240 × 64 × 52 cells)

standard structured grid for a cross comparison in the framework of the EU project FLOMANIA (Haase et al., 2006). The grid is sufficiently fine to resolve the immediate wall vicinity. The averaged y^+ value of the wall-nearest computational nodes over the entire wing surface corresponds approximately to 1. The solution domain was extended to 7.5 chord lengths in the x and y directions and 9.5 chord lengths in the z direction. Similar as in the previous case, the far-field boundary conditions were applied at the solution domain boundaries. The symmetry boundary conditions were imposed at the x-y ($z = 0$) plane, Fig. 11-right. The wall pressure contours displayed in Fig. 11-left as well as the Mach-number contours (Fig. 12) and the pressure coefficient distribution (Fig. 13) at selected spanwise locations illustrate the region of alternating favourable and adverse pressure gradients causing local acceleration and

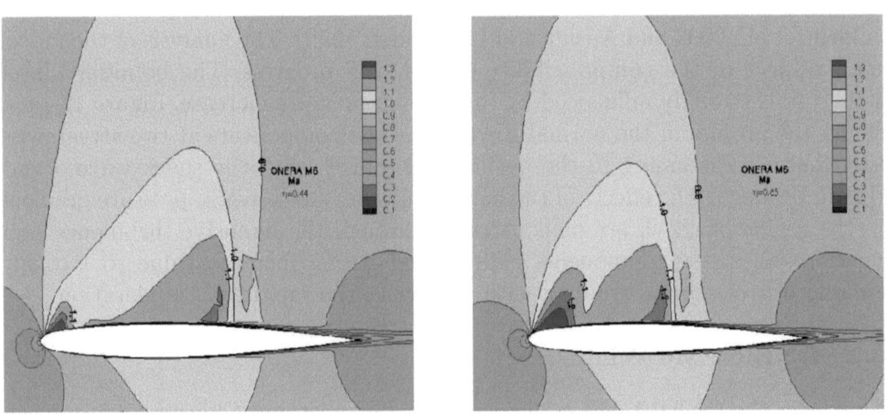

Fig. 12. Mach-number contours in the transonic flow around ONERA M6 wing at 44% and 65% span

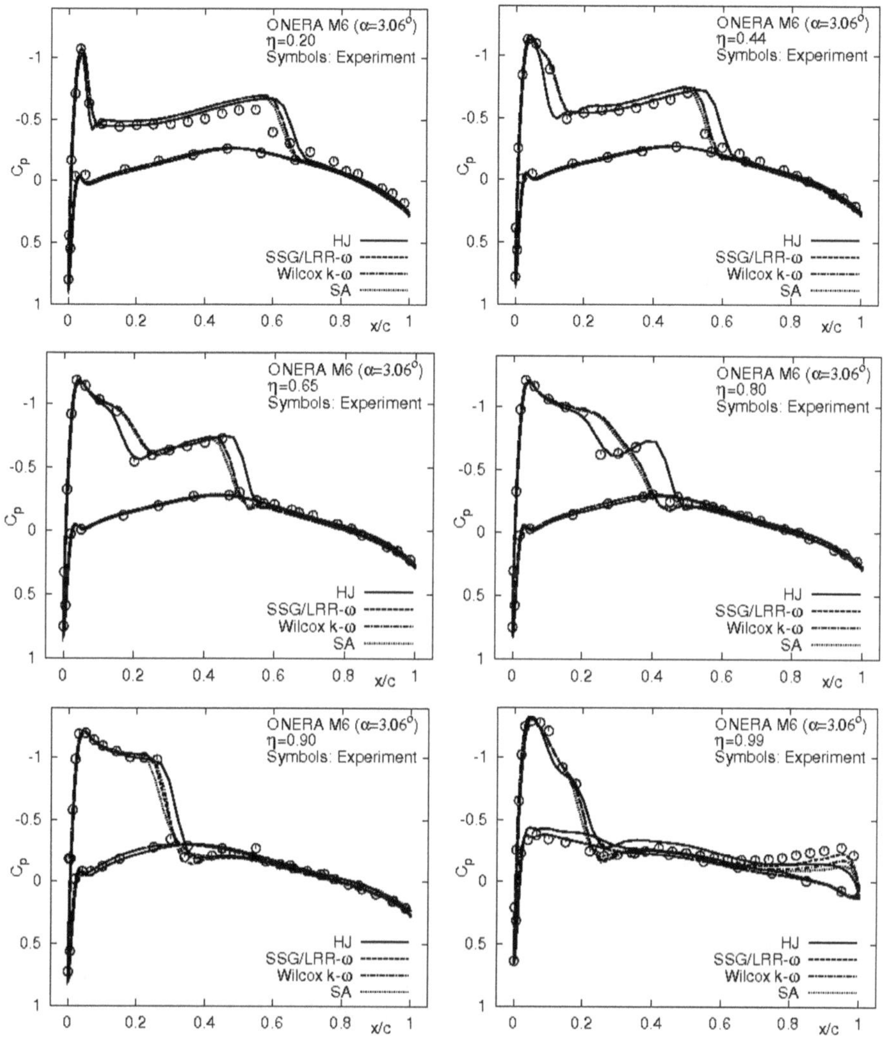

Fig. 13. ONERA M6 wing: pressure coefficient distribution at selected spanwise location

deceleration within the supersonic part of the flow. The consequence of such flow behaviour is its double-shock-like structure representing a double expansion into the supersonic regime. The flow experiences a rapid acceleration in the leading edge region of the wing ($x/c \in [0.1, 0.2]$) up to Ma $\approx$ 1.4, along with the steep rise of the pressure level, Figs. 12 and 13. A prompt deceleration of the flow down to Ma $\approx$ 1.1 follows creating consequently a weak shock, whose front propagates within the supersonic region towards the second shock of considerably increased strength. The second shock changes its position depending on the span location. The gradual coalescence of the shocks occurs in the final 15%–20% of the wing span in the wing-tip region, Fig. 13. The varying positions of both shocks with respect to the spanwise location as well as the wall pressure magnitude over the entire wing surface are reasonably well captured by all models. The pressure coefficient distribution illustrates the equality of the wall pressure values at both wing surfaces over the whole wing span corresponding to the second half of the chord. The only exception is the last spanwise position situated directly at the wing tip, $\eta = 0.99$. Here, the influence of the tip vortex created by the flow from the lower wing surface towards the upper wing surface around the tip is clearly visible in the pressure distribution. This feature is reasonably well captured by the present Reynolds stress transport models. All model predictions are of comparable quality, the only significant difference is documented at the location corresponding to 80% of the span. Here, the HJ model returned the double-shock structure in correct agreement with experimental results, whereas the other models applied predicted a premature coalescence of two shock fronts. While the position and strength of the first shock at the spanwise positions $\eta = 0.44$, 0.65 and 0.80 obtained by applying the HJ model agree well with the experimental data, the predicted position of the second shock exhibits a certain delay compared to the measurements and other model results. The results obtained by LRR/SSG and eddy-viscosity models (positions $\eta = 0.44$ and 0.65) display correct onset of the first shock. However, the course of its front (milder slope), whose foot is shifted downstream, indicates the pressure relaxation of a somewhat lower intensity compared to a steeper pressure gradient obtained by the HJ model. The final outcome of such a situation reflects a gradual shortening of the region between two shocks (characterized by a mild pressure increase) until their premature merging at $\eta = 0.80$.

3.3 DLR-ALVAST

The DLR-ALVAST is a generic transport aircraft model that can be equipped with turbine-driven engine simulators (Kiock, 1996). It has been used, e.g. in the EU-project ENIFAIR, for studying engine interference effects (Burgsmüller and Hoheisel, 2000). In the present work numerical simulations have been performed for the clean wing configuration at a free stream Mach number of $\mathrm{Ma}_\infty = 0.75$ and a Reynolds number of $\mathrm{Re} = 4.3 \times 10^6$ based on the mean chord length. The incidence has been varied to achieve a lift coefficient of $C_\mathrm{L} = 0.5$, which is the design point.

The computations with the SSG/LRR-ω model have been carried out on a grid consisting of 4.1×10^6 cells (this grid was adopted from the study of Fassbender, 2004). Particular care has been taken in the grid generation process, focussing especially on properly resolving the boundary layer over the entire aircraft surface (with 20–24 grid cells accommodated in the normal-to-the-wall direction). The near-wall grid lines have a wall distance well below $y^+ = 1$ over the entire geometry, except a small portion at the fuselage tail. The artificial dissipation coefficients have been set to $k^{(2)} = 1/2$ for first-order accuracy at shocks and to $k^{(4)} = 1/32$ for second order in regions where the solution is smooth. The

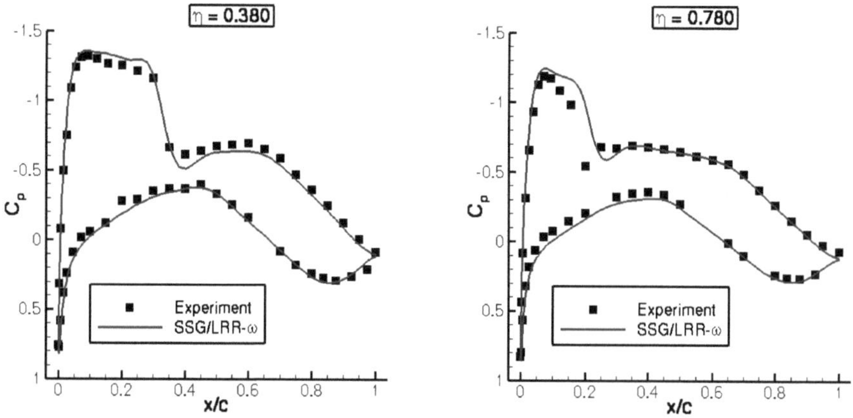

Fig. 14. DLR-ALVAST generic aircraft: experimental and numerical pressure distribution at 38% and 78% span

relatively high value of the latter coefficient has been necessary in order to damp oscillations that are generated by a separation zone at the wing-body junction. The settings of these coefficients influence mainly the strength of the shock (not the position) with respect to the pressure difference at the shock itself (immediately before and after the shock) in the streamwise direction (the higher the values of the dissipation coefficients, the weaker, i.e. the more smeared out, the shock), as it was investigated by Kroll and Jain (1987). Nevertheless a rather low value of the Martinelli coefficient (Martinelli and Jameson, 1988) of $\zeta = 0.3$ could have been used, avoiding excessive artificial dissipation in highly stretched cells. The computations with the HJ model experienced some stability problems, whose debugging is in progress. On the other hand, the flow structure over the DLR-ALVAST wing is characterized by a single shock similar to the RAE2822 cases. Accordingly, it was not expected that the HJ model results would have been substantially different compared to those obtained with the LRR/SSG-ω model. In the following only the latter results will be presented. Figure 14 shows the pressure distributions in two spanwise wing sections obtained with the SSG/LRR-ω Second-Moment Closure in comparison with experimental data from the EU-project ENIFAIR in the ONERA S1MA wind tunnel (Burgsmüller and Hoheisel, 2000). While the overall agreement is fairly good, some deviations occur with respect to the shock position. It is observed that in the inboard part of the wing the shock is predicted upstream of the measured position, while outboard it is predicted downstream of the measured position towards the wing tip. These deviations can to some extent also be attributed to the fact that the aircraft model (made of plastics) heated up, hence deformed during the experiments (see e.g., the work of Hoheisel, 1999).

3.4 Numerical Issues

The following statements can be made with respect to the numerical robustness of the Reynolds stress models. The SSG/LRR-ω model requires about 100% more CPU time per iteration compared to the Wilcox k-ω model. Generally, the number of iterations needed depends on the flow problem and is not necessarily higher when using a Reynolds stress model. In the case of the ONERA M6 wing 1500 iterations have been used with the Wilcox k-ω model and 2500 with the SSG/LRR-ω model. Hereby, the first 700 iterations have been performed on coarse grids (full multi-grid). The iteration number for the HJ model is somewhat higher, but only with respect to the generation of the initial fields of the turbulent quantities in the immediate wall vicinity. These circumstances relate to the initial field (the HJ model computations started from the LRR/SSG-ω model results) adjustment to the HJ model equations.

4 Results and Discussion

The performance of a near-wall, Reynolds-stress turbulence model accounting separately for the viscous effects and the effects of Reynolds-stress and dissipation anisotropy are investigated under the transonic flow conditions characterized by the shock formation and strong shock/boundary layer interaction in several two-dimensional and three-dimensional flow configurations. The overall agreement with the available experimental data for the surface pressure distribution with respect to the shock position is satisfactory in all flow configurations computed. A double-shock-like structure of the flow over the ONERA-M6 wing was captured, in good agreement with the experimental findings. Important departures are revealed concerning the shock-induced separation at the RAE2822/10 airfoil. Despite a correctly predicted shock position, the departure from the experimentally obtained post-shock pressure level caused the flow to remain attached. Further analysis of these phenomena is necessary, especially with respect to the specific model features that require consideration concerning the compressibility effects and the conditions of strong property variation. Computations were also performed by using another near-wall, Reynolds-stress transport closure with constant model coefficients (LRR/SSG-ω) as well as with two widely used eddy-viscosity-based transport model schemes, one-equation ν_t-transport model (Spalart and Allmaras, 1994) and the k-ω due to Wilcox (1988). With the exception of the premature shock waves coalescence in the ONERA-M6 case, the LRR/SSG-ω model predictions follow reasonably the HJ model results with respect to the pressure distribution in all configurations considered, despite its inferiority in capturing the near-wall Reynolds stress anisotropy.

Acknowledgments

The present work was performed in the framework of the German project MEGADESIGN dealing with the aerodynamic simulation and optimisation in the aircraft design. The financial support of the German Ministry for Economy and Technology (BMWi) through the grant 20A0302K for R. Jester-Zürker is gratefully acknowledged.

References

1. Apsley, D.D., Leschziner, M.A.: Advanced Turbulence Modelling of Separated Flow in a Diffuser. Flow, Turbulence and Combustion 63, 81–112 (1999)
2. Barakos, G., Drikakis, D.: Numerical simulation of transonic buffet using various turbulence closures. Int. J. Heat and Fluid Flow 21, 620–626 (2000)
3. Bardina, J.E., Huang, P.G., Coakley, T.J.: Turbulence modelling validation, testing and development. NASA Technical Memorandum 110446 (1997)
4. Batten, P., Craft, T.J., Leschziner, M.A., Loyau, H.: Reynolds-Stress-Transport Modelling for Compressible Aerodynamics Application. AIAA Journal 37(7), 785–797 (1999)
5. Burgsmüller, W., Hoheisel, H.: ENIFAIR – EU Research into Engine Integration on Future Transport Aircraft. Air and Space Europe 2(2), 81–85 (2000)
6. Cook, P.H., McDonald, M.A., Firmin, M.C.P.: Aerofoil RAE 2822 – Pressure Distributions, and Boundary Layer and Wake Measurements: Experimental Data Base for Computer Program Assessment. Report of the Fluid Dynamics Panel Working Group 04, AGARD AR 138 (1979)
7. Eisfeld, B., Brodersen, O.: Advanced Turbulence Modelling and Stress Analysis for the DLR-F6 Configuration. AIAA Paper 2005-4727 (2005)
8. Eisfeld, B.: Numerical simulation of aerodynamic problems with a Reynolds stress turbulence model. In: Rath, et al. (eds.). Notes on Numerical Fluid Mechanics, vol. 92, pp. 413–421. Springer, Heidelberg (2006)
9. Fassbender, J.K.: Improved Robustness for Numerical Simulation of Turbulent Flows around Civil Transport Aircraft at Flight Reynolds Numbers. DLR-Forschungsbericht 2003-09 (2004)
10. Gibson, M.M., Launder, B.E.: Grounds Effects on Pressure Fluctuations in the Atmospheric Boundary Layer. J. Fluid Mech. 86, 491–511 (1978)
11. Gerolymos, G.A., Vallet, I.: Wall-normal-free near-wall Reynolds-stress model for three-dimensional turbomachinery flows. AIAA Journal 40(2), 199–208 (2002)
12. Hanjalić, K., Jakirlić, S.: Contribution towards the Second-Moment Closure Modelling of Separating Turbulent Flows. Computers and Fluids 22(2), 137–156 (1998)
13. Hanjalić, K., Jakirlić, S.: Second-Moment Turbulence Closure Modelling. In: Launder, B.E., Sandham, N.H. (eds.) Closure Strategies for Turbulent and Transitional Flows, pp. 47–101. Cambridge University Press, Cambridge (2002)
14. Haase, W., Brandsma, F., Elsholz, E., Leschziner, M.A., Schwamborn, D.: EUROVAL – A European Initiative on Validation CFD Codes. Notes on Numerical Fluid Mechanics, vol. 42. Vieweg Verlag, Braunschweig (1993)
15. Haase, W., Aupoix, B., Bunge, U., Schwamborn, D.: FLOMANIA – A European Initiative on Flow Physics Modelling. Notes on Numerical Fluid Mechanics and Multidisciplinary Design, vol. 94. Springer, Heidelberg (2006)
16. Hoheisel, H.: Data Evaluation of the ENIFAIR High Speed Tests. DLR-Report, DLR-IB 129-99/11 (1999)
17. Jakirlić, S., Hanjalić, K.: A Second-Moment Closure for Non-Equilibrium and Separating High- and Low-Re-Number Flows. In: Proc. 10th Symposium on Turbulent Shear Flows, The Pennsylvania State University, USA, August 14-16 (1995)

18. Jakirlić, S., Hadzić, I., Djugum, A., Tropea, C.: Boundary-Layer Separation Computed by Second-Moment Closure Models. In: Wagner, et al. (eds.). Notes on Numerical Fluid Mechanics, vol. 77, pp. 215–222. Springer, Heidelberg (2002)
19. Jakirlić, S.: DNS-based scrutiny of RANS-approaches and their potential for predicting turbulent flows. Habilitation Thesis, Darmstadt University of Technology (2004)
20. Jakirlić, S., Eisfeld, B., Jester-Zürker, R., Kroll, N.: Near-wall, Reynolds-stress model calculations of transonic flow configurations relevant to aircraft aerodynamics. Int. J. Heat and Fluid Flow 28(4), 602–615 (2007)
21. Jameson, A., Schmidt, W., Turkel, E.: Numerical solutions of the Euler equations by finite-volume methods using Runge–Kutta time-stepping schemes. AIAA Paper 81-1259 (1981)
22. Kiock, R.: The ALVAST Model of DLR. DLR-Institutsbericht 129-96/22 (1996)
23. Kroll, N., Jain, R.K.: Solution of Two-Dimensional Euler Equations – Experience with a Finite-Volume Code. DFVLR-Forschungsbericht, DFVLR-FB 87-41 (1987)
24. Kroll, N., Radespiel, R., Rossow, C.-C.: Accurate and Efficient Flow Solvers for 3D Applications on Structured Meshes. Special Course on Parallel Computing in CFD. AGARD-Report R-807, ch. 4 (1995)
25. Kroll, N., Rossow, C.-C., Schwamborn, D., Becker, K., Heller, G.: MEGAFLOW – A Numerical Simulation Tool for Transport Aircraft Design. ICAS Congress 2002, Paper No. 1.10.5 (2002)
26. Le, H., Moin, P., Kim, J.: Direct Numerical Simulation of Turbulent Flow over a Backward-Facing Step. J. Fluid Mech. 330, 349–374 (1997)
27. Leschziner, M.A., Batten, P., Loyau, H.: Modelling shock-affected near-wall flows with anisotropy-resolving turbulence closures. Int. J. Heat Fluid Flow 21, 239–251 (2000)
28. Martinelli, L., Jameson, A.: Validation of a Multi-grid Method for the Reynolds Averaged Equations. AIAA Paper 88-0414 (1988)
29. Menter, F.R.: Two-equation eddy viscosity turbulence models for engineering applications. AIAA Journal 32, 1598–1605 (1994)
30. Moser, R.D., Kim, J., Mansour, N.N.: Direct numerical simulation of turbulent channel flow up to $\mathrm{Re}_\tau = 590$. Physics of Fluids 11(4), 943–945 (1999)
31. Radespiel, R., Rossow, C.-C., Swanson, R.: Efficient Cell-Vertex Multi-grid Scheme for the Three-Dimensional Navier–Stokes Equations. AIAA Journal 28(8), 1464–1472 (1990)
32. Schmitt, V., Charpin, F.: Pressure Distributions on the ONERA-M6-Wing at Transonic Mach Numbers: Experimental Data Base for Computer Program Assessment. Report of the Fluid Dynamics Panel Working Group 04, AGARD AR 138 (1979)
33. Spalart, P.R., Allmaras, S.R.: A one-equation turbulence model for aerodynamic flows. La Recherche Aérospatiale 1, 5–21 (1994)
34. Speziale, C.G., Sarkar, S., Gatski, T.B.: Modelling the Pressure-Strain Correlation of Turbulence: an Invariant Dynamical Systems Approach. J. Fluid Mech. 227, 245–272 (1991)
35. Wilcox, D.C.: Reassessment of the scale-determining equation for advanced turbulence models. AIAA Journal 26, 1299–1310 (1988)
36. Wilcox, D.C.: Turbulence modelling for CFD, 2nd edn. DCW Industries, La Canada (1998)

Transition Prediction for Three-Dimensional Configurations

N. Krimmelbein[1] and R. Radespiel[2]

[1] University of Braunschweig, Institute for Fluid Mechanics, Bienroder Weg 3, 38106 Braunschweig, Germany
n.krimmelbein@tu-braunschweig.de
[2] University of Braunschweig, Institute for Fluid Mechanics, Bienroder Weg 3, 38106 Braunschweig, Germany
r.radespiel@tu-braunschweig.de

Summary. A computational method for automatic transition prediction for general three-dimensional configurations is presented. The method consists of a coupled program system including a 3D Navier–Stokes solver, a transition module, a boundary layer code and a stability code. The transition module has been adapted to be used with parallel computation to account for the high computational demand of predicting flows around three-dimensional configurations. A comprehensive investigation on general computational and parallel performance identifies the numerical effort for the transition prediction method. The procedure has been validated comparing numerical results with experiments for the flow around an inclined prolate spheroid. Feasibility studies on generic transport aircraft show the code's capability to predict transition lines on general complex geometries.

1 Introduction

Predicting and modelling the laminar-turbulent transition in Reynolds-averaged Navier–Stokes (RANS) solvers is a necessary requirement for the computation of flows around complex, three-dimensional geometries. Neglecting or using inaccurate transition locations may lead to significant errors of the predicted aerodynamic performance.

Transition from laminar to turbulent flow is a complex phenomenon and can occur through very different mechanisms depending e.g. on on-flow parameters such as Reynolds and Mach number or free-stream turbulence, on surface properties or on the detailed development of the laminar boundary layer.

Existing transition prediction methods vary in their approaches from purely empirical transition criteria over physically based stability equations that take into account non-local and non-linear effects of disturbance growth to direct numerical simulations (DNS). The state-of-the-art transition prediction method for thin aircraft boundary layers is the e^N-method, based on local, linear stability theory. e^N-methods were thoroughly calibrated and used in many applications of wing flows that are unstable due to Tollmien–Schlichting and cross flow instabilities [17, 19].

Recent advances in predicting transition for complex flows address the prediction of unsteady transition on moving airfoils [14] and the application to 2D flows

N. Kroll et al. (Eds.): MEGADESIGN and MegaOpt, NNFM 107, pp. 93–106.
springerlink.com

with laminar separation bubbles [25, 26]. Increasing focus is placed on the prediction of transition for three-dimensional boundary layers on high aspect ratio wings and high lift configurations [10, 11], flows around bodies of revolution [24] and general three-dimensional aircraft configurations [3, 8].

The present work on transition prediction addresses flows around 3D complex configurations. For automatic transition prediction in Navier–Stokes computations a coupled program system is developed and extensively tested. The transition prediction module uses the 3D Navier–Stokes solutions to apply empirical transition criteria or linear stability theory and is capable to be used with parallel computation for general three-dimensional components and configurations.

2 Description of Methods

2.1 Linear Stability Theory

The classical linear stability theory evaluates the stability of a laminar boundary layer by examining the development of small disturbances and is used for transition prediction in form of the e^N-method. The principal approach of the theory is to superimpose harmonic waves as disturbance onto the mean flow quantities, which leads to a system of second-order differential equations for the amplitude function. For incompressible media, the system can be rearranged and written as one fourth-order differential equation, the Orr–Sommerfeld equation [16].

If the system of second-order differential equations is considered, a generalized eigenvalue problem for the complex disturbance frequency ω can be formulated, which can be transformed analytically into a standard eigenvalue problem. For the temporal stability problem this eigenvalue problem is linear. The parameters of the eigenvalue problem depend on the basic flow properties (Ma, Re), on the local velocity and temperature profiles of the laminar boundary layer and on their first and second derivatives [16].

The solution of the eigenvalue problem results in local amplification rates, which are integrated along appropriate integration paths [16]. For temporal theory, the Gaster transformation [1] can be applied by assuming reasonably small amplification rates, to transform temporal growth into spatial growth. The maximum amplification direction is related to the direction of the group velocity [1] and the trajectory of the group velocity may be approximated by a streamline at the boundary layer edge [1, 15]. The exponent of the integrated amplifications yield the so called N-factor that can be used for transition prediction with the e^N-method by comparing the values of N along the integration path with experimentally determined critical N-factors.

2.2 Numerical Methods

Navier–Stokes Solver TAU

The DLR Navier–Stokes solver TAU [21, 22] is a program for the prediction of viscous and inviscid flows around general complex geometries. The solver is based on the finite-volume method and uses a dual-grid approach where the flow variables are associated with the vertices of the original grid.

TAU can be used on structured and unstructured (hybrid) grids. Generally, a semi-structured grid layer above surfaces is used to resolve boundary layers, whereas the rest of the computational domain is filled with an unstructured grid. The solver computes the fluxes with a second-order central scheme or one of various upwind schemes with linear reconstruction for second-order accuracy. Time integration is performed by either applying an explicit, multistage Runge–Kutta scheme or an implicit, lower-upper symmetric Gauss–Seidel (LU-SGS) scheme. Turbulent flows are modelled using different Spalart–Allmaras or k-ω turbulence models, for transitional flows the turbulent production terms are suppressed in the laminar flow area [22]. For convergence acceleration residual smoothing, local time stepping and a multi-grid approach can be applied. To improve the solver capability for incompressible flows, a low-Mach-number preconditioning method is implemented [22].

For parallel computations, a domain decomposition approach is used and the message passing concept using MPI [27] is applied. In a preprocessing step, the grid is divided into a certain number of subdomains using a bisection algorithm [22]. After partitioning the grid, the solver computes the flow data on a single domain per process. Data is regularly exchanged for grid points lying at the edge of the subdomains with the data of the corresponding neighbour domains.

Transition Prediction Module

For automatic transition prediction in Navier–Stokes computations a transition prediction module, implemented directly into the DLR Navier–Stokes solver TAU, has been developed [8, 10] (Fig. 1). The transition prediction module includes certain external programs, e.g. the boundary layer code Coco [18] for swept, tapered wings and the linear stability equations solver Lilo [20]. Both, Coco and Lilo are sequential programs, which are accessible from the transition module via file I/O and system calls. The transition prediction module has been developed with special focus on predicting transition for flows around general complex, three-dimensional geometries and consequently supports parallel computing.

Although the development was focussed on using linear stability theory in form of the e^N-method with amplification rates determined by the linear stability solver for the transition prediction, a series of other transition methods are implemented as well. These range from simple empirical criteria (Michel [13], Mayle [12]), empirical criteria based on stability theory (AHD, C1 [2]) to database-based e^N-methods (Stock/Degenhardt [23], Casalis/Arnal [4]).

If the e^N-method using linear stability theory is considered for transition prediction, a suitable integration path to calculate the N-factors has to be found. Since the group velocity direction coincides with the amplification direction [16], and this direction can be approximated by the direction of the flow at the boundary layer edge [8], the integration path can be approximated by an edge streamline. Note that the boundary layer edge is actually not represented by a single edge (or external) streamline. Referring to edge streamline means "the local projections of the loci, where the streamlines cross the boundary layer edge" [5]. Generally, all available transition criteria of the transition prediction module are applied along edge streamlines using a "locally monoclinic orthogonal external streamline oriented coordinate system" [5]. The edge streamlines are either integrated from the boundary layer edge velocity distribution of the Navier–Stokes solution or are determined from the boundary layer code.

It can be shown, that for general transition prediction, and especially for the prediction of cross flow type transition, a high normal-to-wall mesh density is required to resolve the boundary layer adequately [8]. To overcome this constraint, a boundary layer method is in many cases an efficient alternative to a high-resolution Navier–Stokes computation. However, boundary layer methods, and in particular quasi-2D boundary layer methods, have certain limitations for complex flows, e.g. for flows around complex geometries and low aspect-ratio wings or in the presence of laminar separation bubbles. To deal with both, the use of coarse grids across boundary layers for rapid engineering applications and highly resolved boundary layers for detailed flow analysis, boundary layer data for the transition module can either be extracted directly from the Navier–Stokes solution or the first-order boundary layer code for swept, tapered wings is used. The boundary layer method is applied to line-in-flight cut-planes distributed along the wing span. The surface pressure distribution, the geometrical sweep angles and the on-flow conditions (Re, Ma, etc.) are used as main input data. Since the geometrical sweep angles are extracted automatically from the grid topology at the leading and trailing edges, the application of the boundary layer code along edge streamlines is not suitable. Generally, the streamlines do not necessarily cross the leading or the trailing edge, and hence the automatic determination of the sweep angles would be more difficult. Additionally, an alteration or detailed analysis of the pressure distribution around the attachment line to account for the use of the effective sweep angle concept is simplified by the use of line-in-flight cuts.

In a preprocessing step, right before the RANS solver iterations start, data not depending on the flow solution are processed (Fig. 1), e.g. reading the input parameters of the transition module or to determine the wall-normal lines, the line-in-flight cuts and geometrical sweep angles. During the solution process of the solver, the iteration process is interrupted in certain intervals, and the transition prediction is executed (Fig. 1). A new transition line is calculated, where required with additional calls of the external programs. New laminar and turbulent regions are created according to the new transition locations and passed to the RANS solver where the solution process continues (Fig. 1). The prediction of the transition lines itself is an iterative process. The transition prediction is continuously called from the solver, until the transition lines are converged (Fig. 1).

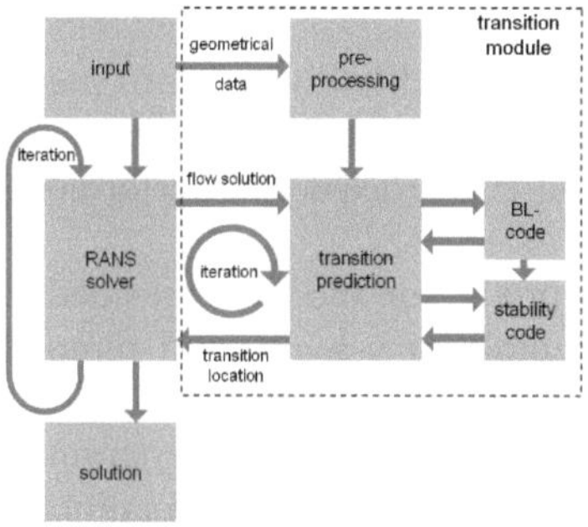

Fig. 1. Coupled program system. Navier–Stokes solver with transition module

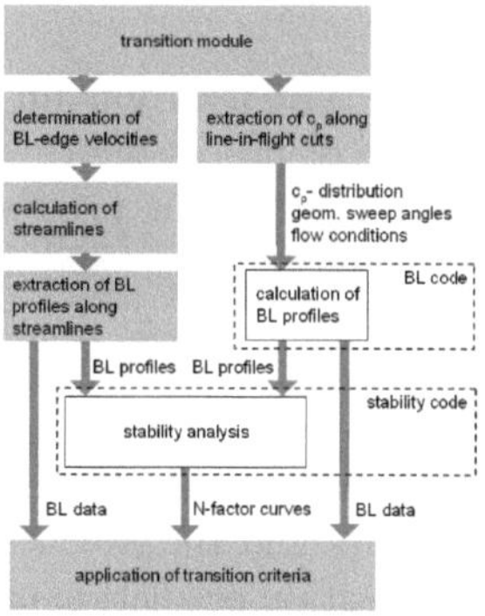

Fig. 2. Determination of new transition locations with the transition module

For each transition prediction step, the following general procedure is executed (Fig. 2). For all surface points of the geometry the boundary layer profiles are assembled along wall-normal lines, integral boundary layer data are calculated and the boundary layer edge velocity is determined. After all edge streamlines have been integrated, the velocity profiles along the streamlines are extracted and passed to the stability solver. The N-factor curves from the stability analysis are then evaluated in the transition module to give new transition locations. Alternatively, if using empirical transition criteria, the boundary layer data may be directly evaluated by the transition module (Fig. 2). The new transition lines are used to define new laminar and turbulent regions in the grid for the Navier–Stokes solver.

For the application of the boundary layer code, the line-in-flight cuts are separated in an upper and lower surface part at the corresponding attachment line and the pressure distribution is passed along with the on-flow conditions and local sweep angles to the boundary layer code. With the locally swept, tapered wing assumption, the velocity profiles along the cutting lines are calculated and are either directly transferred to the stability solver, or are evaluated by the transition module to calculate integral boundary layer data to apply empirical transition criteria (Fig. 2).

The stability analysis in form of the e^N-method yields a series of N-factor curves, which are analysed in the transition module. Limiting, respectively critical, N-factors are applied using the $2N$-factor strategy, treating the N-factors for Tollmien–Schlichting and cross flow instabilities generally independently. The interaction of Tollmien–Schlichting and cross flow waves can not be evaluated by linear local stability theory and is instead modelled by applying a stability diagram, where the critical N-factor of one instability form depends on the local N-factor of the other stability form [24].

3 Implementation and Parallelization Issues

The DLR TAU code uses domain decomposition for parallel computation. For a given number p of processors the computational grid is divided into p subgrids (subdomains). Each of the processors computes on one of the subgrids. A continuous communication of local data between the processes is performed using MPI [27].

Parallelization by means of the transition prediction module is needed for the determination of wall-normal lines, the assembly of velocity and temperature profiles along the normals, the calculation of edge streamlines, the assembly of line-in-flight cuts and the execution of the sequential, external programs. Data within the transition prediction module are not exchanged for grid points placed at the edge of the subdomains. Instead, either global data are communicated, i.e. information not associated with a certain grid point, or non local data (boundary layer profiles) associated with surface grid points near domain boundaries are transferred between different processes.

Calculation of geometrical data in form of lines (i.e. wall-normal lines, edge streamlines) within the transition module is effectively a ordered assembly of a list of grid points. These points are gathered, beginning at starting points (i.e. surface points, user-defined streamline starting points) and ending at geometrically or user-defined endpoints. A possible, geometrically defined limit is a domain boundary. In this case, the endpoints will be communicated to the neighbour domain where they serve as new start points for another loop of the assembly of the line.

Velocity profiles inside the boundary layer are easily accessible, if the surface point and the point associated with the end of the wall-normal line are in the same domain. In this case, only the knowledge of the point list for the wall normal has to be known, and the velocities at each wall-normal point can be interpolated from the Navier–Stokes solution. If a boundary layer profile is cut by a domain boundary, the corresponding data have to be assembled from different domains. For this case, the velocities for each wall-normal part are interpolated from the Navier–Stokes solution. All parts of the velocity profile not lying inside the domain containing the surface point of the wall normal have to be communicated from the neighbour domains.

The transition prediction module uses different external programs for the determination of boundary layer data and the stability analysis. These programs are stand-alone codes, not explicitly written for the utilization by a parallel and automated transition prediction process within a Navier–Stokes solver. They are executed during runtime of the Navier–Stokes solver by interrupting the execution of the transition prediction module by system calls and running start-up scripts.

The external programs Lilo [20] and Coco [18] are designed to process the data of a single streamline at one sequential run. An alternative approach, compared to the domain decomposition principle of the TAU code, is applied for the parallelization. For each process of the Navier–Stokes calculation one external program is executed sequentially. I.e., if the Navier–Stokes calculation is run on p processors, the external programs are started from each of the p processes independently, so that p streamlines can be processed parallel and a quasi-parallel performance of the programs is achieved. After calculating boundary layer data and streamlines by the transition module, all input files for the external programs are written and the streamlines are virtually distributed to the processes. Each process executes for its streamlines the external programs. After that, all substantial output from the external programs is passed to the transition module.

4 Results

4.1 Parallelization Performance

Different calculations have been performed to investigate the parallelization of the transition module and the effect on the computational demand. The conception of the study on the parallelization of the transition prediction module is to apply one single transition prediction step to a fully converged, transitional flow solution.

As test case, the flow around the 6:1 prolate spheroid according to [7] was chosen. The flow parameters, $\alpha = 5.0°$, $\mathrm{Re} = 6.5 \times 10^6$, $\mathrm{Ma} = 0.13$, were set to obtain both, Tollmien–Schlichting and cross flow instabilities for nearly the complete laminar section of the prolate spheroid. An evaluation of both types of instabilities on all streamlines leads to the maximum computational effort for the transition prediction.

The computations were carried out using a computer cluster equipped with AMD opteron 2.2 GHz processors and 1-gigabit-ethernet. The computational grid around the prolate spheroid has an overall number of points of 2.8 million, with a resolution of 128 grid points normal to the wall in the structured grid part. Low-Mach-number preconditioning was applied together with the implicit LU-SGS time integration scheme and a 3-w multi-grid cycle. Turbulent flow was modelled using the standard Spalart–Allmaras turbulence model.

As reference, the computational demand of the TAU code was evaluated in terms of wall clock time (*wct*, Fig. 3). The dependence of the computational time for one multi-grid iteration on the number of processors shows that the TAU solver parallels very well for the numbers of processor investigated. For the computational demand of one transition prediction step, as long as the number of processors p exceeds the number of streamlines n processed by the transition module, a good scaling is achieved (Fig. 4). It can be seen, that most of the computational effort is caused by the linear stability equations solver. Depending on the number of streamlines (6–24), the computational time for the transition prediction is approximately 20–90 times as high as for one multi-grid iteration of the solver.

If the number of domains exceeds the number of streamlines to be processed, no further benefit is gained in terms of computational time from the parallel execution of the transition module. This is due to the type of parallelization

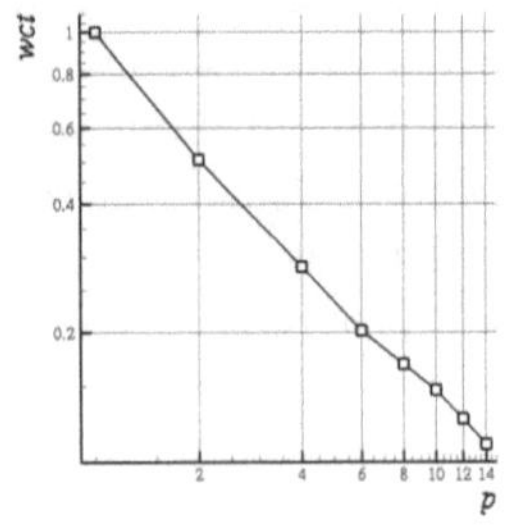

Fig. 3. Computational time *wct* vs. number of process *p*. One TAU multi-grid iteration 3-w cycle, implicit LU-SGS scheme.

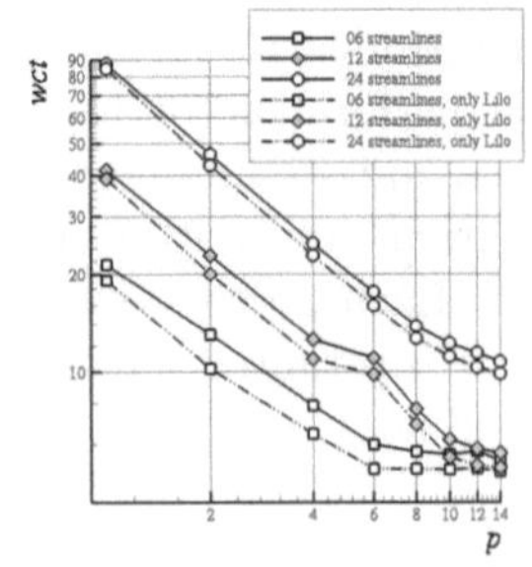

Fig. 4. Computational time *wct* vs. number of process *p*. Transition prediction step with Lilo, 6, 24 and 24 streamlines.

of the execution of the external programs, as described in Sect. 3. If normalizing the computational time needed for one transition prediction step with the corresponding computational time for one multi-grid cycle of the solver for a certain number of process p, it can be seen, that the computational effort of one transition prediction step compared to one solver multi-grid cycle is rather constant (Fig. 5). For the here investigated case, the computational demand of processing 6 (12, 24) streamlines is approximately 30 (45, 90) times as high as the computational demand of one solver iteration.

4.2 Code Validation

First validation investigations have been performed in predicting the transition for the fully three-dimensional flow around a 6:1 prolate spheroid. For this case, at certain on-flow conditions transition is characterized to change from pure

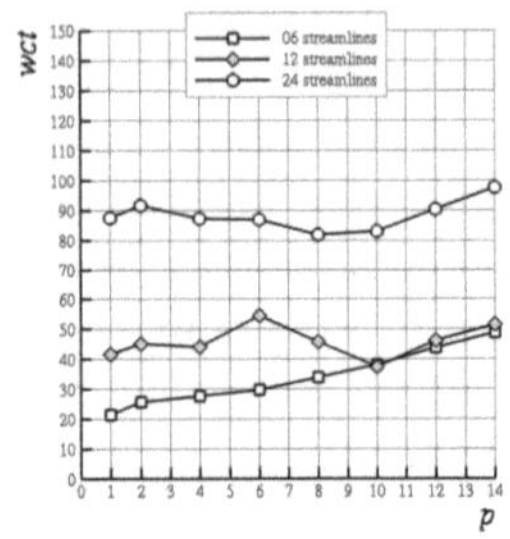

Fig. 5. Computational time *wct* vs. number of process *p*. Transition prediction step, normalized with *wct* for one multi-grid cycle of the solver.

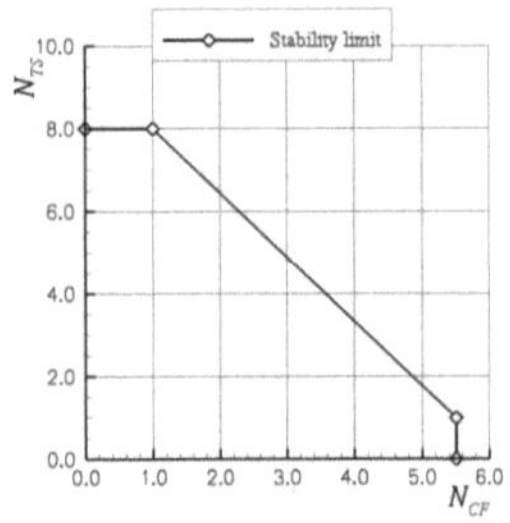

Fig. 6. Stability diagram. N_{TS} vs. N_{CF}.

Tollmien–Schlichting transition to pure cross flow transition with regions where both types of waves interact and lead to transition.

Comprehensive measurements of the flow behaviour around an inclined prolate spheroid were accomplished at the DFVLR (now DLR) 3 m × 3 m low speed wind tunnel, Göttingen. Surface hot film probes measuring the local wall shear stress were applied for the investigation of the three-dimensional boundary layer [6, 7]. 12 measuring stations in streamwies direction and 30-80 measuring stations in circumferencal direction were used. The evaluation of the local wall shear stress provides detailed information of the laminar-turbulent transition of the boundary layer.

Local, linear stability theory can not be used to analyse the interaction of Tollmien–Schlichting and cross flow waves from first principles. However, an empirical approach to overcome this deficiency is to reduce the critical N-values in the N_{TS}-N_{CF}-space for simultaneously excited Tollmien–Schlichting and cross flow waves by assuming that the critical N_{TS}-factor decays linearly with increasing N_{CF}. Numerical investigations of the transition for the flow around the inclined prolate spheroid in [24] yield the diagram of Fig. 6 which was applied during the validation calculations to account for the interaction of the two wave types.

The calculations were carried out using low-Mach-number preconditioning and the implicit LU-SGS time integration scheme. Turbulent flow was modelled using the standard Spalart–Allmaras turbulence model. In the structured grid part, a normal-to-wall resolution of 128 points was used, i.e. 60–100 points resolve the laminar boundary layer of the prolate spheroid, the overall number of grid points is 2.8 million. On-flow conditions varied from Ma $= 0.03$ to Ma $= 0.13$, $\alpha = 5.0°$ to $\alpha = 15.0°$ and Re $= 1.5 \times 10^6$ to Re $= 6.5 \times 10^6$.

For the low Reynolds number (Re $= 1.5 \times 10^6$, Ma $= 0.03$, $\alpha = 5.0°$ and $10.0°$), transition is triggered purely by Tollmien–Schlichting instabilities and is very well predicted by the transition module (Figs. 7 and 8). Although for $\alpha = 10°$ cross flow amplifications are present during the iterative transition prediction, the final, converged transition is purely caused by Tollmien–Schlichting instabilities.

Turning to the higher Reynolds number (Re $= 6.5 \times 10^6$, Ma $= 0.13$, $\alpha \in \{5.0°, 10.0°, 15.0°\}$), transition is now caused by Tollmien–Schlichting and cross flow amplifications (Figs. 9–11). Generally, for these cases the transition is caused by Tollmien–Schlichting waves near the windward and leeward symmetry lines, whereas for the remaining part of the geometry cross flow instabilities play a growing role with increasing angle of attack. For $\alpha = 5.0°$, nearly the whole transition line is represented by simultaneously excited Tollmien–Schlichting and cross flow waves. Increasing the angle of attack to $10°$ leads to a development of a region with pure cross flow transition, that is even enlarged for $\alpha = 15.0°$.

In conclusion, all transition lines for the test cases investigated are predicted in good agreement with the experimental results. For the higher Reynolds number, transition is predicted slightly too upstream, but the general qualitative shape is represented fairly well (Figs. 7–11).

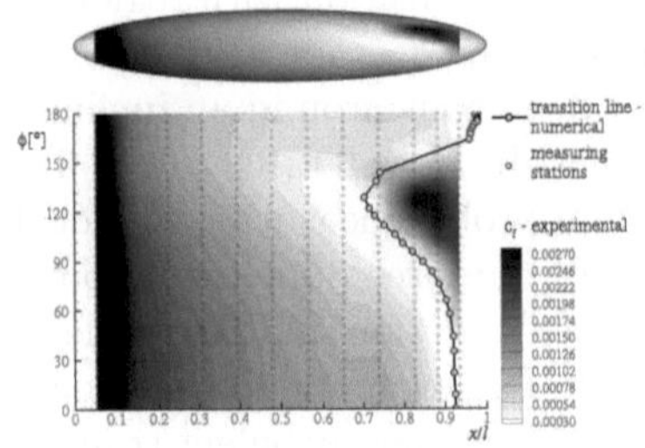

Fig. 7. Comparison of computed and measured transition location. 6:1 prolate spheroid, $\alpha = 5.0°$, $\mathrm{Re} = 1.5 \times 10^6$, $\mathrm{Ma} = 0.03$.

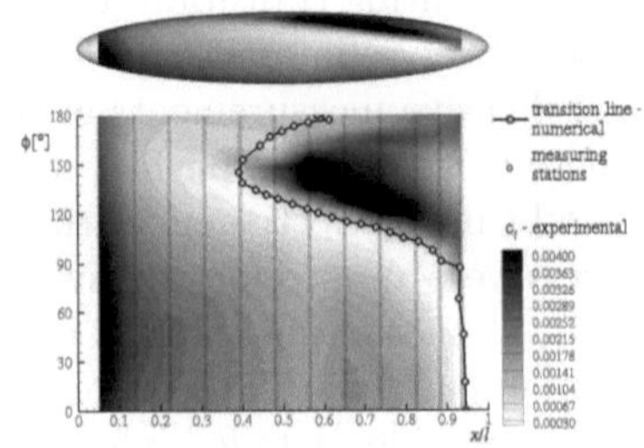

Fig. 8. Comparison of computed and measured transition location. 6:1 prolate spheroid, $\alpha = 10.0°$, $\mathrm{Re} = 1.5 \times 10^6$, $\mathrm{Ma} = 0.03$.

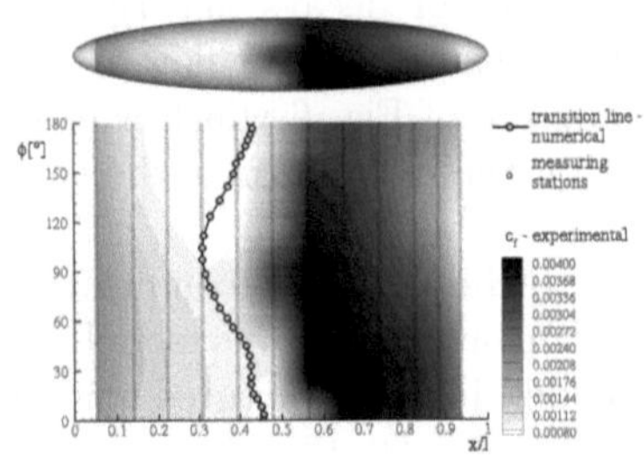

Fig. 9. Comparison of computed and measured transition location. 6:1 prolate spheroid, $\alpha = 5.0°$, $\mathrm{Re} = 6.5 \times 10^6$, $\mathrm{Ma} = 0.13$.

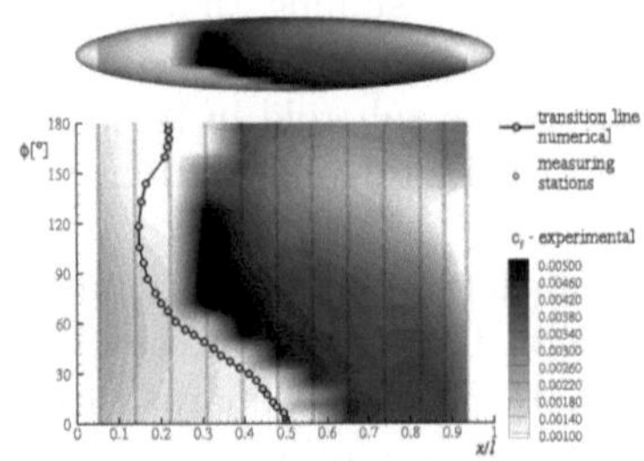

Fig. 10. Comparison of computed and measured transition location. 6:1 prolate spheroid, $\alpha = 10.0°$, $\mathrm{Re} = 6.5 \times 10^6$, $\mathrm{Ma} = 0.13$.

4.3 Feasibility Study

Generic Transport Aircraft

The first case of the feasibility study displays the flow around a generic, complex three-dimensional aircraft configuration. The objective was to predict transition simultaneously on all relevant surfaces of the configuration, i.e. body, vertical tail plane and upper and lower surfaces of main wing and horizontal tail plane. A very moderate resolution of the structured part of the grid was applied, with 32 grid points normal to the wall, except for the horizontal tail plane with 48 points normal to the wall. The overall number of grid points was 12 million. For faster convergence, low-Mach-number preconditioning was applied, together with the implicit LU-SGS time integration scheme. Turbulent flow was modelled using the standard Spalart–Allmaras turbulence model.

The flow conditions were chosen to ensure attached flow over nearly all surfaces ($\alpha = -4.0°$, $i_\mathrm{H} = 4.0°$, $\mathrm{Re} = 2.3 \times 10^6$, $\mathrm{Ma} = 0.2$). With regard to the moderate resolution of the boundary layers, only Tollmien–Schlichting instabilities

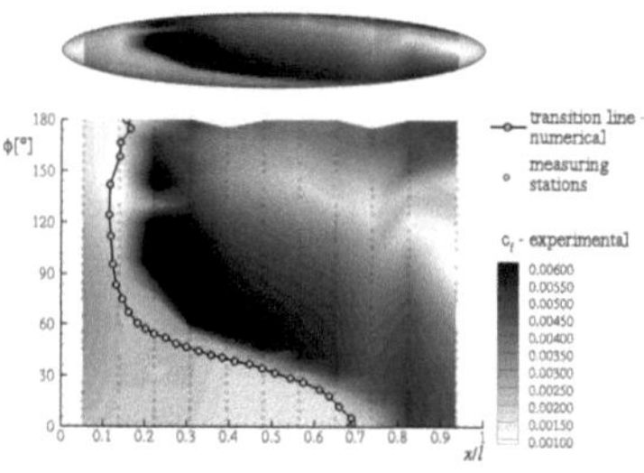

Fig. 11. Comparison of computed and measured transition location. 6:1 prolate spheroid, $\alpha = 15.0°$, $\mathrm{Re} = 6.5 \times 10^6$, $\mathrm{Ma} = 0.13$.

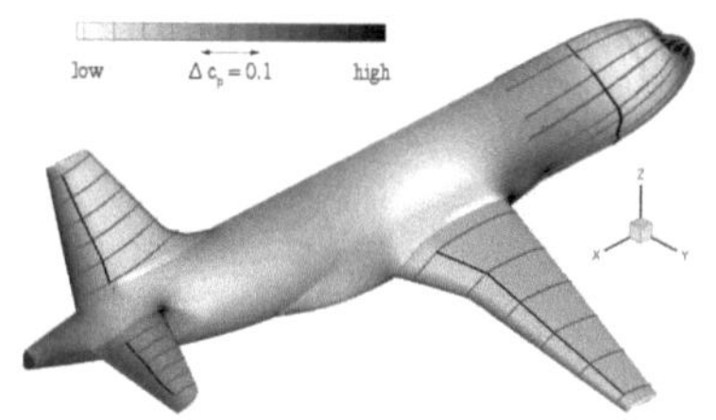

Fig. 12. Edge streamlines and transition lines. Generic transport aircraft, $\alpha = -4.0°$, $\mathrm{Re} = 2.3 \times 10^6$, $\mathrm{Ma} = 0.2$, $i_{\mathrm{H}} = 4.0°$.

were considered. On the body surface and the upper surface of the main wing two simple transition criteria were applied. On the body, a modified $c_{\mathrm{p,min}}$-criterion was used, setting the transition a short distance downstream of the pressure minimum. The main wing is equipped with a deflected flap, leading to large separated areas before transition would have been predicted by linear stability theory. Here, the laminar separation point was used as transition point instead, in order to avoid convergence problems of the Navier–Stokes iterations. For all other surfaces the laminar stability theory in form of the e^N-method was applied, with a critical N-factor of 7.5. Figure 12 shows the calculated edge streamlines together with the converged transition lines on the upper surfaces of the configuration. The predicted transition lines are located in the adverse pressure gradient region, as expected for the investigated transition scenario, and are physically plausible.

3D High Lift Configuration

The second test case of the feasibility study is the flow around a generic, three-dimensional high lift configuration. The grid for this configuration does not allow a good resolution of the laminar boundary layer, since it has only 20 points normal to the wall. Therefore this test case served as a platform to apply the boundary layer method for infinite, swept and tapered wings.

The high lift configuration is equipped with slat and flap. The overall number of grid points for this case is 8 million. Low-Mach-number preconditioning was applied, together with the implicit LU-SGS time integration scheme. Turbulent flow was modelled using the standard Spalart–Allmaras turbulence model. The flow conditions are set to match flow conditions that lead to the beginning of local separation ($\alpha = 12.0°$, $\mathrm{Ma} = 0.174$, $\mathrm{Re} = 1.34 \times 10^6$).

Transition has been predicted on all wing parts (except for the lower surface of the slat), using two different methods to determine the main flow properties for the linear stability analysis: i) extraction of the velocity profiles directly from the Navier–Stokes solution, ii) using the pressure distribution and local sweep angles

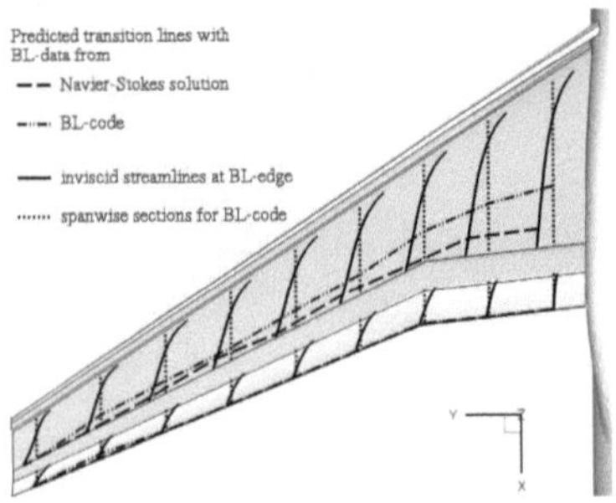

Fig. 13. Edge streamlines, line-in-flight cuts and transition lines. 3D high lift configuration, lower surfaces, $\alpha = 12.0°$, $\mathrm{Re} = 1.34 \times 10^6$, $\mathrm{Ma} = 0.174$.

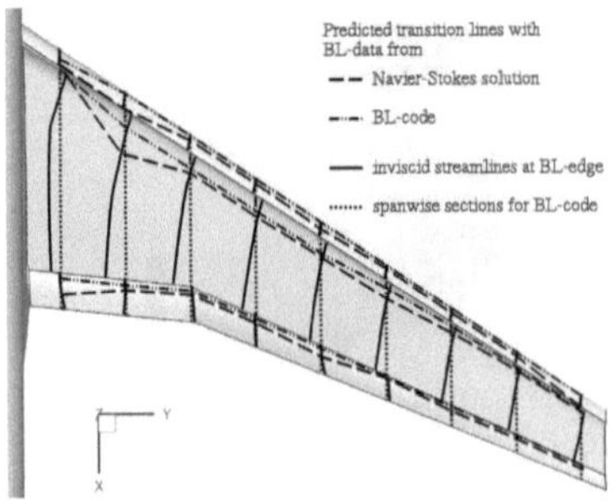

Fig. 14. Edge streamlines, line-in-flight cuts and transition lines. 3D high lift configuration, upper surfaces, $\alpha = 12.0°$, $\mathrm{Re} = 1.34 \times 10^6$, $\mathrm{Ma} = 0.174$.

from line-in-flight cuts, and applying the boundary layer code to generate the velocity profiles. At the lower surface of the deflected flap no transition was found with both prediction methods for almost the whole span. This is in accordance with the appearing pressure distribution on this element, which has a favourable gradient from the attachment line to the trailing edge. On the main wing lower surface, the predicted transition locations with the two methods differ significantly. While the predicted transition with boundary layer data from the boundary layer code is close to the local pressure minimum, the transition lines with boundary layer data directly from the Navier–Stokes solution are computed further downstream, close to the trailing edge (Fig. 13). The poor agreement between the two transition lines is caused by the insufficient resolution of the boundary layer in the Navier–Stokes solution, both in wall-normal and in streamwise directions. The same problem applies at the upper surfaces of the configuration (Fig. 14). While both prediction methods yield smooth transition lines located shortly after the local pressure minimum, the transition positions from the prediction method with the boundary layer code are well converged after a few iterations of the transition prediction module. In contrast, the transition lines from the method with boundary layer data from the Navier–Stokes solution are somewhat oscillating as the result of laminar separation before transition.

Generally speaking, the application of the transition prediction module shows the capability of this method to predict transitional flows on general 3D high lift configurations. However, validation against experimental data for 3D high lift configurations has not yet been possible due to the lack of reliable experimental data.

5 Conclusions

The general approach of the present work is to demonstrate the applicability of transition to complex configurations and fully three-dimensional boundary layers. The development of a transition prediction module attached to a

Navier–Stokes solver is presented, with emphasis on the parallelization of the transition prediction procedure for the application to domain decomposed Navier–Stokes solutions. A study on the effect of the parallelization is accomplished and detailed results on the computational demand are given.

The flow around three different configurations was investigated, where two cases served to demonstrate the feasibility of our approach, and one test case was successfully used for the validation of the transition module. It is shown, that the presented method is generally capable of predicting transition on general aircraft components, however comparisons with experimental data for 3D wing flows still need to be performed in the future.

Acknowledgements

The present work was performed as part of the German research project MEGADESIGN [9]. The authors thank Andreas Krumbein of DLR, German Aerospace Center, Göttingen, Germany, for general discussions of automatic transition prediction, testing of the transition module and his feedback as a user, Geza Schrauf of Airbus Deutschland GmbH, Bremen, Germany, for providing the codes Lilo and Coco and for guidance in their application, Hans Peter Kreplin of DLR, German Aerospace Center, Göttingen, Germany, for providing experimental data of the prolate spheroid, Thomas Gerhold of DLR, German Aerospace Center, Göttingen, Germany, for discussions of the parallelization of the TAU code and Norbert Kroll of DLR, German Aerospace Center, Braunschweig, Germany, for general support of the project.

References

1. Arnal, D.: Boundary Layer Transition: Prediction Based on Linear Theory. AGARD-R-793 (1994)
2. Arnal, D., Habiballah, M., Coustols, E.: Laminar stability theory and transition criteria in two- and three-dimensional flow. Rech. Aerosp. 2 (1984)
3. Arthur, M.T., Atkin, C.J.: Transition modelling for viscous flow prediction. In: 36th AIAA Fluid Dynamics Conference and Exhibit, San Francisco, AIAA-2006-3052 (2006)
4. Casalis, G., Arnal, D.: Database method – Development and validation of the simplified method for pure crossflow instability at low speed. ELFIN II technical report No. 145, ONERA-CERT (1996)
5. Hirschel, E.H.: Basics of Aerothermodynamics, p. 201. Springer, Heidelberg (2005)
6. Kreplin, H.-P., Meier, H.U., Maier, A.: Wind Tunnel Model and Measuring Techniques for the Investigation of Three-Dimensional Turbulent Boundary Layers. In: 10th AIAA Aerodynamic Testing Conference, San Diego, AIAA-78-781 (1978)
7. Kreplin, H.-P., Vollmers, H., Meier, H.U.: Wall Shear Stress Measurements on an Inclined Prolate Spheroid in the DFVLR 3m × 3m Low Speed Wind Tunnel, Göttingen. DFVLR-AVA, Report IB 22-84 A 33 (1985)
8. Krimmelbein, N., Radespiel, R., Nebel, C.: Numerical Aspects of Transition Prediction for Three-Dimensional Configurations. In: 35th AIAA Fluid Dynamics Conference and Exhibit, Toronto, AIAA-2005-4764 (2005)

9. Kroll, N., Becker, K., Rieger, H., Thiele, F.: Ongoing Activities in Flow Simulation and Shape Optimization within the German MEGADESIGN project. In: 25th International Congress of the Aeronautical Sciences, ICAS 2006 (2006)
10. Krumbein, A.: Automatic Transition Prediction for High-Lift Systems Using a Hybrid Flow Solver. Journal of Aircraft 42, 1362–1366 (2005)
11. Krumbein, A.: Automatic Transition Prediction and Application to 3D Wing configurations. In: 44th AIAA Aerospace Sciences Meeting and Exhibit, Reno, AIAA-2006-914 (2006)
12. Mayle, R.E.: The Role of Laminar-Turbulent Transition in Gas Turbine Engines. J. Turbomachinery 113 (1991)
13. Michel, R.: Determination du point de transition et calcul de la trainee des profiles d'ailes en incompressible. ONERA (1952)
14. Radespiel, R., Windte, J., Scholz, U.: Numerical and Experimental Flow Analysis of Moving Airfoils with Laminar Separation Bubbles. AIAA Journal 45(6), 1346–1356 (2006)
15. Schlichting, H., Gersten, K.: Boundary Layer Theory, 8th edn. Springer, Heidelberg (2000)
16. Schrauf, G.: Transition prediction using different linear stability analysis strategies. In: 12th AIAA Applied Aerodynamics Conference, Colorado Springs, AIAA-1994-1848 (1994)
17. Schrauf, G., Perraud, J., Lam, F.: Comparison of Boundary-Layer Transition Predictions Using Flight Test Data. Journal of Aircraft 35(6) (1998)
18. Schrauf, G.: COCO – A program to compurte velocity and temperature profiles for local and nonlocal stability analysis of compressible, conical boundary layers with suction. ZARM Technik Report (November 1998)
19. Schrauf, G.: Large-Scale Laminar-Flow Tests Evaluated with Linear Stability Theory. Journal of Aircraft 41(2) (2004)
20. Schrauf, G.: LILO 2.1. User's Guide and Tutorial. GSSC Technical Report 6, originally issued September 2004, modified for Version 2.1 (2006)
21. Schwamborn, D., Gerhold, T., Hannemann, V.: On the Validation of the DLR-TAU Code. In: Nitsche, W., Heinemann, H.-J., Hilbig, R. (eds.) New Results in Numerical and Experimental Fluid Mechanics. Notes on Numerical Fluid Mechanics, vol. 72, pp. 426–433. Vieweg (1999)
22. Schwamborn, D., Gerhold, T., Heinrich, R.: The DLR TAU-Code: Recent Applications in Research and Industry. In: Wesseling, P., Onate, E., Periaux, J. (eds.) European Conference on Computational Fluid Dynamics, ECCOMAS CFD (2006)
23. Stock, H.W., Degenhart, E.: A Simplified e^N Method for Transition Prediction in Two-Dimensional, Incompressible Boundary Layers. Z. Flugwiss. Weltraumforsch 13 (1989)
24. Stock, H.W.: e^N Transition Prediction in Three-Dimensional Boundary Layers on Inclined Prolate Spheroids. AIAA Journal 44(1), 108–118 (2006)
25. Windte, J., Scholz, U., Radespiel, R.: Validation of the RANS-Simulation of Laminar Separation Bubbles on Airfoils. Aerospace Science and Technology 10(6), 484–494 (2006)
26. Wokoeck, R., Krimmelbein, N., Ortmanns, J., Ciobaca, V., Radespiel, R., Krumbein, A.: RANS Simulation and Experiments on the Stall Behaviour of an Airfoil with Laminar Separation Bubbles. In: 44th AIAA Aerospace Sciences Meeting and Exhibit, Reno, AIAA-2006-244 (2006)
27. Message Passing Interface Forum. MPI: A message-passing interface standard 1.0. Technical report (1994), http://www.mpi-forum.org

Application of Transition Prediction

Andreas Krumbein

Deutsches Zentrum für Luft- und Raumfahrt e.V.,
Institut für Aerodynamik und Strömungstechnik, Numerische Verfahren,
Bunsenstraße 10, 37073 Göttingen, Germany
andreas.krumbein@dlr.de

Summary. A Reynolds-averaged Navier–Stokes solver, a laminar boundary-layer code and different transition prediction methods for the prediction of Tollmien–Schlichting and cross flow instabilities were coupled for the automatic prediction of laminar-turbulent transition on general three-dimensional aircraft configurations during the ongoing flow computation. In this article, the procedure is applied to a two-dimensional three-element high-lift airfoil configuration which is characterized by the existence of laminar separation bubbles. The automatic transition prediction procedure is applied using different operation modes and different transition prediction strategies.

1 Introduction

The modelling of laminar-turbulent transition in Reynolds-averaged Navier–Stokes (RANS) solvers is a crucial issue when high quality simulation results for aircraft shall be produced. Especially the simulation of flows around high-lift systems of aircraft may result in significant errors when the transition points are of insufficient accuracy or are not taken into account at all. High-lift systems very often involve multi-component wings (e.g. slat, main wing, and flaps) and may have very high levels of total circulation. Because all components of the high-lift system are in close interaction with one another the total circulation and the complete flow field is affected by the transition line on any of the components.

Although the overall lift value may be predicted with satisfactory accuracy slight deviations between the real and the computed pressures can lead to large errors in the computed overall drag value. It could be shown that the overall pressure drag of a high-lift configuration, which dominates the drag value of the configuration as a whole as well as the drag of every single element, is composed of a balance of very large positive and negative contributions, such as the suction forces at the noses or the resistance forces in the coves and the trailing edge regions. The contribution of one single element may be one order of magnitude larger than the resulting overall drag of the complete configuration. Thus, a relative error of 5% of the computed drag on the slat upper side may result in a change of 50% for the overall drag value [1].

Another aspect of taking into account transition is that in many cases the high potential of higher order turbulence models can be made use of only when the

N. Kroll et al. (Eds.): MEGADESIGN and MegaOpt, NNFM 107, pp. 107–120.
springerlink.com

areas of laminar-turbulent transition are known and deployed in the computational procedures with sufficiently high accuracy. Thus, in modern computational fluid dynamics (CFD) tools a robust transition modelling must be established together with reliable and effective turbulence models. Only if the transition locations are taken into account with sufficient accuracy all physical characteristics of the flow field can be reproduced in such a way that the demanding quality requirements are satisfied.

For the design process of wings, there exists the demand for a RANS-based CFD tool that is able to handle flows automatically and autonomously with laminar-turbulent transition. Existing transition prediction methods vary from empirical transition criteria via the local, linear stability equations based on small disturbance theory or non-local, linear and non-local, non-linear stability methods using the parabolized stability equations over large eddy simulations to direct numerical simulations of the Navier–Stokes equations. Empirical transition criteria and the e^N-method [2,3] based on local, linear stability theory and the parallel flow assumption represent *state-of-the-art* methods for the prediction of transition onset in many industrial applications. Although they do not account for a number of fundamental aspects in the transition process e^N-methods are used in aircraft industry most frequently for design purposes covering transition due to Tollmien–Schlichting (TS) and cross flow (CF) instabilities. Because there are no other practical methods presently available for industrial applications e^N-methods together with the two-N-factor method and empirical criteria for transition mechanisms which are not covered by the e^N approach (e.g. bypass and attachment line transition) are going to be used further on for the design of aircraft wings and wing systems even for a future laminar wing of transport type aircraft.

Recently the unstructured/hybrid RANS solver TAU [4] of the Deutsches Zentrum für Luft- und Raumfahrt, German Aerospace Center (DLR) has been provided with a general transition prediction functionality which can be applied to general three-dimensional aircraft configurations. The developments and first technical validation steps have been carried out at the Institute of Fluid Mechanics (ISM) of the Technical University of Braunschweig (TU-BS), [5,6]. The TAU code is used together with the laminar boundary-layer method in [7] and the local linear stability code in [8]. These two codes and an infrastructure part of the TAU code are components of a so called "transition prediction module" that is coupled to the RANS solver and that interacts with the RANS solver during the computation in a very similar way as it is documented in [9].

For a long time it was necessary to use transition database methods in order to apply the e^N-method for transition prediction in a fully automatic way so that the transition location iteration could be executed without intervention (automatic) by the user of the RANS code and without *a priori* knowledge of the transition characteristics of the specific flow problem (autonomous). Now the fully automated local, linear stability solver in [8] is available using a frequency estimator for the detection of the relevant regions of amplified disturbances for TS instabilities and a wave length estimator for CF instabilities.

In this article, the coupling structure between the TAU code and the transition prediction module is outlined and the transition prediction strategy is described together with the different operation modes of the transition prediction module which can be selected by the user. The main objective is to demonstrate the different characteristics of the different operation modes and the different transition prediction strategies and their impact on the computational results which are obtained for a two-dimensional three-element high-lift airfoil configuration which is characterized by the existence of laminar separation bubbles for the flow case presented. The computational results are compared to experimental findings.

2 Transition Prediction Coupling

On the one hand, the transition prediction module consists of an infrastructure part inside the flow solver which performs pre-processing operations necessary for each step of the transition prediction procedure, for example, the extraction of the surface pressure distribution from a wing section. On the other hand, the module contains a number of additional components which basically execute the transition prediction. These additional components are a laminar boundary-layer (BL) code for swept, tapered wings [7], two e^N-database methods, one for TS and the other for CF instabilities [10, 11] and a local, linear stability code [8].

With respect to the calculation of the laminar BL parameters the coupled system can be run in two different modes: Either the TAU code communicates the surface pressure distribution of the configuration to the laminar BL method, the laminar BL method then computes all of the laminar BL parameters that are needed for a selected transition prediction method and the transition prediction method determines new transition locations that are given back to the RANS solver (*BL mode 1*). Or the TAU code computes the necessary BL parameters internally and communicates them directly to the transition prediction method (*BL mode 2*).

Also with respect to the transition prediction (*PD*) method, the system can be run in two different modes: Either the two e^N-database methods (*PD mode 1*) or the local, linear stability code (*PD mode 2*) can be used for the determination of transition points due to TS or CF waves.

This coupled structure results in an iteration procedure for the transition locations within the iterations of the RANS equations. The structure of the approaches using the two different *BL modes* is outlined graphically in Fig. 1.

During the computation, the RANS solver is stopped after a certain number of iteration cycles usually when the lift has sufficiently converged, that is when pressure oscillations have been damped to a sufficiently low degree. Then the transition module is called, geometrical data are processed and all laminar viscous data – basically the velocity profiles in streamwise and crossflow direction and their 1st and 2nd derivatives – are calculated either by the BL code or by the TAU code itself. Then, either the two e^N-database methods or the stability code analyze the laminar boundary layer and try to determine a transition point. For

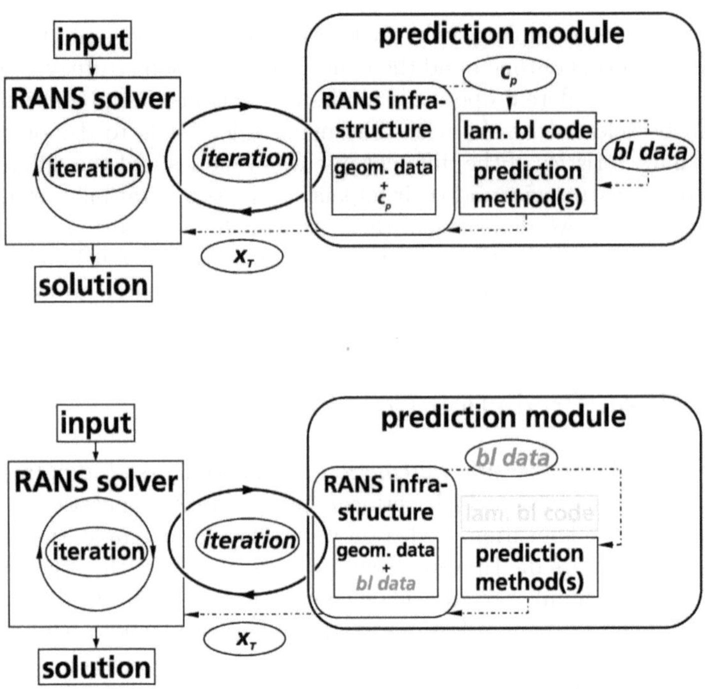

Fig. 1. Coupling structure: *BL mode 1* (above) and *BL mode 2* (below)

BL mode 1 this is possible only when the transition point is located upstream of the separation point predicted by the BL code because the BL code terminates when a separation is detected. If a transition point due to TS or CF instabilities was found it is communicated back to the RANS solver. If no transition point due to TS or CF instabilities upstream of the laminar separation point could be found the laminar separation point is used as approximation of the real transition point. This is an attempt to predict transition due strictly to the presence of separation bubbles. This approach often yields a good approximation of the real transition point when transition does not occur before the laminar boundary layer separates, particularly for low Reynolds number flows. For *BL mode 2* the laminar BL data are calculated beyond the point of laminar separation which is detected by the TAU code inside the RANS computational grid. Thus, transition inside laminar separation bubbles can be detected without relying on an approximation. Practically, the determination of transition inside laminar separation bubbles is only realizable using *PD mode 2* because the e^N-database methods lack parameters for the base flow profiles in laminar separation bubbles.

These steps are done for the upper and lower sides of all specified wing sections. When all new transition locations have been communicated back to the RANS solver, each transition location is slightly underrelaxed to damp oscillations in the convergence history of the transition locations. Then, all underrelaxed transition points – they represent a transition line on the upper or lower surface of a wing element in form of a polygonial line – are mapped onto the surface grid of the configuration applying a transition setting algorithm subdividing the surface of the geometry into laminar and turbulent regions, and the computation is continued. In so doing, the determination of the transition locations becomes an iteration process itself. With each transition location iteration step the underrelaxation factor is reduced until a converged state of all transition points has been obtained.

3 Computational Results

The system was applied to the A310 take-off configuration [12] consisting of slat, main airfoil and flap defined by Ma $= 0.221$, Re $= 6.11 \times 10^6$ and $\alpha = 21.40°$. According to [13], as value for the limiting N-factor for TS instabilities $N_{\mathrm{T}} = 9$ was used. In the experiments [12] the following transition locations were determined on the upper sides of the slat, $(x^{\mathrm{T}}_{\mathrm{upp}}/c)_{\mathrm{slat}} = 0.15$, and the flap, $(x^{\mathrm{T}}_{\mathrm{upp}}/c)_{\mathrm{flap}} = 0.345$. On the main airfoil upper side the transition location was not measured, but the location of the upper side kink – the point where the slat trailing edge is located when the configuration is undeflected – is useful as a point of orientation, $(x^{\mathrm{kink}}_{\mathrm{upp}}/c)_{\mathrm{main}} = 0.19$. On the lower sides, the transition points were not measured. In the computations a standard one-equation turbulence model was applied. The computations were started with free stream initialization and were carried out for two different computational grids [14] exhibiting different grid densities. In the fine grid, the grid resolution was highly increased compared to the coarse grid in streamwise as well as in wall normal direction in the structured parts which resolve the boundary layers. While the coarse grid (grid 1) consists of about 22,000 primary grid points the fine grid (grid 2) has about 122,000 points, Fig. 2.

Because in the experiments a laminar separation bubble on the slat upper side caused transition, the computations using grid 2 were intended to resolve the laminar separation bubble and to investigate the impact on the transition locations caused by the different modes of the transition prediction module.

The three different combinations of modes which are currently available in the TAU code for two-dimensional cases are the following: a) *BL mode 1 & PD mode 1*, b) *BL mode 1 & PD mode 2* and c) *BL mode 2 & PD mode 2*. For the combinations b) and c) an extrapolation of the N-factor curves computed by the linear stability code can be applied. The extrapolation is usually switched off for case b), so that a laminar separation point from the BL code approximates transition when transition due to TS or CF instabilities does not occur. For case c), the extrapolation is usually switched on, because the numerical dissipation in the RANS code affects the quality of the laminar BL data which are calculated inside the RANS grid when case c) is applied. The effect of the numerical dissipation in the RANS code is an upstream influence of the turbulent quantities which start to develop downstream of the current transition point. Nevertheless, they have an influence also on the laminar BL profiles directly upstream of the transition point and give them a slightly bellied shape. Thus, because these influenced BL profiles are interpreted as fully laminar ones by the linear stability

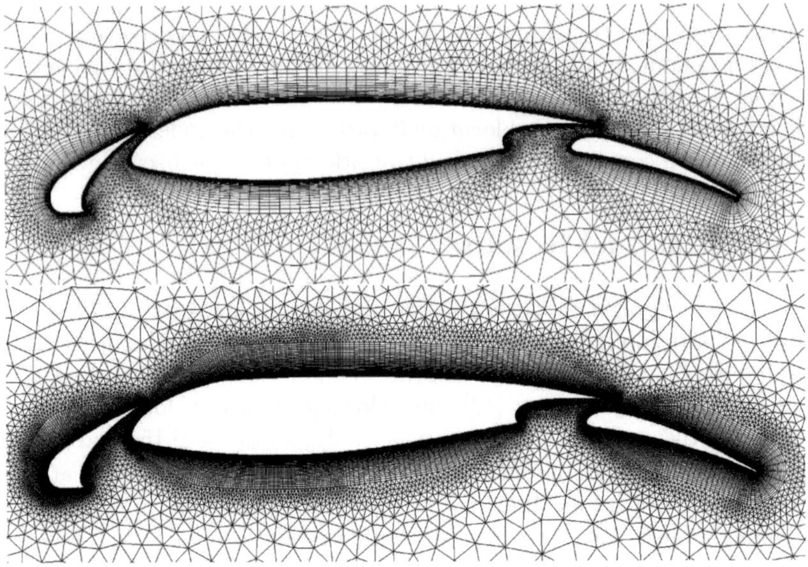

Fig. 2. Computational grids: grid 1 (above), grid 2 (below)

code, they are considered as stable or at least stabilized BL profiles. For those surface stations which are influenced in such a way, the values of the N-factor curves tend to decrease, so that it can happen that the critical N-factor is not reached by the envelope of the computed N-factor curves. This disadvantageous effect can the significantly reduced, when the envelope of the N-factor curve is extrapolated from a point where it is not yet affected by this upstream effect. In the TAU code, a linear extrapolation is started from that point on where the steepest slope of the envelope of the N-factor curves is found. This extrapolation is then used to determine the new transition point, which is found where the extrapolation curve reaches the critical N-factor.

In this article, results obtained for case b) without extrapolation, for case c) with extrapolation, and for case b) with extrapolation are shown, where the latter combination yields a new prediction strategy that keeps the computational speed of the BL code while getting rid of the approximation of the transition point by a laminar separation point. This new approach can yield much better results when it is favoured or necessary to use the laminar BL code.

The transition prediction procedure which was started with initial transition points located almost at the upper side trailing edges of the particular elements (on the lower sides, fully laminar flow was assumed up to the trailing edges) was run with a pre-prediction phase of 1000 iteration cycles where the laminar separation points which occur in the RANS grid are used as transition points in order to stabilize the computation. The pre-prediction interval was 20 iteration cycles. Then, the transition prediction iteration was started using a prediction interval of 500 iteration cycles.

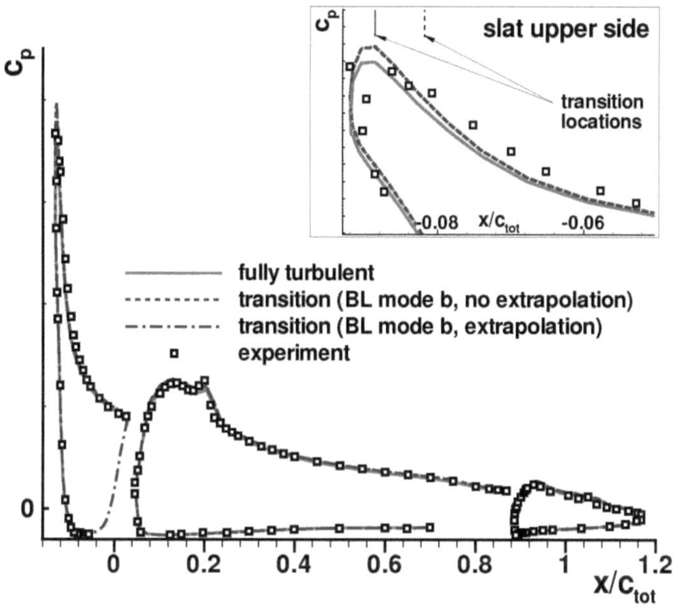

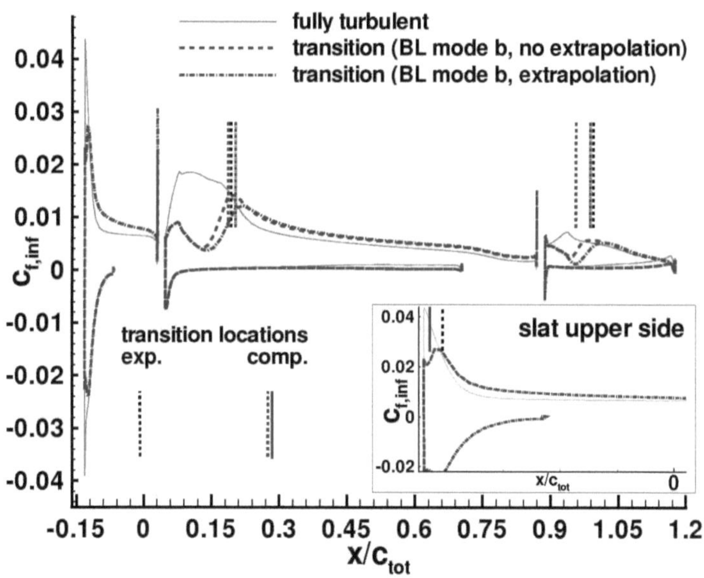

Fig. 3. c_p- (above) and c_f-distributions (below) for grid 1

In Fig. 3, the c_p- and c_f-distributions for grid 1 are shown. As expected, the results from the computations with predicted transition (PD) yield more

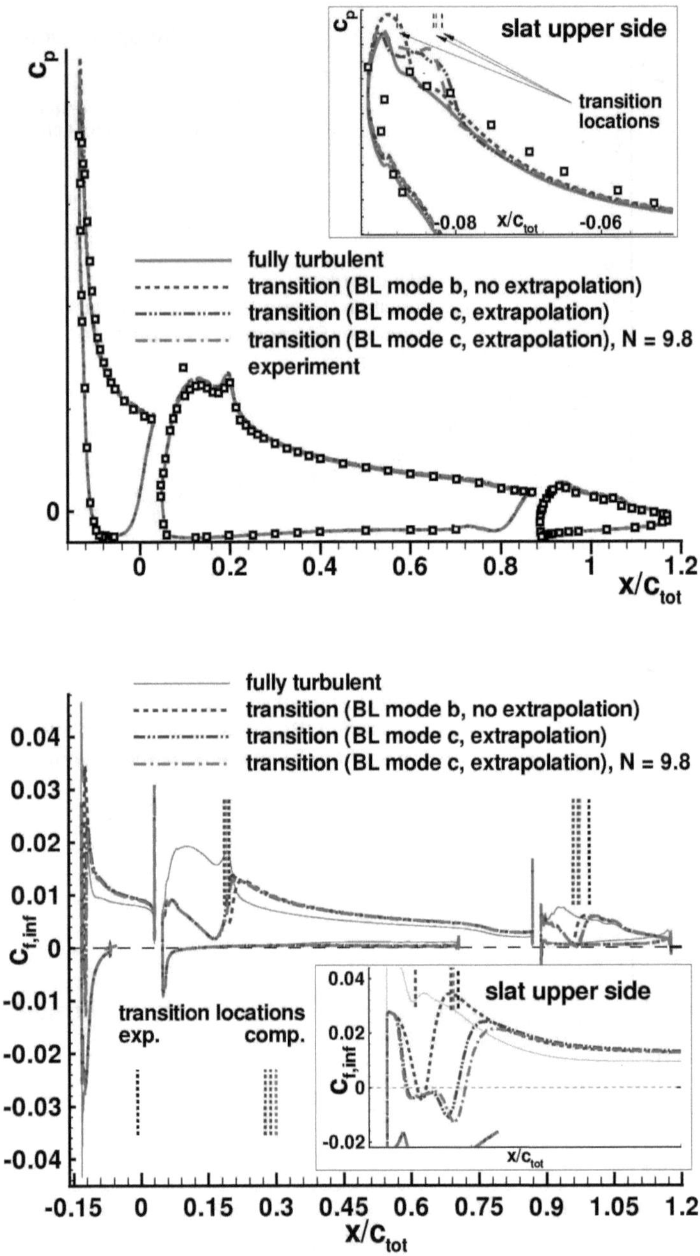

Fig. 4. c_p- (above) and c_f-distributions (below) for grid 2

negative pressure levels on the upper sides of all elements than the fully turbulent (FT) results. This effect is pronounced in the suction peak areas. In terms of the pressure no significant difference between case b) without extrapolation and case b) with extrapolation can be recognised. The comparison of the c_f-distributions clearly shows the transition from laminar to turbulent flow on all elements. On the main wing element, transition occurs directly upstream of the kink for case b) without extrapolation, on slat and flap upper side the predicted transition points are located clearly upstream of the experimentally determined locations. Especially, on the flap the deviation is remarkable. While there is no effect on the slat when case b) with extrapolation is applied – on the slat, the flow is still very stable, so that no amplified modes can be detected by the linear stability code and the expected laminar separation bubble can not evolve because grid 1 is not fine enough –, significantly different transition locations are predicted for the main element and the flap. The transition point on the main element is now located downstream of the upper side kink, the transition location on the flap almost coincides with the experimentally determined transition point. Although this result is based only on a first test case example, it shows that significant improvements of the predicted transition locations can be obtained using the BL code and the extrapolation technique of the envelope of the N-factor curves.

Figure 4 shows the results for grid 2. For grid 2, the mode combination b) without extrapolation and c) yield converged results. For the computations with transition, a separation bubble on the slat upper side is reproduced. For b) without extrapolation, where the transition point inside the separation bubble was approximated using the laminar separation point from the BL code, the resulting bubble is of too small extent and of too small strength due to the fact that through this kind of approximation the turbulence production starts too far upstream, so that the separation bubble can not fully develop. For mode combination c), where the stability analysis is carried out inside the laminar separation bubble, the extent and strength of the bubble show a good qualitative agreement with the experimental pressures. Whereas for b) without extrapolation the predicted transition location on the slat upper side is not very different from that in grid 1, for mode combination c) the measured transition point now is reproduced with excellent accuracy. The transition point on the main wing element is determined downstream of the kink using b) without extrapolation and more upstream of the kink than it was the case for grid 1 when combination c) is used. In the transition region on the main element, the interaction between transition and the influence of the kink lead to a relatively strong change between the c_f-distributions from mode combination b) without extrapolation on the one hand and from c) on the other hand. On the flap upper side the transition point resulting from b) without extrapolation shows almost the same deviation from the measured value as for grid 1.

For mode combination c) however a visible downstream shift of the transition point occurs decreasing the gap between the former computational results and the experiment significantly. The error reduction between the predicted and measured flap upper side transition point is about 40%.

In addition to the results using the critical N-factor $N_T = 9$ for the TS instabilities, more results using the value $N_T = 9.8$ were produced for mode

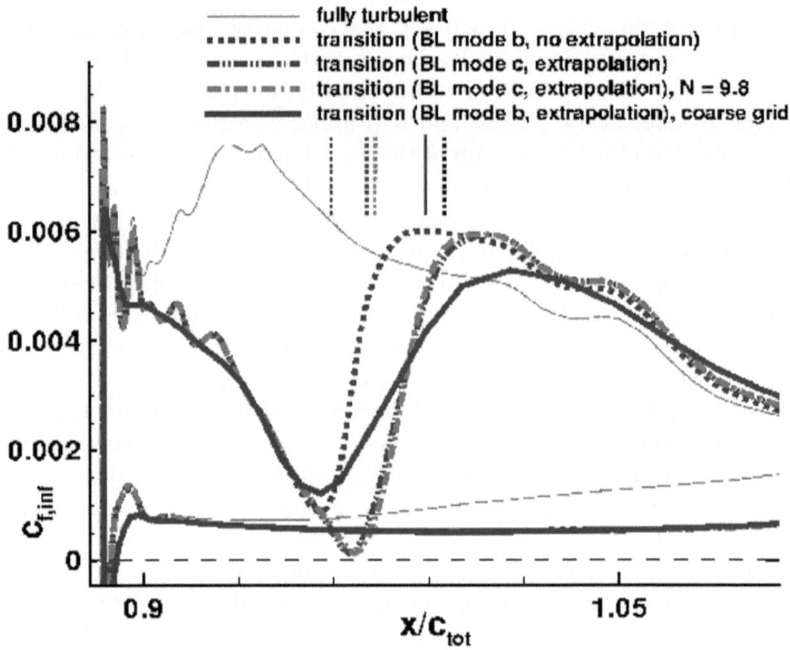

Fig. 5. c_f-distributions on the flap for grid 2

combination c). The value $N_T = 9.8$ can be found in the literature [15] for the same wind tunnel where the measurements reported in [12] were carried out. As can bee seen in Fig. 4, a further slight improvement of the predicted transition locations on the slat and flap upper sides can be achieved. On the slat, the effect is an increase of the laminar separation bubble in strength and extent. A comparison of the c_p-distributions of the results from mode combination c) with $N_T = 9$ and $N_T = 9.8$ shows that only a very small shift of the transition location inside the bubble can have a visible effect on the pressure in that area and, thus, can influence easily the shape and size of the bubble and the strength of the separation in the simulation.

An inspection of the skin friction distribution in the transition area of the flap upper side of the results obtained with mode combination c) reveals that the boundary layer is shortly before separation as shown in Fig. 5. In the test report [12] it is documented that transition on the slat upper side was provoked by a laminar separation bubble in which the transition occurs. For the flap transition location, there is no indication in the test report that this could be the case too. In the test report a clear distinction is made between transition locations which occur inside a separation bubble and other transition locations for which separation could not be observed. Thus, for the upper side flap it was always assumed that transition occurred due to a TS instability. If this assumption is true the simulation using the prediction approach based on mode combination

b) without extrapolation is expected to yield a much better accuracy of the transition location as is obtained in the result presented in this article or in other publications. But this is not the case. The prediction approach which uses a laminar separation point as an approximation of the transition point fails and predicts a transition point which is located significantly too far upstream. It is commonly accepted that a laminar boundary layer code is better suited than a RANS code to predict the point of the onset of a laminar separation. This is due to the fact that, on the one hand, the parabolic character of the boundary layer equations prevents any upstream influence, an effect which is always disadvantageous in a RANS simulation with transition when a laminar separation point is located in the direct vicinity of a transition point. On the other hand, the boundary layer codes used for transition prediction purposes use many points in the wall normal direction to resolve the laminar boundary layers – the BL code in the TAU transition prediction module uses 100 points – and apply a fourth-order accurate finite difference scheme in wall normal direction for the discretisation of the BL profiles, in contrast to a nominally second-order accurate discretisation scheme in the RANS codes. These properties of the laminar BL codes and the experience that laminar separation predictions using BL codes very often give clearly better results than those using RANS codes – normally, a laminar separation point predicted by a laminar BL code is located further upstream than that of a RANS code when the same point distribution on the surface is used – have provided much confidence in the correctness of these results. Also the author is not in doubt about that.

Now it turns out that if one does not approximate the transition point by a laminar separation point from a laminar BL code, the results can be significantly improved. In the simulations using mode combination c) no separation occurs on the flap upper side and laminar boundary layer profiles can be computed up to a point where the boundary layer almost separates. Here, again the situation occurs that a laminar BL code predicts laminar separation further upstream than the RANS code does. The upstream influence in the RANS code prevents the separation of the boundary layer and the evolution of a separation bubble. Because the upstream influence is an effect which is highly dependent on the numerical dissipation it is very probable that a further refinement of the grid – in streamwise as well as in wall normal direction – will improve the results once more and will lead to an evolution of a separation bubble. When this happens, the transition module computes laminar boundary-layer profiles beyond the laminar separation point and will detect transition inside the bubble in the same way as it was done for the slat upper side. If this supposition is true must still be verified.

In addition to the results for grid 2 (fine), the results for grid 1 (coarse) from the computation with mode combination b) with extrapolation are shown in Fig. 5. Here, the upstream effect becomes particularly apparent and is accompanied by an error reduction between the predicted and measured flap upper side transition point of about 80%.

In Fig. 6, the velocity profiles in the area of the transition point on the flap for grid 1 and mode combination b) with extrapolation and grid 2 and mode

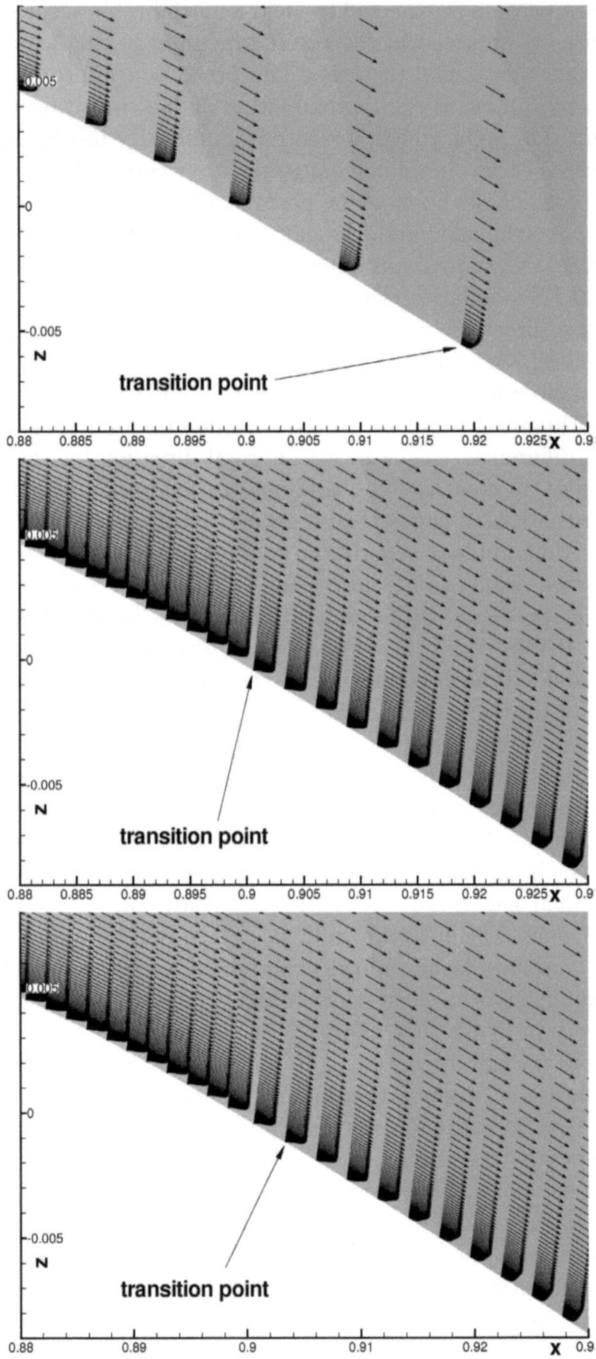

Fig. 6. Velocity profiles in the flap transition region: grid 1 (above), grid 2 and $N_{\mathrm{T}} = 9$ (middle), grid 2 and $N_{\mathrm{T}} = 9.8$ (below), x- and z-scales are independent

combination c) for $N_T = 9$ and $N_T = 9.8$ are shown. The corresponding transition locations are marked. The result for grid 1 explicitly demonstrates the upstream effect which gives the laminar boundary-layer profiles a slightly bellied shape that they would not have without this effect.

A close inspection of the results for grid 2 shows that about three profiles upstream of the respective transition point are affected by the upstream effect with increasing absolute values of the velocity vectors at the first grid point apart from the wall.

In addition to the verification that a finer grid resolution leads to the expected laminar separation bubble on the flap upper side two more things have to be done in order to fully conclude this investigation: The first is the application of mode combination b) with extrapolation in a computation using grid 2 (fine). The second is a test of a mixture of the mode combinations where the external (BL code) and the internal (from the RANS grid) computation of the boundary-layer data with and without extrapolation are applied in different combinations to the three different elements of the configuration.

4 Conclusion

The TAU code coupled to a newly developed transition prediction module was applied to a two-dimensional three-element high-lift airfoil configuration which is characterized by the existence of laminar separation bubbles. The prediction of the transition location was carried out in a fully automatic manner during the ongoing RANS computation so that no intervention of the user is needed. It could be shown that when the computational grid is fine enough transition locations inside laminar separation bubbles can be predicted with high accuracy while the separation bubble itself can be reproduced well with respect to its extent and strength.

Different operation modes of the transition prediction module were applied in the investigations. It was shown that the approximation of a transition point by a laminar separation point from a boundary-layer code can result in a poor result for the predicted transition location. This shortcoming can be overcome when the extrapolation of the N-factor curves is used. The extrapolation of the N-factor curves can always be used when the laminar boundary layers are computed from the RANS computational grid which requires a grid resolution which is fine enough in streamwise as well as in wall normal direction. In principle, the extrapolation can also be used when the laminar boundary layers are computed by the laminar boundary-layer code. In this case, however, it can happen that up to the point of laminar separation which is detected by the boundary-layer code the perturbation modes are not amplified so that no N-factor curve exists. In such a case, the internal computation of the boundary layers is the only way to solve the problem.

Acknowledgements

The development work of the transition prediction module within the TAU code was carried out by the Institute of Fluid Mechanics of the Technical University of Braunschweig within the German Research initiative MEGADESIGN.

References

1. Rudnik, R., Ronzheimer, A., Schenk, M.: Berechnung von zwei- und dreidimensionalen Hochauftriebskonfigurationen durch Lösung der Navier–Stokes Gleichungen. Jahrbuch 1996 – Bd. II der DGLR – JT 96-104, Deutscher Luft- und Raumfahrtkongress, DGLR-Jahrestagung, pp. 717–726 (September 1996)
2. Smith, A.M.O., Gamberoni, N.: Transition, Pressure Gradient and Stability Theory. Douglas Aircraft Company, Long Beach, Calif. Rep. ES 26388 (1956)
3. van Ingen, J.L.: A suggested Semi-Empirical Method for the Calculation of the Boundary Layer Transition Region. University of Delft, Dept. of Aerospace Engineering, Delft, The Netherlands, Rep. VTH-74 (1956)
4. Kroll, N., Rossow, C.-C., Schwamborn, D., Becker, K., Heller, G.: MEGAFLOW – A Numerical Flow Simulation Tool For Transport Aircraft Design. In: ICAS Congress 2002 [CD-Rom], ICAS, Toronto, Canada, pp. 1.105.1–1.105.20 (2002)
5. Nebel, C., Radespiel, R., Wolf, T.: Transition Prediction for 3D Flows Using a Reynolds-Averaged Navier–Stokes Code and N-Factor Methods. AIAA-2003-3593
6. Krimmelbein, N., Radespiel, R., Nebel, C.: Numerical Aspects of Transition Prediction for Three-Dimensional Configurations. AIAA-2005-4764
7. COCO – A Program to compute Velocity and Temperature Profiles for Local and Nonlocal Stability Analysis of Compressible, Conical Boundary Layers with Suction, ZARM Technik Report (November 1998)
8. Schrauf, G.: LILO 2.1 User's Guide and Tutorial. Bremen, Germany, GSSC Technical Report 6, originally issued September 2004, modified for Version 2.1 (July 2006)
9. Krumbein, A.: Automatic Transition Prediction and Application to Three-Dimensional Wing Configurations. Journal of Aircraft 44(1), 119–133 (2007); also AIAA Paper 2006-914 (June 2006)
10. Stock, H.W., Degenhardt, E.: A simplified e^N-method for transition prediction in two-dimensional, incompressible boundary layers. Zeitung für Flugwissenschaft und Weltraumforschung 13, 16–30 (1989)
11. Casalis, G., Arnal, D.: ELFIN II Subtask 2.3: Database method – Development and validation of the simplified method for pure crossflow instability at low speed. ELFIN II – European Laminar Flow Investigation, Technical Report n° 145, ONERA–CERT, Département d'Études et de Recherches en Aérothermodynamique (DERAT), R.T. DERAT n° 119/5618.16 (December 1996)
12. Manie, F., Piccin, O., Ray, J.P.: Test Report of the 2D Model M1 in the ONERA F1 Wind Tunnel, GARTEUR AD(AG-08), TP-041 (1989)
13. Arthur, M.T., Dol, H., Krumbein, A., Houdeville, R., Ponsin, J.: Application of Transition Criteria in Navier–Stokes Computations, GARTEUR AD(AG-35), TP-137 (2003)
14. Wild, J.: Private communications, DLR, Institute of Aerodynamics and Flow Technology. Braunschweig, Germany (November 2003/2005)
15. Cliquet, J., Houdeville, R., Arnal, D.: Application of Laminar-Turbulent Transition Criteria in Navier–Stokes Computation, AIAA-2007-515

Numerical Simulation Quality Assessment for Transport Aircraft

Klaus Becker[1] and Jochem Häuser[2]

[1] Airbus, Airbusallee 1, 28199 Bremen
klaus.becker@airbus.com
[2] HPCC-Space GmbH, Karl-Scharfenberg-Straße 55, 38229 Salzgitter

Summary. The MEGADESIGN project was supposed to investigate the status that hybrid CFD prediction has on accuracy, in a relative as well as absolute sense. Different test cases have been analyzed and their results were compared to requirements specified by aerodynamic engineers. Produced solutions stayed significantly below requested error limits, however, the maximum lift area with flow separation still poses a challenging task. Proposals have been elaborated to improve the physical modeling of the tests, which seem to be the major source of remaining errors.

1 Introduction

Within the MEGADESIGN project, one of the topics was quality assessment of numerical flow simulation. On the background of ever-running activities in the field of simulation accuracy, MEGADESIGN has provided a status for transport aircraft, which helps to increase reliability of aerodynamic design and data work. While the prediction of aerodynamic data of an aircraft in principle requires high accuracy absolute aerodynamic data, design engineers more often need relative data that describe the aerodynamic progress on their design path, thus giving hints on further improvement potential. Because predictive accuracy plays an important role in this business, MEGADESIGN activities in that area were very welcome. One of the tasks was to improve and assess existing capabilities and demonstrate the state that has been reached with hybrid RANS CFD.

2 Aspects of Accuracy

In general, there are two views on accuracy – the relative and the absolute one. During the design phase it is of high importance to reliably judge on the design progress. This leads to some requirements concerning the predictive accuracy of differences in aerodynamic data between

- different configurations at given flight parameters or
- different flight parameters at a given configuration.

Accuracy related to such comparisons is called relative accuracy.

N. Kroll et al. (Eds.): MEGADESIGN and MegaOpt, NNFM 107, pp. 121–131.
springerlink.com

Absolute accuracy is related to the quality of numerical simulation concerning the prediction of absolute aerodynamic data for a given configuration at given flight parameters.

Activities within the MEGADESIGN project dealt with both inputs and defined major milestones in that field against which the developments should be assessed. These milestones were defined for high lift configurations although a lot of assessment work was also done on cruise configuration analysis.

- Requirements on relative accuracy
  "Differences between high lift configurations should be predicted with a maximum error of 5 lift counts in the linear range, 10 lift counts and 50 drag counts around maximum lift."
- Requirements on absolute accuracy
  "The maximum error for a landing configuration should be less than 5 lift counts and 100 drag counts for a C_{L} between 1 and 2. Around maximum lift an error of 10 lift counts and ± 1 degree for the maximum lift angle could be accepted."

3 Status on Accuracy

3.1 Cruise Configuration Analysis

For the project it was interesting to analyze the level of accuracy that could be reached concerning various types of aerodynamic data. To start with, global aerodynamic coefficients like lift and drag values were considered. In order to eliminate complex sources of uncertainties in the simulation, flow around a simple wing/body configuration was chosen to be investigated. Numerical results were compared to experimental values provided by a test in ARA wind tunnel. It was interesting to see the global coefficients vary with angle of attack. Both lift and drag polars were produced for both a transonic cruise and a dive Mach number (Fig. 1).

Computations were done on a 5.8×10^6 grid point hybrid mesh which was produced from scratch, providing acceptable y^+ values in the order of 1. The comparison includes an assessment of a list of turbulence models available in the TAU code. It is good to see that for this example of a modern type aircraft the choice of the turbulence model has a rather small effect. For both Mach numbers, the numerical polars show excellent agreement with the experimental data.

Transferring these results into aircraft application, it has been demonstrated that numerical simulation is able to predict aerodynamic performance within the same error bounds as experiments. Based on both approaches, lift over drag ratio of A380 has been predicted within 1% accuracy around design point – according to flight test results (Fig. 2).

Looking into details of the flow, MEGADESIGN project performed some activities to compare results based on different tools. Both, the TAU code from DLR [1] and the CFD^{++} code from Chakravarthy [2] were engaged to investigate the DLR F6 wing/body/pylon/nacelle-configuration. This configuration

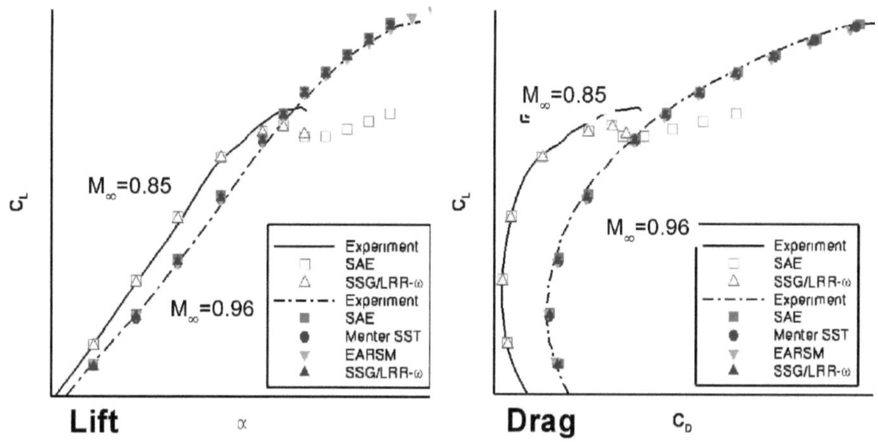

Fig. 1. Lift and drag coefficient comparison at high transonic Mach numbers for a set of turbulence models

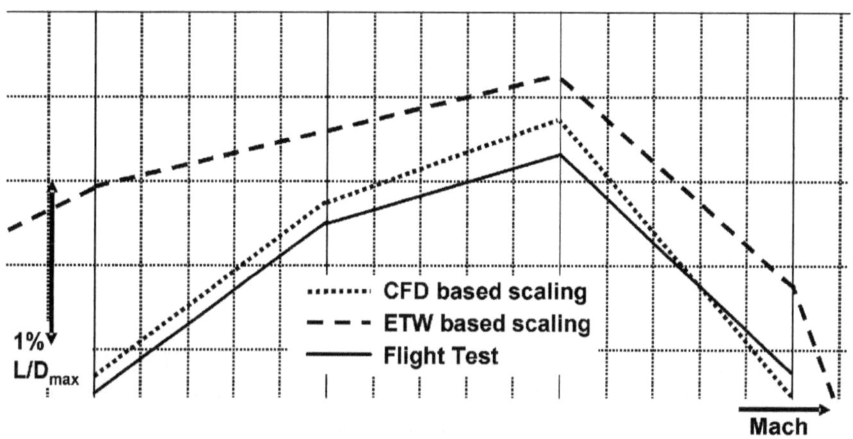

Fig. 2. Pre-flight prediction of aircraft aerodynamic efficiency

exhibits some sensitive flow behavior at wing/body junction, namely an area of flow separation and horseshoe vortex near the wing trailing edge (see Fig. 4).

Comparison of codes is not easy because there are essential differences in approach and details. CFD^{++} is a cell-centered code while TAU is using cell-vertex discretization techniques. Therefore two meshes were chosen that produced a comparable number of discrete equations. Mesh density and point distribution were done according to best practice knowledge, taking into account Y+ and other well-known recommendations.

In general, it is quite astonishing how small the differences in solution are. Remembering well the widespread results presented by some GARTEUR group [3]

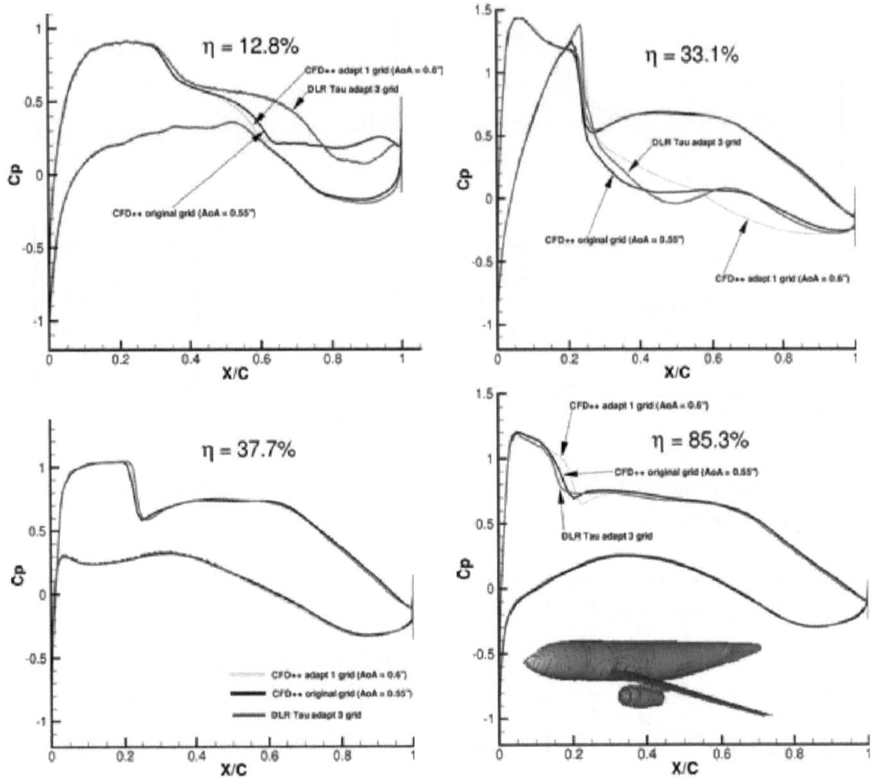

Fig. 3. DLR F6 computations using TAU and CFD^{++} codes with two-equation turbulence models

activities 20 years ago, CFD has made tremendous progress. Here the comparison was done at equal global lift coefficient, which caused some differences in angle of attack because the codes tried to compensate for some local differences in the prediction of separation, for example.

In detail, for most parts of the wing surface, pressure distributions are quite similar, except the representation of local phenomena. Pressure on the lower side of the wing is predicted almost exactly the same, except directly inboard of the pylon.

On the outer wing we can find a deviation in shock location and strength (Fig. 3, bottom right). This could be an effect of mesh resolution in the shock and trailing edge region as well as of a difference in global lift. Going further inboard, these differences completely vanish (Fig. 3, bottom left). Inboard near to the pylon (Fig. 3, top right) we see some larger differences on the lower wing side, which may be addressed partly to a resolution problem of the geometry. However, main reason for these differences seems to be the sensitivity of the flow. It is likely that some local separation may occur in this area. For the most

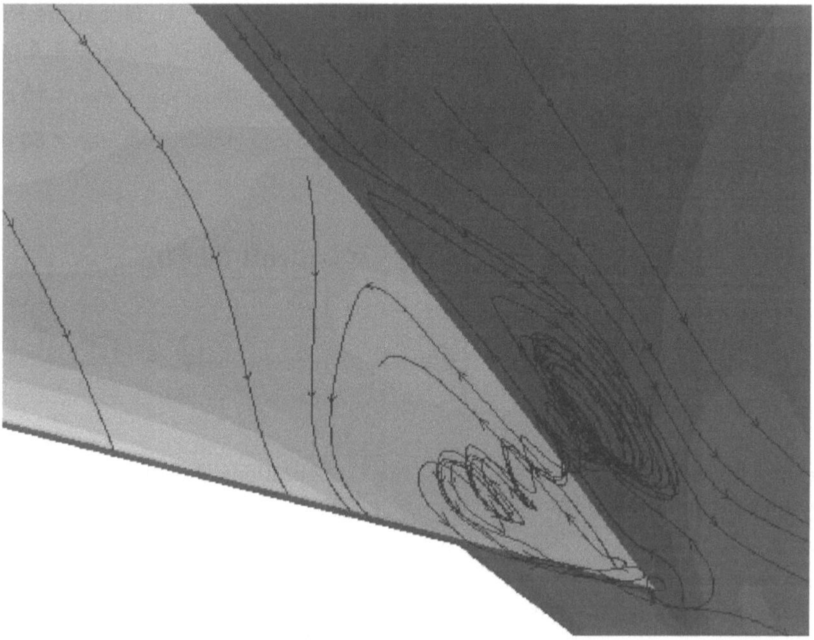

Fig. 4. Separation/vortex area at wing body junction trailing edge

inboard section (Fig. 3, top left) it is obvious that CFD^{++} code predicts a larger separation zone (Fig. 4) than TAU for the selected turbulence models.

In our opinion, this investigation provided another valuable contribution to increase confidence in numerical modeling and related CFD solutions.

3.2 High Lift Configuration Analysis

High lift configuration analysis is more complex than for cruise configurations because of geometrical and physical reasons. Although the hybrid grid approach helps to overcome the difficulties of structured meshing, the flow phenomena appearing are difficult to model. This is especially true because high lift investigations for aerodynamic design as well as aerodynamic data are looking for lift or maximum lift optimization. Flow separation from onset up to massive separation is in the major focus of simulation, which poses tough requirements on turbulence modeling or flow modeling in general.

Looking at relative accuracy, validation exercises showed a quite satisfying outcome. All requirements defined in the milestone could be fulfilled. After some improvements of the TAU code with respect to robustness a best practice meshing in context with conventional application of TAU led to satisfying results (Fig. 5).

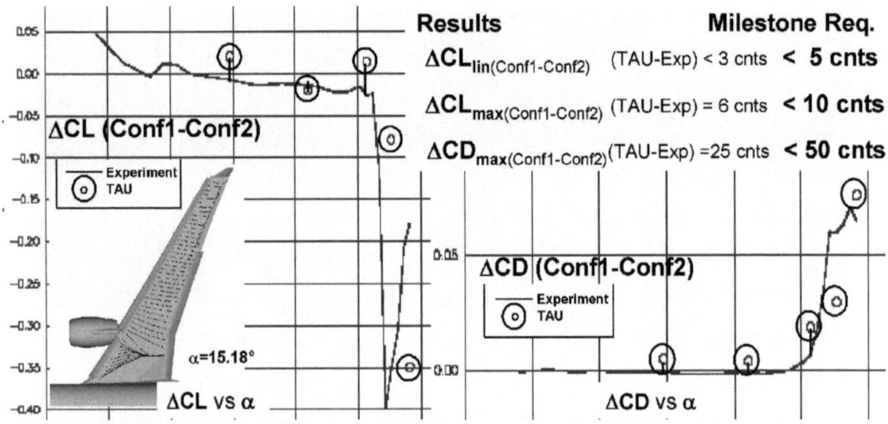

Fig. 5. Lift and drag differences between two configurations at equal flow conditions, experimental and computational values

To compare results of two configurations is not straightforward in the high lift case, it depends on what the objective of the task is. To check whether one configuration produces more lift than another will require comparison at the same angle of attack. So we recommend to compare differences in experimental ΔC_{L} prediction – between both configurations – with computational ΔC_{L} at the same angle of attack. Fig. 5 (left side) shows a ΔC_{L} difference taken from experiment as a line whereas computational ΔC_{L}s are given by the circles. In the linear range both predictions differ by 3 lift counts only which is below the requested 5 counts level. Considering maximum lift we have to take into account that C_{Lmax} will be achieved at different angle of attack for the single configurations – and also at angles which are different between experiment and computation. Thus we recommend to compare just the differences in C_{Lmax} and check the angle of attack afterwards. From the results one could derive that ΔC_{L} between both configurations at maximum lift is predicted at only 6 counts difference to the experimental prediction, and ΔC_{Dmax} only about 25 drag counts. Both values are below the requested bounds.

Looking into the flow at more detail may give some hints on locally unsatisfactory results. For a specific high lift configuration pressure values have been compared, i.e. TAU results against ETW test (Fig. 6). Obviously, for most of the sections there is an excellent agreement of surface pressure values. The only exception is near the wing tip where some separation occurs in the numerical simulation. TAU code predicts separation on the flap whereas experimental data do not indicate any separation. This directly points to one of the major remaining problems of numerical simulation, i.e. separation prediction.

Concerning absolute accuracy in low speed some problems become obvious. While the linear area of the lift polar fits very well with the experimental data

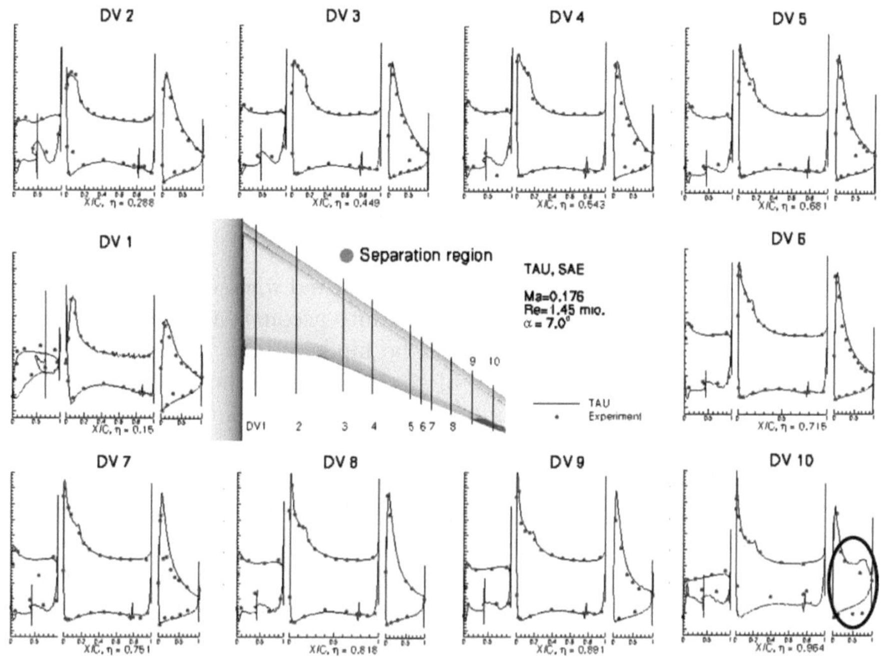

Fig. 6. Pressure distributions on high lift wing, $Ma_\infty = 0.176$, $Re = 1.45 \times 10^6$, $\alpha = 7°$

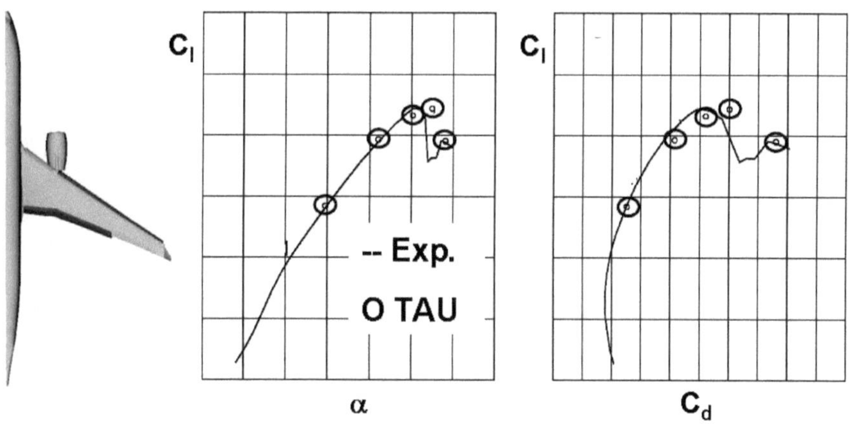

Fig. 7. Comparison of lift and drag values – CFD vs. experiment

the maximum lift area is not so well represented by numerical simulation (Fig. 7). Although maximum lift value is predicted with an error below 10 lift counts the deviation of the respective angle of attack is more than 1 degree. And the difference in drag values shows 240 counts, which is much more than the requested 100 counts.

With respect to these results one should keep in mind that flow separation is rather sensitive to the test arrangement. Therefore a deeper analysis of both would be recommended.

4 Means to Improve Accuracy

There are a number of points that have to be reflected when doing investigations for accuracy checking. Without taking them into account, an effective comparison between theory and experiment will be hindered and the results may be open to interpretation. The objective of any comparison should therefore be to make the numerical and experimental simulation, i.e. the physical model set-up, as similar as possible.

MEGADESIGN dealt with the transition effect issue of flow modeling. The integrated process of RANS simulation with transition prediction based on boundary layer stability analysis has been tested in order to judge on the modeling effect. Seeing the picture provided for four wing sections on the configuration presented in Fig. 8 an immediate improvement of the main wing trailing edge and flap pressure could be observed for the most outboard section.

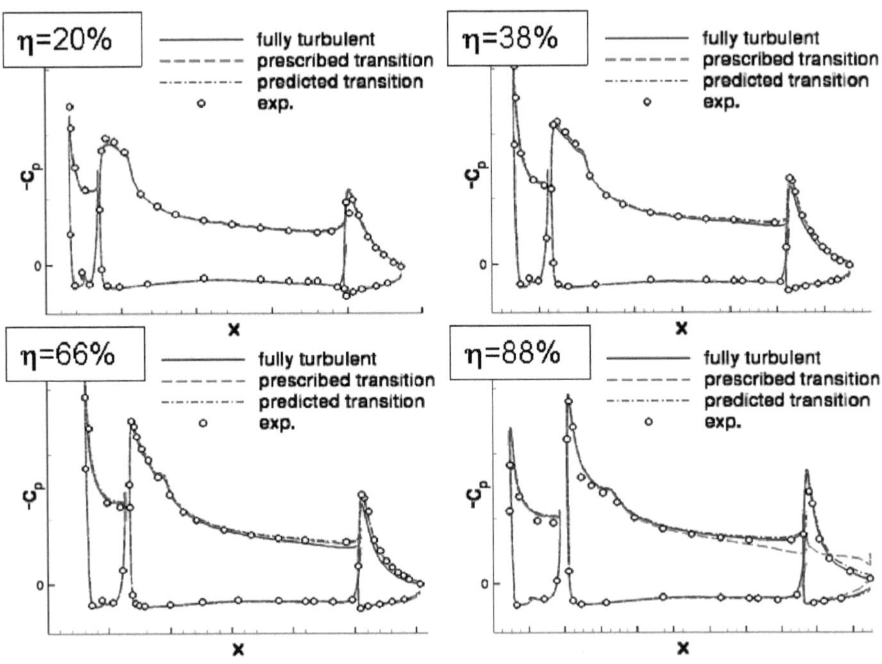

Fig. 8. CFD results using integrated transition prediction – compared to experimental values, $Ma_\infty = 0.174$, $Re = 1.35 \times 10^6$, $\alpha = 14°$

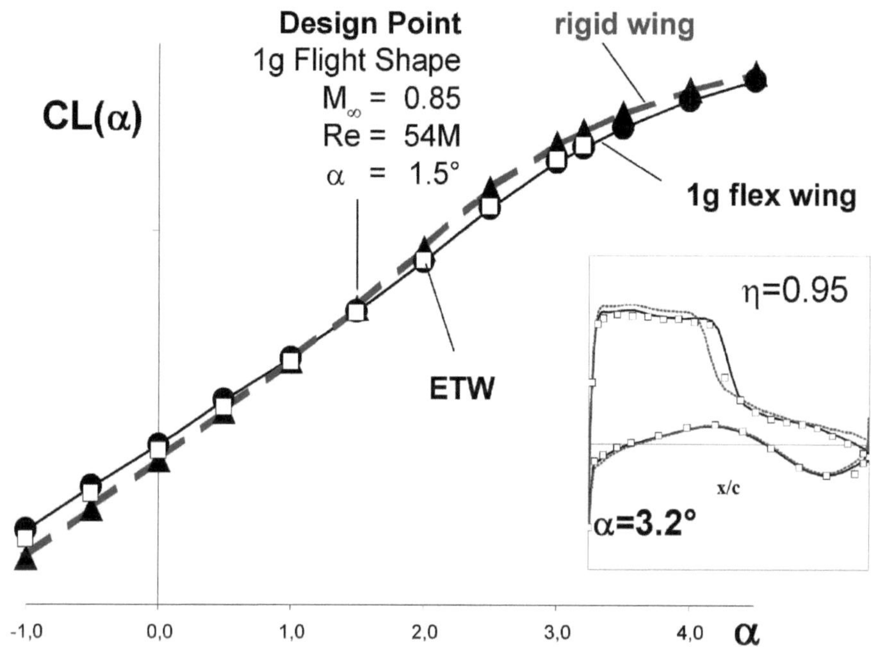

Fig. 9. Static deformation effect on global lift coefficient in highspeed, related pressure distributions at 95% wing span section

While fully turbulent simulation seems to deliver good results for the 88% section, is reveals some deficiencies for the 66% wing cut. Use of integrated transition prediction clearly improves the predictive capability in all sections, which indicates the high sensitivity of the numerical flow to completeness of the physical flow model.

Another source of error is modeling of the test situation. Instead of comparing numerical and test results corrected for free flight it turns out that proper modeling of the whole geometric environment of the test may lead to improved results. Several projects already dealt with proper simulation of wind tunnel tests [6, 7, 8, 4, 5]. The results show for all cases that it is essential to model the geometrical environment as correct as possible. This includes a variety of entries like wind tunnel walls, sting or peniche for half model and tunnel inlet/outlet sections. Concerning the high lift test mentioned above (Fig. 6) this would finally mean to even model test equipment like pressure tube bundles which are exposed to the air. The resulting effects are specifically relevant for drag analysis but also influence (local) separation behavior of the flow.

Another important source of deviation between experimental and numerical flow data is static deformation. Usually, even a very stiff wind tunnel model deforms under the air pressure and thus the flow is along shapes which are different to those provided by the construction CAD models. These effects can by

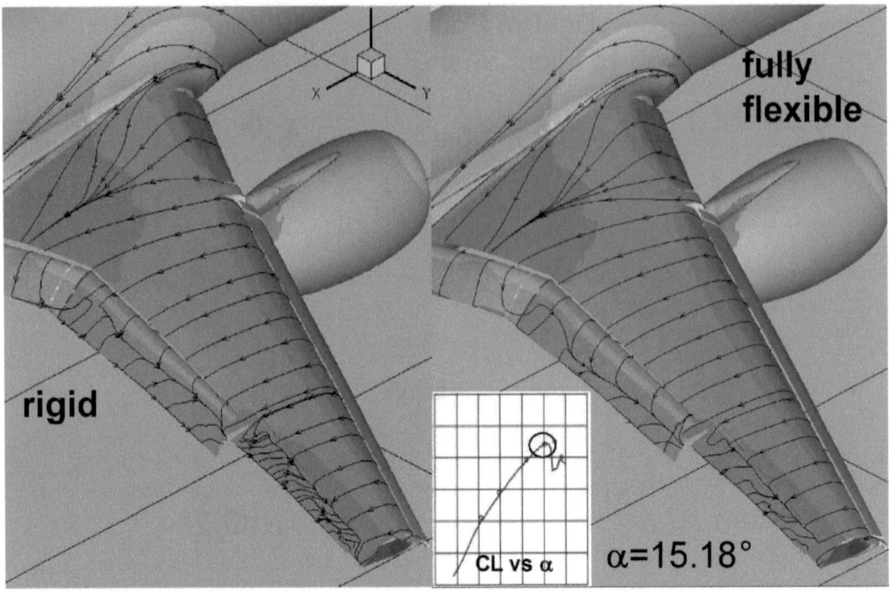

Fig. 10. High lift component deformation effect on flow separation – see separation line on outer flap

no means be neglected, specifically in the case of separation onset or boundary layer investigations. This means that static deformation in the loop with CFD has to become the usual way of simulating experiments in aerodynamics. Investigations done in the high speed regime showed a significant improvement in global coefficient (C_L) as well as local C_p prediction (Fig. 9).

But also for low speed computations static deformation has a considerable effect. The method developed in MEGADESIGN is to simulate – on the basis of a simplified structural model – the deformation of the main wing and on top the individual relative change of slats and flaps. As will be described later in the Fluid Structure Coupling section of this book, mean deformation of the high lift elements alone does not change flow characteristics, however, only the individual superimposed deformations yield a change in separation behavior on the flaps, for example (Fig. 10). This is due to the fact that local effective gap and overlap determine the flow separation behavior.

All in all it has to be stated that any deep validation of CFD results vs. experiment has to be build on a proper modeling. Without these measures it could be hard to judge on the source of simulation errors. But providing CFD results and comparisons with this type of comprehensive modeling could essentially contribute to further strengthen the reliability and reputation of numerical simulation for aircraft aerodynamic development.

5 Conclusions

MEGADESIGN has performed quite a number of activities to assess and improve numerical simulation accuracy. Best practice guidelines have been developed for the optimum use of RANS CFD in complex configuration simulation. It could be demonstrated that relative accuracy of TAU simulations can be judged to be within the limits expected by aerodynamic engineers. That means differences between high lift configurations can be predicted with a maximum error of 5 lift counts in the linear range and 10 lift counts as well as 50 drag counts around maximum lift. Absolute accuracy still remains an issue for RANS simulation. Although lift prediction seems to be within desired bounds, i.e. 5 lift counts deviation in linear range and 10 lift counts around maximum lift, accurate drag prediction is not really possible for complex high lift cases. Also the angle of attack where maximum lift occurs could only be predicted with an error larger than the requested 1 degree.

These deficiencies in simulation capabilities point out the urgent need for further development, be it on physical models, discretization or numerical algorithms to provide robust and reliable CFD solutions. Investigations are under way – within related projects – to improve the numerical model, i.e. to investigate the influence of the wind tunnel environment, intrinsic transition prediction and, perhaps most important, structural static deformation.

References

1. Gerhold, T.: Overview of the Hybrid RANS Code TAU. In: Kroll, N., Fassbender, J.K. (eds.) Notes on Numerical Fluid Mechanics and Multi-disciplinary Design, vol. 89, pp. 81–92. Springer, Heidelberg (2005)
2. Chakravarthy, S., et al.: The CFD^{++} Computational Fuid Dynamics Software Suite, Metacomp Technologies (2000)
3. GARTEUR AD(AG-05): Accuracy Study of Transonic Flow Computations for Three Dimensional Wings, GARTEUR/TP-30 (March 1988)
4. Melber-Wilkending, S., Wichmann, G.: Project ForMEx – A new CFD Approach for transposition of Wind Tunnel data towards Flight Conditions. In: 15th AG STAB/DGLR Symposium, Darmstadt, 29-November - 1-December (2006)
5. Melber-Wilkending, S., Heidebrecht, A., Wichmann, G.: A new approach in CFD supported Wind Tunnel testing. In: 25th International Congress of the Aeronautical Sciences, ICAS, September 3-8 (2006)
6. van der Burg, J.W.: Final Task Report, EUROLIFT II, Technical Report D1.1-1 (August 2007)
7. Jakob, H.: Method for Determination of Wing Deformation from Measured Pressure Distributions, EUROLIFT II, Technical Report D1.1.1-7 (December 2004)
8. Jakob, H.: Numerical Investigation of the Influence of Pressure Tube Bundles at a Wind Tunnel Model in Landing Configuration, EUROLIFT II, Technical Report D1.1.2-4 (November 2006)

6 Conclusion

[illegible] numerical simulation [illegible] the [illegible] have been [illegible] the [illegible] age of RANS CFD [illegible] damaged [illegible] [illegible] that [illegible]

Part III

Fluid Structure Coupling

Computational Methods for Aero-Structural Analysis and Optimisation of Aircrafts Based on Reduced-Order Structural Models

L. Reimer, G. Wellmer, C. Braun, and J. Ballmann

Mechanics Department (Lehr- und Forschungsgebiet für Mechanik)
RWTH Aachen, Templergraben 64, 52064 Aachen
lars@lufmech.rwth-aachen.de

Summary. In this part of the MEGADESIGN project, aeroelastic effects are introduced into the aerodynamic analysis of aircrafts by coupling DLR's flow solvers TAU and FLOWer to a Timoshenko-beam solver. The emerging aeroelastic solvers and a method for the automatic identification of Timoshenko-beam models for wing-box structures were integrated into a simulation environment enabling the combined optimisation of aerodynamic wing shape and structure.

1 Introduction

In designing an aircraft one strives to achieve an optimal result, taking into account technical and economical constraints. To this end the application of CSD (Computational Structural Dynamics) codes for dimensioning the structure and the application of RANS-based CFD (Computational Fluid Dynamics) solvers for aerodynamic analysis have become industrial standard. However, structural and aerodynamic properties of an aircraft are generally still determined and optimised independently, often even by different departments of the manufacturer. Since this approach does not duly consider the existing mutual influence between fluid motion and structural deformation, a wing design following aforementioned process is always suboptimal. This becomes even more momentous with the increasing size and flexibility of modern transport aircraft wings [1].

To enhance industrial design processes towards consideration of fluid-structure interaction, one of the main goals within MEGADESIGN was the creation of a design environment which enables simultaneous optimisation of aerodynamic shape and structure of wings. Following this demand a scheme for multidisciplinary optimisation (MDO) of aircraft wings was defined, as shown in Fig. 1. LFM is specialised in the development of computational methods for aeroelastic analysis based on reduced-order structural (ROS) models, in particular beam models based on the Timoshenko approach. For this reason, LFM contributed the following two software modules to the MDO scheme of MEGADESIGN: the *Aeroelastic Coupling Module (ACM)* and the *Timoshenko-Beam Generator (TBG)*.

The concept of the ACM involves a partitioned approach distinguishing four modules: an advanced CFD solver, preferably RANS-based, a CFD volume grid

N. Kroll et al. (Eds.): MEGADESIGN and MegaOpt, NNFM 107, pp. 135–150.
springerlink.com

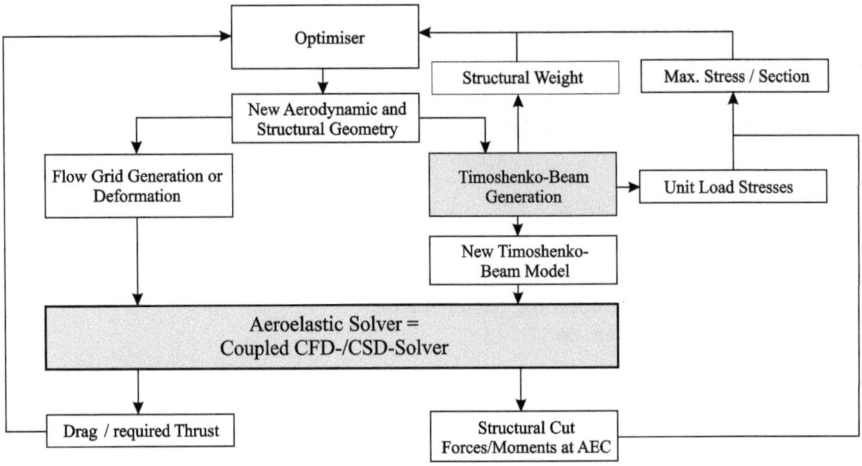

Fig. 1. The MEGADESIGN scheme for simultaneous optimisation of aerodynamic shape and structure of wings. The *Timoshenko-Beam Generation* emcompasses the TBG and the CSD code FEAFA.

deformation solver included in and adapted to the CFD solver, a CSD code and the ACM itself. An aeroelastic multi-field solver for computational aeroelasticity (CAE) arises from coupling of the respective modules through the ACM. Accordingly to the MDO scheme this is supposed to compute iteratively the aeroelastic equilibrium configuration (AEC) of the configuration to be optimised under the constraint of a prescribed constant lift (see Fig. 1). In the course of MEGADESIGN the both CFD solvers FLOWer and TAU, already developed in the MEGAFLOW project [2], were coupled to LFM's inhouse CSD code FEAFA (Finite Element Analysis for Aeroelasticity) through the ACM. Apart from several other types of finite elements (FE) FEAFA comprises a Timoshenko-beam element which was enhanced and is favoured for determining aerodynamic loads in aeroelastic applications [3,4]. The advantage of the chosen beam model lies in the quite accurate determination of the AEC for wings or even complete aircraft configurations at only small additional computational costs compared to flow computations about rigid configurations disregarding the model deformation.

In addition to bygone validations and succesfull applications of FLOWer-ACM in other projects [5,6,7], both couplings, FLOWer-ACM and TAU-ACM, were validated in MEGADESIGN against measurement data obtained from aeroelastic experiments in the HIRENASD project [8,9,10]. The concepts followed through in ACM, its functionalities and aforementioned validations are presented.

The second program, the TBG, addresses the automatic identification of beam models for thin-walled hollow wing box structures. It is thereby an important component of the projected MDO scheme. In regard to MDO, the structural properties can be described efficiently by only a few design parameters when using beam modelling. Moreover, it leads to a reduction in overall computational optimisation effort. The accuracy of the beam models computed by the TBG for generic wing box structures is presented in this paper by comparing computed deformation fields with those resulting from shell model representations in a commercial CSD code. Additionally, the suitability of the TBG for its intended use in the MDO scheme is proven and the influence of structural design parameters is argued.

2 The *Aeroelastic Coupling Module* – Solver-Independent Coupling of Computational Fluid and Structural Dynamics Codes

The realisation of the following functionalities were most prominent in the development of the ACM:

All sequences which are relevant for the aeroelastic coupling, concerning both spacial and temporal issues, should be enclosed in and controlled by the ACM. This way code changes in the single field solvers for CFD and CSD to be coupled by the ACM for assembling a CAE solver can be widely avoided.

The coupling with the MEGAFLOW CFD software, i.e. the unstructured CFD solver TAU and the block-structured code FLOWer, required the data transfer in regard to spacial coupling to be independent of the cell geometry of the CFD surface mesh. Therefore, the input data to the ACM are defined as pointwise representations of the CFD surface mesh and the corresponding loads consistenly assigned to the CFD nodes from the aerodynamic stress distribution [3]. The only output data is the pointwise representation of the currently deformed wetted aerodynamic surface and the corresponding CFD surface mesh.

The transfer algorithms applied for spacial coupling have to obey mechanical laws, i.e. and conserve force, moment and work transferred through the coupling interface. Moreover, the spacial coupling model and procedure should be applicable to complete aircrafts in both cruise and high-lift configuration. The control and high-lift devices were projected to be considered flexible with their own deformation degrees of freedom. The additional computational effort including data transfer and CSD solution should be far below those demands for pure CFD solutions. This recommends a local data transfer scheme and ROS discretisation.

Although the main emphasis in MEGADESIGN was on the computation of aeroelastic steady states in the MDO scheme, also unsteady aeroelastic coupling schemes are provided by the ACM. It offers computationally efficient loose coupling schemes, which are enhanced by extrapolation of either the state of the structure or the flow field during the coupling step. As an additional option, an imperative tight coupling scheme is also provided by the ACM [3, 11, 12]. The steady aeroelastic solution is advanced in time in the sense of a block Gauss–Seidel algorithm which is improved in regard of convergence with under-relaxation of the structure [3]. The competitiveness of ACM's spacial coupling scheme is demonstrated exemplarily for the configuration depicted in Fig. 2. It shows the unstructured TAU surface mesh of the so-called DLR-F6 high-lift configuration including wing, fuselage, pylon, nacelle, slats, flaps and fairings. To test the spacial coupling ability of the ACM with such a complex configuration, the overall structure of the aircraft was represented by a framework of Timoshenko-beams with fictitious structural properties. A load distribution prescribed on the surface mesh was projected to the beam framework using the load transfer algorithm of the ACM which will be explained later on. Based on the structural deformation computed by FEAFA, the CFD surface mesh is deformed with the deformation transfer algorithm available in the ACM. Although the CSD solution is provided by FEAFA in this case, the applicability of the ACM is not confined to this particular solver.

The load distribution was intentionally chosen to produce large and visible deformations at the main wing and in particular at the flaps, even though these deformations exceeded the limits of linear structural modelling. The structural response and the ensuing large deformation of the CFD surface mesh are shown in Fig. 2. The surface mesh remains smooth, even in crucial regions with intersections of several CFD surface components and high cell density, as is pointed out by the close-up on the vicinity of the flap track fairing.

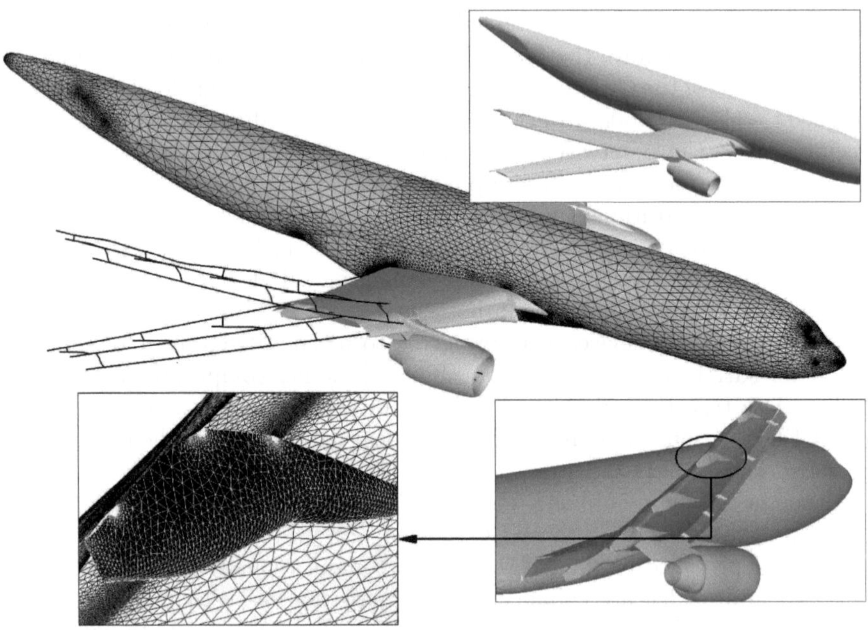

Fig. 2. Exemplary application of the load/deformation transfer scheme of the ACM to the DLR-F6 high-lift configuration (TAU CFD surface mesh, structural representation with a framework of Timoshenko-beams)

2.1 Load/Deformation Projection for Single Beams

The load/deformation transfer scheme for frameworks made of beams is realised in the ACM in two steps. At first the transfer of aerodynamic loads and structural deformations is performed between a single CFD surface mesh point and a single and most nearby beam element (see Fig. 3). This follows the assumption that cross-sections, which are oriented perpendicular to the local beam axis in the reference configuration, remain plane even in the course of deformation.

To transfer the consistent CFD force $\mathbf{F}$ (see Fig. 3), first the projection of the CFD surface point onto the most nearby beam element is determined. The equivalent force-moment group ($\mathbf{F}_\mathrm{P}$, $\mathbf{M}_\mathrm{P}$) is computed for the surface force at the projection point. The interpolation of this force-moment group to equivalent forces and moments at the FE nodes of the respective element is done by using its isoparametric shape functions. The interpolation coefficients are defined by evaluation of the shape functions at the natural coordinate r_P of the projection point.

The transfer of structural deformation data to the CFD surface mesh works vice versa. Having computed the structural deformations (translations and rotations) at the FE nodes within the CSD solver, nodal deformations ($\mathbf{u}$, φ) are interpolated to the projection point by evaluating the beam shape functions in turn at the projetion point. Pursuent to the assumption that beam cross-sections remain plane during deformation, the displacement of the CFD surface point $\mathbf{u}$ is computed from displacements, as given at the natural coordinate of the projection point, i.e. rotation $\varphi_\mathrm{P} \times \mathbf{d}$ and translation $\mathbf{u}_\mathrm{P}$. Because shear deformation is taken into consideration in Timoshenko's beam theory the cross-section previously perpendicular to the beam axis does not remain perpendicular in the deformed configuration.

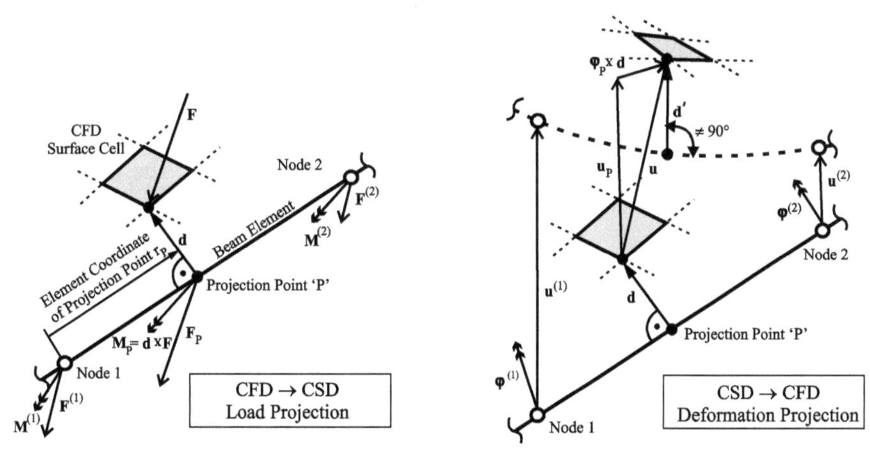

Fig. 3. Load/Deformation transfer between a single point of the CFD surface mesh and the most nearby beam element

For the whole aforementioned spacial transfer procedure the proof can be supplied, that the mechanical work is conserved at virtual displacements of the structure [3], which includes all kinematically compatible displacements. The equivalence of forces and moments holds in the aforementioned procedure by the fact, that the structural shape functions sum up to one.

2.2 Extension of the Load/Deformation Projection to Frameworks of Beams

For structural models composed of a framework of beams, like the one depicted in Fig. 2, non-uniqueness problems arise when a load/deformation transfer scheme is used, which is based solely on relations between a single CFD surface point and its most nearby structural element [3]:

For example, a CFD surface node of the wing close to the trailing edge can be erroneously projected onto those parts of the beam framework which represent the flap at a wing-flap configuration. Uniqueness is preserved by explicit assignment of the CFD surface components to components of the beam framework. Structural parts that do not have any mechanical contact with the aerodynamic surface, like e.g. flap-tracks, can be explicitly excluded from the load/deformation transfer by omitting any assignments [3].

In areas of kinks or junctions of beams another kind of non-uniqueness arises because more than one orthogonal projections is possible. This can result in unphysical dislocations on the CFD surface. To keep or recover the smoothness of the surface after the deformation transfer, the displacement of the affected surface node is interpolated between the displacements of all possible projection points. Therein the interpolation coefficients depend on the deviation of the projection angle from orthogonality and the projection distance [3].

In intersection areas of CFD surface components, gaps can emerge in the CFD surface mesh which are caused by the projection procedure (e.g. wing-fuselage intersection). These can be prevented if the displacement of the currently considered CFD surface node is not only affected by its directly assigned component of the beam framework, but also by all adjacent components, i.e. the indirectly assigned components. In this case the displacement of the CFD point is blended between directly and indirectly projected deformations. Here, the coefficients are weighted with the distances to the directly and indirectly assigned components [3].

The load/deformation transfer remains conservative, if the aerodynamic loads to be transferred are charged with the same weighting coefficients as used for the interpolation and blending of deformations. The transfer rules for loads and deformations are explicitly given without solving any systems of equations and thus can be evaluated in the computation very efficiently. The concept of the aforementioned weighting coefficients retains the required smoothness of the CFD surface and preserves the locality of the load/deformation transfer in applications to complex configurations.

2.3 Validation against Aeroelastic Experiments

The elastic wing model investigated in transonic aeroelastic experiments at high Reynolds numbers within the scope of the HIRENASD project was regarded as one of the test cases for the aeroelastic solvers FLOWer-ACM and TAU-ACM. The planform of the HIRENASD wing complies with wings of modern high speed transport aircrafts in regard of aspect ratio and taper. It has a supercritical profile but is not optimised for real application. To reduce the influence on the flow resulting from the wind tunnel wall boundary layer, a fuselage substitute was designed (Fig. 5), which was fixed to the wind tunnel turntable in the experiments without having any mechanical contact with the wing. A Timoshenko-beam model was defined which identifies the structural properties of the complete wind tunnel assembly [9, 10]. This consists of the wing, its clamping device, the housing of a vibrational excitation mechanism as part of the clamping device and the wind tunnel balance which was especially developed for the HIRENASD experiments [10]. The CFD volume mesh captures the boundary layers developing along the wind tunnel ceiling and the surfaces of the fuselage substitute and the wing.

The consideration of the elasticity of the wing in RANS-based simulations with FLOWer-ACM improves the accuracy of the results significantly towards those values measured in the experiments. This becomes evident by comparing measured and computed lift curves (Fig. 4, left) for the design point of the HIRENASD wing (Ma $= 0.80$). Even at a moderate root angle of attack $\alpha = 2°$, the lift computed without consideration of model deformation (FLOWer) comes out 19% higher than verified by the experiment and predicted by the aeroelastic solver FLOWer-ACM.

The Mach number marking the onset of the lift collapse is shifted to lower Mach numbers when increasing the angle of attack. In contrast to the pure flow simulation with FLOWer, i.e. ignoring deformations, this effect was quite

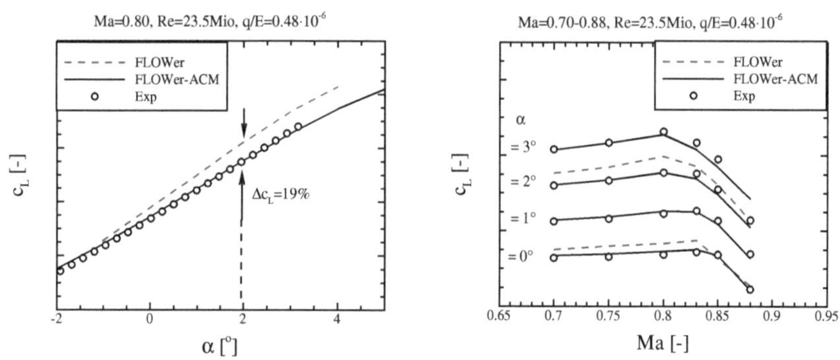

Fig. 4. Lift as a function of the angle of attack at wing root (left) and the Mach number (right), as is generated by the undeformed and the AE configuration (simulated with FLOWer resp. FLOWer-ACM)

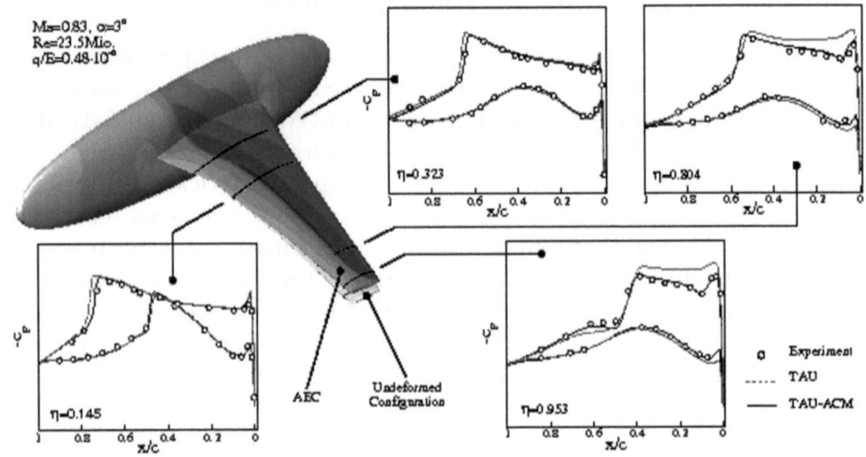

Fig. 5. Computed pressure distributions for the undeformed and the AE configuration (simulated with TAU resp. TAU-ACM) in comparison to measured data

accurately predicted by the aeroelastic simulation with FLOWer-ACM (Fig. 4, right).

This improvement in simulation accuracy is also evident in pressure distributions computed using TAU-ACM which are plotted in some of the experimental measurement sections in Fig. 5 for Ma $= 0.83$ and $\alpha = 3°$. Though the boundary layer at the wind tunnel ceiling is not resolved here by the CFD mesh, the suction niveau itself and especially the suction peaks and the shock positions meet the experimental characteristics, provided that wing elasticity is considered. Even though the wing bending deflection is rather small, about 38.8 mm at the tip, a nose-down twist of about $1.3°(= 0.43\,\alpha)$ arises due to the kinematic coupling of bending and twist deformation caused by the sweep. In the transonic case this significant reduction of local angle of attack must be considered by all means to predict the wing aerodynamics correctly, as is demonstrated in Fig. 5. Both established aeroelastic solvers TAU-ACM and FLOWer-ACM are capable to compute these aeroelastic effects accurately.

3 The *Timoshenko-Beam Generator* – Automatic Identification of Timoshenko-Beam Properties for Multi-Cellular Thin-Walled Wing Structures

In each MDO step the structural data entering the computation of the AEC must be conform with the specifications of the structure defined by the optimiser. The wing structure is assumed to be a multi-cellular thin-walled spar box which is defined to follow predominantly the wetted wing surface. Thus, in the course of optimisation the structural geometry is affected by modifications of the aerodynamic shape of the wing just as by variations of potential structural optimisation parameters, e.g. the number of spars, their chordwise positions, the sheet thicknesses of the spars and of the upper and lower panels (see Fig. 6).

3.1 Beam Identification Methodology

To meet these prerequisites the geometry of the wing box structure is supplied to the TBG in terms of coordinate lists resulting from spanwise sections through the current aerodynamic wing surface. Each multi-cellular hollow section is created by inserting spars at the chordwise positions defined by the user. All coordinates in front of the foremost spar and behind the rearmost spar are discarded and the sheet thicknesses defined by the user are assigned to the respective spars and panels. The idealisation of the structure by means of beam properties comprises the computation of the following data in each section:

- the flexural and torsional stiffnesses,
- the torsional, flexural, and mass centres,
- the moments of inertia, as well as
- the structural mass.

These are computed analytically applying the formulae for multi-cellular thin-walled hollow sections. Therein it is assumed that across the thickness of the sheets the shear stress is constant and can be treated in terms of shear flow. Also, warping stresses are neglected. In slender structures, to which beam theory is limited, the majority of sections is not subject to warping stresses such that this approach should cause only small errors in flow and aerodynamic load computations. The flexural stiffness, the total mass, and its centre can be obtained via straightforward summation over the contour elements. However, the calculation of the torsional stiffness and the torsional centre of a multi-cell section represents a statically indeterminate problem. Each cell has to be cut open

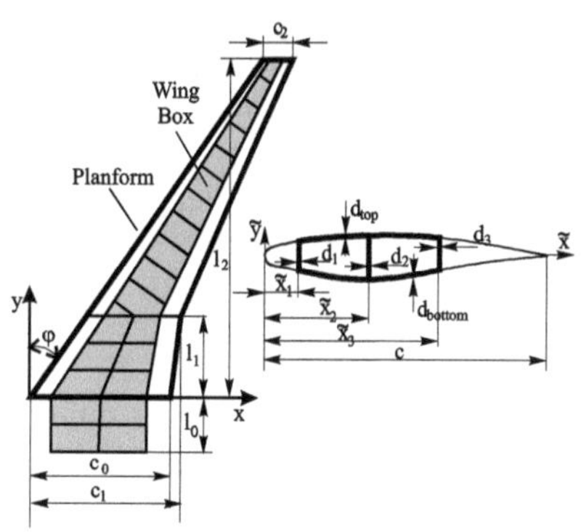

Fig. 6. Generic wing box with spar positions and sheet thicknesses as parameters

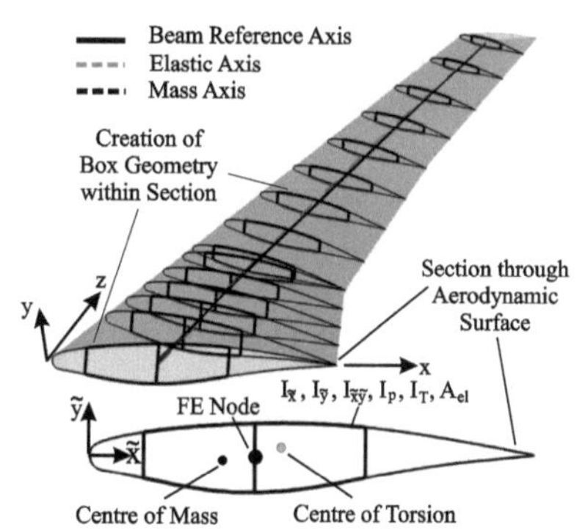

Fig. 7. Equivalent beam model with corresponding beam properties

and additional shear flows have to be introduced to cancel any relative displacement at the cut edges. Both desired quantities, stiffness and torsion centre, have to be derived from the additional shear flows, which result from linear systems of equations [13,14]. The beam properties are computed separately for each read-in section and can be interpolated at intermediate positions.

Next the location of a reference axis is for the beam is defined, which serves only as a geometrical reference line for the coordinates of the torsional, flexural, and mass centres. Of course, this line should be prescribed in some meaningful manner.

The structural data characterising the originated Timoshenko-beam model of the wing box is stored in a manner compatible with FEAFA, which in turn supplies the structural side of the aeroelastic solution process with the stiffness and mass matrices.

3.2 Preparation of Structural Optimisation Constraints

To limit the MDO process to feasible designs, boundary conditions must be imposed to the aerodynamic shape and the structure. A suitable formulation should confine the design space to structures in which the actual von-Mises stress does not exceed the yield stress of the material. Although the beam modelling gives very accurate aerodynamic load distributions for wings and aircrafts in AEC, as was verified by comparisons of aeroelastic simulations and experiments [3,5,6,7], stress values must be derived by analysing the stresses in the real supporting structure, i.e. the actual wing box in this case. For this evaluation the TBG provides, before deriving the Timoshenko-beam model, first stress analyses of the real structure under unit loads.

For now normal and shear stresses per unit force and moment are considered by the TBG. Normal stresses resulting from longitudinal forces and bending moments are functions of the cross-sectional area data. The shear stresses due to transverse forces and torsional moments depend on the shear flows needed for the calculation of the torsional stiffness and torsional centre.

The estimation of the stresses in the wing is then performed in a follow-up to the computation of the AEC. The loads acting upon the structural nodes in the AEC are added up to cut loads. Then the resulting shear and normal stresses in the read-in sections can be evaluated by multiplying the stresses derived by the TBG under unit loads with the actual cut loads (see Fig. 1).

3.3 Quality of Beam Identifications

For the validation of the TBG a generic multi-cellular wing box was defined. Two aspects were considered: the accuracy of the computed deformations based on the beam model and the accuracy of the stress approximation.

Both aspects were investigated by using the commercial FE package MARC to create a thick shell model of the reference geometry. It is similar in plan view (see Fig. 6), size, and inner layout to the structure of a transport aircraft-type wing [15]. However, the configuration of the shell model was simplified by specifying hexagonal cross-sections instead of using cross-sections with curvilinear top and bottom shells.

Inboards of the kink the structure has three spars, whereas outboards it possesses only two. Ribs are introduced at regular intervals to ensure that structural sections keep their profile during deformation so that beam theory can be applied. The shell model and the equivalent beam model were both subjected to point loads at the elastic axis of the wing tip section. Special attention was paid to inhibiting local load transmission effects in the shell model. A comparison of the deformation distributions is shown in Fig. 8 for a vertical force of $F_y = 80\,\text{kN}$ and for a torsional moment of $\|\mathbf{M}_\text{T}\| = 300\,\text{kNm}$. For visualisation purposes the deformations of the beam model were projected onto the shell surface. To this end the same algorithm was used as is implemented in the ACM for the deformation transfer.

The isolines of the bending deflection in Fig. 8 prove that beam theory is indeed applicable to get a reasonable approximation of the deformation. The contour plots match well, with only some differences visible at the root and around the kink. At the points denoted $\mathbf{x}_1$ the local error of the beam model relative to the shell model is $\triangle u_y/u_{y,\,\text{MARC}} = -0.23\%$, but it reaches locally about 10% close to the kink. In this region of load redirection beam theory is, strictly speaking, not valid. Also the kinematic coupling of sections on both sides of the kink is not properly captured by the beam model.

The contour plots of the vertical deflection due to a torsional moment applied at the wing tip do not exhibit the same amount of agreement as for transverse force loading (see Fig. 8). Here the relative error in torsional angles at $\mathbf{x}_1$ is $\triangle\varphi_\text{T}/\varphi_{\text{T},\,\text{MARC}} = -2.8\%$. As before, the discrepancies are concentrated around

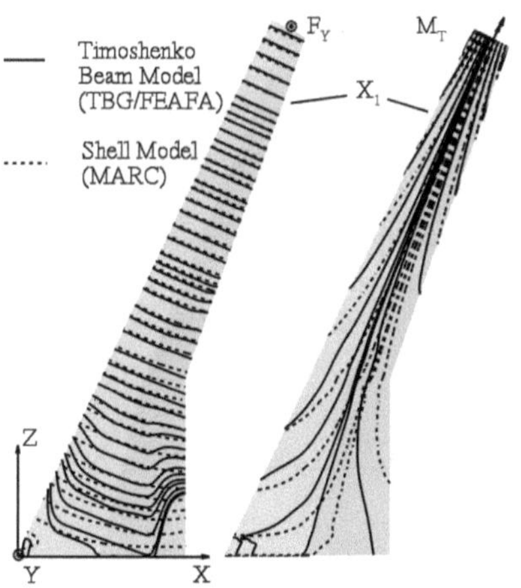

Fig. 8. Isolines of the deformation in y-direction due to a tip load of $F_y = 80\,\text{kN}$ (left) and $\|\mathbf{M}_\text{T}\| = 300\,\text{kNm}$ (right)

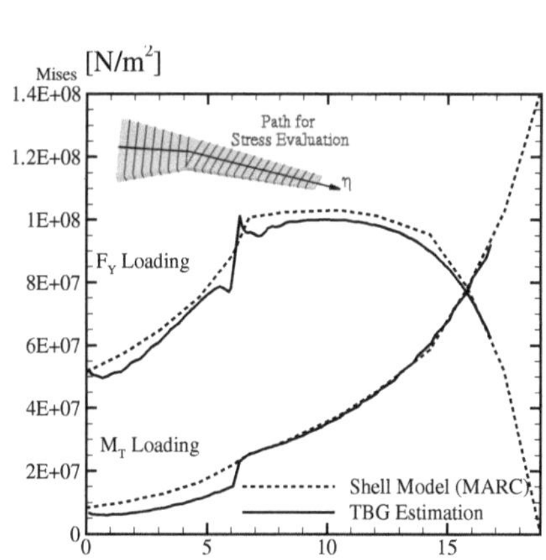

Fig. 9. Von-Mises stress distributions due to a tip load of $F_y = 80\,\text{kN}$ and $\|\mathbf{M}_\text{T}\| = 300\,\text{kNm}$ obtained from MARC and from stress estimations by the TBG

the kink. Apart from the reasons already given, for torsional loads the disregard of warping effects in the formulation of the beam model may play an additional role.

For the validation of the stress prediction generated with the data provided by the TBG, the von-Mises stresses were extracted from the shell model along a path in the top shell part. A comparison of the stress distributions along this path is shown in Fig. 9 for the same loadings as applied for the deformation validation. In both cases the stress approximation from the geometrical information in the TBG is conservative and follows the distributions given by MARC, apart from the stress peak in the vicinity of the kink in case of the loading with the vertical force. Also the TBG strongly overestimates the effective stress close to the root under loading with the torsional moment.

Nonetheless, the accuracy of the stress prediction obtained with the BG is sufficient for defining a boundary condition to the optimisation problem. Once the optimisation process will have delivered a structural design, this would have to be discretised with higher-order models. With the loads acting in the AEC a more detailed investigation of the stresses would have to be carried out, paying special attention to kinks and corners for examination of stress peaks.

3.4 Evaluating the Structural Design Space in Consideration of Fluid-Structure Interaction

After the completion of the TBG and its associated software modules, a process scheme comparable to the MEGADESIGN MDO scheme (see Fig. 1) was set up in a command line script. Apart from proving the suitability of the TBG for its intended application, the process scheme was built to evaluate the influence of different design parameters having regard to fluid-structure interaction.

Because LFM had no optimisation tools at hand and no prior experience in MDO, sweeps over the structural design space were performed instead. The script controlled the computations necessary to obtain the value of the objective function at discrete combinations of structural design parameters and thus created a hyperplane of the objective function.

The formulation of the problem was adapted from the goals of the MEGADESIGN project [15]: With a scaled version of the HIRENASD wing [8,9] as an exemplary geometry, it was aimed to maximise the achievable range R [16],

$$R = \frac{2}{b_{\mathrm{f}}} \sqrt{\frac{2\,g}{\varrho_\infty A}} \frac{\sqrt{c_{\mathrm{L}}}}{c_{\mathrm{D}}} \left(\sqrt{m_0} - \sqrt{m_1}\right) = k \frac{\sqrt{m_0} - \sqrt{m_1}}{c_{\mathrm{D}}} \,. \tag{1}$$

The total mass m_0 was defined as constant, the structural weight influencing the amount of fuel available and thus the zero-fuel mass m_1. The lift c_{L} was accordingly required to remain constant:

$$c_{\mathrm{L}} - \frac{mg}{q_\infty A} = 0 \,. \tag{2}$$

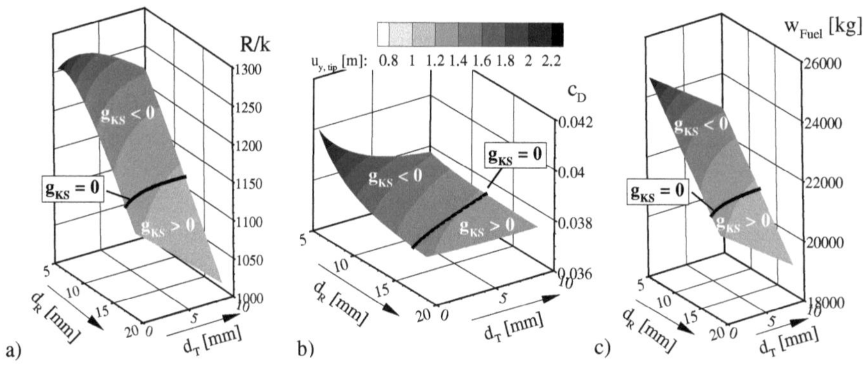

Fig. 10. a) Reduced range R/k, b) the fuel weight w_{Fuel} and c) the drag c_{D} as functions of the sheet thickness at wing root d_{R} and tip d_{T}. The valid design space $g_{\mathrm{KS}} \geq 0$ is limited by the thick black line.

Furthermore the yield stress boundary condition was taken into account:

$$1 - \frac{\sigma_{\mathrm{Mises}}}{\sigma_{\mathrm{F}}/S} \geq 0\,. \tag{3}$$

The yield stress σ_{F} was reduced here by a suitable safety factor S [17].

The procedure described above involved a large number of simulations, making the use of a RANS-based flow solver for the determination of the AEC for each parameter combination impractical within the scope of this sub-project. Rather, a simplified aerodynamic model was implemented in the stand-alone version of the ACM. It is based on the Weissinger subsonic lifting-line theory [18], which was expanded by 2-D polar data to approximately consider viscous profile drag. The method was validated against rigid and aeroelastic simulations using the FLOWer code.

For the investigation of the design space the Weissinger-ACM was embedded in a target-lift loop to fulfill (2). The values of the yield-stress boundary condition were determined with the data supplied by the TBG in all read-in sections of the wing. To gain a single value g_{KS} of all boundary conditions, the Kreisselmeier–Steinhauser (KS) function [1,19] was employed. It represents a continuously differentiable conservative envelope to a set of boundary conditions. The hyperplane investigations resulted in diagrams of the type depicted in Fig. 10. In this particular investigation the sheet thicknesses at wing root d_{R} and at the tip d_{T} were varied within the limits $5\,\mathrm{mm} \leq d_{\mathrm{R}} \leq 18\,\mathrm{mm}$ and $0.5\,\mathrm{mm} \leq d_{\mathrm{S}} \leq 10\,\mathrm{mm}$. Common values were prescribed for the sheet thicknesses of the spars as well as the top and bottom surface in each wing-box section.

The shape of the hyperplane for the objective function R/k in Fig. 10a indicates an absolute maximum at $d_{\mathrm{R}} < 5\,\mathrm{mm}$ and $d_{\mathrm{S}} \approx 3\,\mathrm{mm}$. But this is situated by far beyond the valid design space which is limited by the KS function to $g_{\mathrm{KS}} \geq 0$.

According to (1), an improvement in range can be achieved by two mechanisms: reducing the drag c_D and reducing the wing structural weight. The latter depends linearly on the sheet thicknesses, so does the fuel weight w_{Fuel}. Within the zone of valid designs at $g_{KS} \geq 0$ the range is influenced by the fuel weight (see Fig. 10c) in excess of the drag which is roughly constant in this zone (see Fig. 10b). As the sheet thicknesses of wing root and tip move towards very thin values, the influence of the deformation on the drag gains more and more significance. The absolute maximum of the objective function is reached when the increase of drag obtained from further reductions of the sheet thicknesses is no longer compensated by the increase of fuel weight.

The boundary optimum at about sheet thicknesses $d_R \approx 16\,\mathrm{mm}$ and $d_S \approx 2\,\mathrm{mm}$ is thus predetermined by the yield-stress boundary condition. Hence, structural wing optimisation allowing for fluid-structure interaction is only worthwhile when aerodynamic shape optimisation is combined.

Upon each hyperplane a contour plot of the vertical tip displacement is superimposed. The relative orientation of the deflection isolines to the line $g_{KS} = 0$ reveals that the often used deformation boundary conditions are not suited so well for this kind of optimisation problem and might indeed lead to structural layouts in which the yield stress is exceeded.

4 Summary

In the MEGADESIGN project an *Aeroelastic Coupling Module (ACM)* and a *Timoshenko-Beam Generator (TBG)* were developed at LFM.

The ACM enables the aeroelastic coupling of any kind of CFD and CSD solver regardless of their temporal and spacial discretisation schemes. The arising aeroelastic solver is capable to compute the aeroelastic equilibrium configuration and the dynamic response behaviour for complete transport aircraft. Existing control and high-lift devices can be considered as flexible as well. The ACM comprises the required temporal aeroelastic coupling schemes, different loose and, optionally, a tight one to advance the aeroelastic solution in time.

Both CFD solvers FLOWer and TAU were coupled through the ACM to a CSD code existing at LFM. Due to the accurate direct consideration of fluid-structure interaction in both developed aeroelastic solvers the aerodynamic solution at aeroelastic equilibrium matches well with measurement data obtained from aeroelastic HIRENASD experiments for a wing-fuselage configuration.

The purpose of the TBG is to automatically create beam models of wing structures intended for MDO of wing shapes and structures. The program creates a multi-cellular, thin-walled spar box from cross-sections through the wing's wetted surface with prescribed sheet thicknesses, numbers and positions of spars. The computed beam properties are stored in a manner suitable for the assembly of the stiffness and mass matrices by LFM's in-house CSD code. Also, weight information and data needed for the estimation of stresses in the sections are determined.

The accuracy of the resulting beam model was validated against a shell model created with the commercial FE code MARC, resulting in sufficient accuracy for the intended application.

The TBG was included in a simple optimisation process scheme built with a command line script. This was used to prove the suitability of the BG and to investigate the influence of structural design parameters.

References

1. Martins, J.R.R.A.: A Coupled-Adjoint Method for High-Fidelity Aero-Structural Optimization. Doctoral thesis, Stanford University, Stanford, USA (2002)
2. Kroll, N., Fassbender, J.K.: MEGAFLOW – Numerical flow simulation for aircraft design. In: Notes on Numerical Fluid Mechanics and Multidisciplinary Design, vol. 89. Springer, Heidelberg (2005)
3. Braun, C.: Ein modulares Verfahren für die numerische aeroelastische Analyse von Luftfahrzeugen. Doctoral thesis, RWTH Aachen, Germany (2007)
4. Boucke, A.: Kopplungswerkzeuge für aeroelastische Simulationen. Doctoral thesis, RWTH Aachen, Germany (2003)
5. Braun, C., Boucke, A., Ballmann, J.: Numerical Prediction of the Wing Deformation of a High-Speed Transport Aircraft Type Windtunnel Model by Direct Aeroelastic Simulation. In: International Forum on Aeroelasticity and Structural Dynamics (IFASD) 2005, Munich, Germany, paper IF-147 (2005)
6. Reimer, L., Braun, C., Ballmann, J.: Analysis of the Static and Dynamic Aero-Structural Response of an Elastic Swept Wing Model by Direct Aeroelastic Simulation. In: International Council of the Aeronautical Sciences (ICAS) 2006, Hamburg, Germany, paper ICAS 2006-10.3.3 (2006)
7. Wellmer, G., Chen, B.H., Braun, C., Reimer, L., Ballmann, J.: Numerical Prediction of Aeroelastic Effects on Twin-Sting-Rig Mounted Models for Rear Fuselage and Empennage Flow Investigation in Transonic Windtunnel. In: International Forum on Elasticity and Structural Dynamics (IFASD) 2007, Stockholm, Sweden, paper IF-071 (2007)
8. Ballmann, J.: Transonic High Reynolds Number Aero-Structural Dynamics Experiments in the European Transonic Windtunnel (ETW). In: International Forum on Aeroelasticity and Structural Dynamics (IFASD) 2007, Plenary Lecture, Stockholm, Sweden (2007)
9. Reimer, L., Braun, C., Chen, B.H., Ballmann, J.: Computational Aeroelastic Design and Analysis of the HIRENASD Wind Tunnel Wing Model and Tests. In: International Forum on Aeroelasticity and Structural Dynamics (IFASD) 2007, Stockholm, Sweden, paper IF-077 (2007)
10. Korsch, H., Dafnis, A., Reimerdes, H.G., Ballmann, J., Braun, C., Brakhage, K.H., Olivier, H.: Dynamic Qualification of the HIRENASD Elastic Wing Model. In: Annual Meeting of German Aerospace Association (DGLR) 2006, Braunschweig, Germany, Paper DGLR-2006-045 (2006)
11. Hurka, J., Ballmann, J.: Elastic Panels in Transonic Flow. AIAA Paper 2001-2722 (2001)
12. Massjung, R.: Discrete conservation and coupling strategies in nonlinear aeroelasticity. Computer Methods in Applied Mechanics and Engineering 190, 91–102 (2006)
13. Neuber, H.: Technische Mechanik, zweiter Teil. Springer, Heidelberg (1971)

14. Czerwenka, G., Schnell, W.: Einführung in die Rechenmethoden des Leichtbaus I. BI Hochschultaschenbücher (1967)
15. Heinrich, R.: Spezifikation des Testfalls für den Hauptmeilenstein M8.1 im Verbundvorhaben MEGADESIGN. Deutsches Zentrum für Luft- und Raumfahrt, Institut für Aerodynamik und Strömungstechnik. Braunschweig, Germany (2005)
16. Pamadi, B.N.: Performance, Stability, Dynamics, and Control of Airplanes. AIAA Education Series (1998)
17. Raymer, D.P.: Aircraft Design: A Conceptual Approach. AIAA Education Series (1992)
18. Weissinger, J.: The Lift Distribution of Swept-Back Wings. NACA Technical Memorandum No. 1120 (1947)
19. Klimmek, T.: Multidisziplinäre Optimierung von Flugzeugflügeln. ASIM Fachgruppe 1, Grundlagen und Methoden der Modellbildung und Simulation. Siegen, Germany (2004)

Development and Application of TAU-ANSYS Coupling Procedure

Ralf Heinrich

DLR, Institute of Aerodynamics and Flow Technology,
Lilienthalplatz 7, 38108 Braunschweig, Germany
Ralf.Heinrich@dlr.de

1 Introduction

Aeroelastic effects can play a significant role in wind-tunnel testing under high Reynolds number conditions, as shown for example within the European project HIRETT [12]. Due to the high static pressure in the wind tunnel the deformations can reach a magnitude which cannot be neglected, as shown for example in [3]. And also within the project EUROLIFT I a discrepancy has been found between the computed polar and the polar measured in the ETW wind tunnel. The discrepancy could be attributed to either model deformation, a non-uniform onflow due to the presence of wind tunnel walls or the influence of specific geometry installation effects. The objective of the work of DLR in subtask 1.1.1 of EUROLIFT II has been to investigate numerically the effect of the wind tunnel model deformation under ETW conditions. This can only be done, if the interaction between fluid and structure is taken into account properly. Therefore a process chain has been developed. It allows a coupling between CFD and computational structure mechanics (CSM). Before the results of the numerical study are discussed in Sect. 3, the chain itself and the components of the chain are described. On the CFD side the user has the choice between two DLR codes. The block-structured FLOWer code [7] can be used as well as the hybrid (unstructured) TAU code [13]. Both codes have been mainly developed within the projects MEGAFLOW I–II [7] and MEGADESIGN [8]. The adaptations of the codes needed to use them in the coupling environment will be described in Sect. 2. Because of the deformations on the structure side the volume mesh has to be deformed as well. The TAU code contains a powerful mesh deformation tool for unstructured meshes [4], while a mesh deformation tool was still missing for block-structured meshes. A new tool based on radial basis functions has been developed, which is able to deform block-structured as well as hybrid meshes. This tool has also been used for the numerical study and the validation discussed in this report. It is described in subsection 2.4. On the structure side the commercial finite element (FEM) code ANSYS [16] is currently used. Since no source code of this tool is available, the code itself cannot be adapted for the needs of the coupling environment. However, within ANSYS the user has

N. Kroll et al. (Eds.): MEGADESIGN and MegaOpt, NNFM 107, pp. 151–167.
springerlink.com

the possibility to use a scripting language called APDL (ANSYS parametric design language). This scripting language enables the development of an interface for the communication between the components of the process chain. For the coupling of a CFD and a CSM code, data has to be exchanged between both disciplines across the so called contact surface of the fluid and structure domain. The CSM code needs the aerodynamic loads (forces or pressure and skin-friction) provided by the CFD code to calculate the deformations of the structure. Based on the surface deflections of the contact (or coupling) surface, the CFD mesh has to be deformed in order to provide corrected aerodynamic loads. If the surface discretization on CFD and CSM side is the same, this is an easy task. But if the discretization is different, as is usually the case, the data has to be interpolated properly. For this purpose a new interpolation module has been developed, as described in subsection 2.3.

2 The Process-Chain and Its Components

In Fig. 1 left a flow chart shows the sequence of steps followed during a coupled computation to achieve the aeroelastic equilibrium. Figure 1 right shows a corresponding convergence history of the density residual and the lift of the CFD code for a generic wing body configuration in the transonic regime. Usually in a first step a CFD computation is performed using the undeformed geometry. Output of the CFD computation is the aerodynamic load on the coupling surface (pressure and friction coefficient or force distribution). The aerodynamic loads have to be mapped from the nodes of the CFD surface to the nodes of the CSM surface. Therefore the CFD and CSM code have to export the coupling surfaces. For the data transfer the interpolation module is required. It imports the coupling surfaces, the data to be interpolated and performs the mapping from CFD nodes to CSM nodes. Depending on the quantity to be interpolated different interpolation techniques can be selected. After mapping the aerodynamic loads from the CFD to CSM side the structure code is started. Output of the structure code is the deformation of each structural node of the CSM coupling surface. These have to be interpolated to the nodes of the CFD coupling surface. Because of the change of the CFD surface the volume mesh has to be deformed as well. The input for the volume mesh deformation are the undeformed CFD mesh and the deflections of the surface nodes. Now the second loop of the coupling procedure can be done starting again with a CFD computation. The new (deformed) volume mesh and the last CFD solution are used as input for the second CFD computation. In the convergence history (Fig. 1, right) we see that the density residual is increased again and the lift coefficient changes its value. After a number of iterations a converged state is reached again. The aerodynamic loads are transferred again to the CSM surface using the interpolation module. Then a CSM solution is computed and so on. The process is repeated until the equilibrium state is reached or a user specified number of iterations have been performed. To determine if the equilibrium state is reached the solution of the actual and the previous coupling cycle is compared. If the change is smaller than

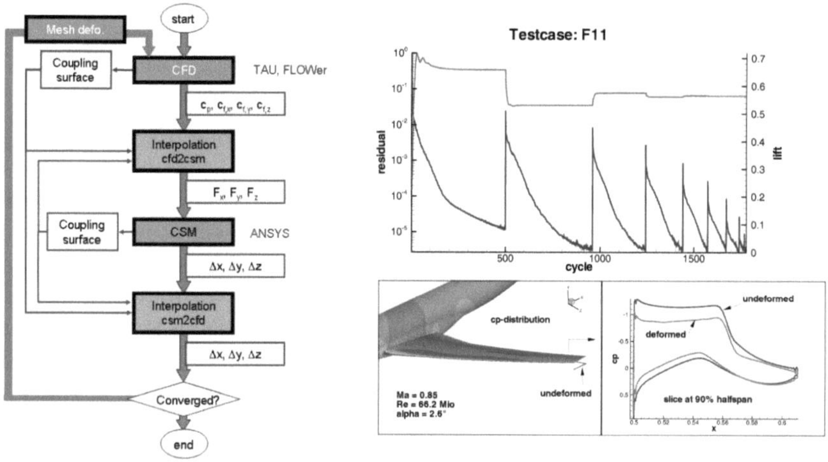

Fig. 1. Left: Flowchart of a coupled computation. Right: Convergence history of a coupled computation for the generic DLR F11 configuration under cruise conditions.

a prescribed value the coupled process is stopped and the aeroelastic equilibrium is assumed to be reached. As reference quantity for the convergence criterion the lift or the maximum deflection can be used, for example.

2.1 Aerodynamic Codes

As previously mentioned the user has the choice between the block-structured FLOWer code and the hybrid (unstructured) TAU code. Both codes are well established simulation tools for solving the Reynolds averaged Navier–Stokes (RANS) equations. They are in daily use at DLR, universities and aerospace industry. For details the reader is referred to [7, 8, 13]. Within subtask 1.1.1 of EUROLIFT II only TAU has been used. For the communication with the interpolation module, an interface is needed to transfer the relevant data. To specify the coupling surface the xyz-coordinates are needed and a connectivity list. Each xyz-coordinate is associated with a node index. These indices are used in the connectivity list to specify the surface elements of the coupling surface. Triangles as well as quadrilaterals can be used as element type. Each surface element is associated with a "boundary marker". The boundary marker can be used to group a selection of surface elements to components like e.g. wing, slat or flap. This is very useful while setting up a coupled computation (see subsection 2.3). For each surface node variables can be ex- or imported. Export values for the CFD code are currently pressure and friction coefficients in xyz-direction or forces. Import values are the deflections in xyz-direction. In the current implementation the transfer of data is realized with file IO. As data format the AMIF (aerodynamic mesh interface format) specification of MSC [18] is used. However, it should be emphasized that a change of the data format or a switch to direct communication with the interpolation module can be done with limited effort.

2.2 Structure Code

On the structure side the well known commercial FEM code ANSYS is used. For details the reader is referred to [16]. No source code is available, so interfaces cannot be plugged into ANSYS directly. However, within ANSYS a powerful scripting language called APDL can be used. In our case we make use of two ANSYS scripts. One script is used to export the coupling surface in the same format as the CFD code. This is done only once at the beginning of a coupled computation. A second script is used to import the structural loads (which have been interpolated from the CFD nodes to the CSM nodes), to perform a FEM analysis and to export the resulting deflections of the CSM nodes in AMIF.

2.3 Interpolation Module

The task of the interpolation module is the mapping of aerodynamic loads from the CFD to the CSM coupling surface and the mapping of deflections from the CSM coupling surface back to the CFD coupling surface. Depending on the quantity to be interpolated different interpolation techniques can be selected. Main input of the interpolation module is:

- CFD and CSM coupling surfaces
- Quantities to be interpolated or calculated
- For each quantity the interpolation technique and direction has to be specified (source and destination coupling surface)

In many cases different coordinate systems and scalings are used on the CFD and CSM side. Therefore the user can specify (per input parameters) a number of transformations for each coupling surface resulting in a common coordinate system. Translation, rotation (around user specified axis and angle) and scaling can be selected as transformation type.

Interpolation of Aerodynamic Loads

Figure 2 left shows the outer wing region of the CFD and the CSM coupling surfaces of a generic wing. On both sides a similar mesh resolution is used. The shape of both meshes matches well and a linear interpolation of the pressure or pressure coefficient is sufficient for the mapping of the aerodynamic loads. Based on the interpolated pressure corresponding forces can be calculated for each surface node. In Fig. 2 right the pressure distribution on the CFD side is compared to the interpolated values on the CSM side. A good agreement of input and interpolated values could be achieved. In some cases the agreement of the coupling surfaces is not as good as in the previous example or no connectivity on the CSM side is given (scattered data). Then a linear interpolation of the pressure or other quantities is not possible. Instead the user can select a nearest neighbor search for the forces, see Fig. 3. For a given point i on the

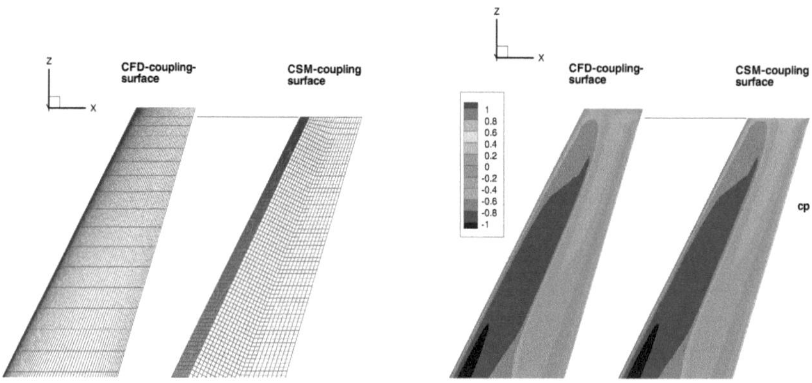

Fig. 2. Left: CFD and CSM coupling surface for generic LANN wing [14] Right: Comparison of calculated c_p-distribution on CFD side and linear interpolated c_p-distribution on CSM side (generic LANN wing, Ma= 0.82, $\alpha = 0.6°$, tip region, inviscid)

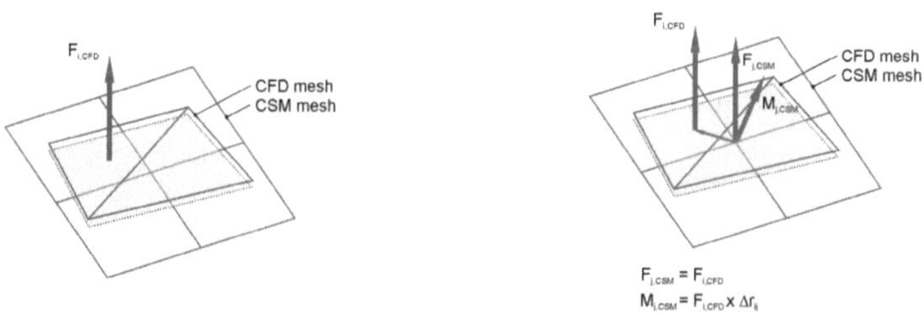

Fig. 3. Mapping of a force from CFD side to nearest neighbor on CSM side

CFD side the nearest neighbor on the CSM side is searched. If a connectivity on the CSM side is given (like in the figure), the nearest point with the same normal orientation is selected. If the nearest point j is found, $F_{i,\text{CFD}}$ is added to the actual value of $F_{j,\text{CSM}}$ (more than one CFD force can be mapped to the same CSM node). Additionally a moment $F_{i,\text{CFD}} \times \Delta r_{ij}$ is mapped to the node j, because the force is moved. This ensuress a conservative interpolation scheme with respect to the force and moment balance on the CFD and CSM side. Figure 4 shows the influence of the interpolation scheme for the previous example. On the left we see the CFD surface mesh and a slice through the surface mesh at 90% half wing span. The symbols belong to the resulting profile and pressure coefficient distribution without coupling (on the undeformed surface). Three coupled computations have been made, all until the equilibrium state is reached. The solid thick line corresponds to the result using linear interpolation of the pressure coefficient, the dashed line corresponds to the results achieved with the conservative force interpolation described before and the dash-dotted line is computed using the nearest neighbor search for c_p. There are no large

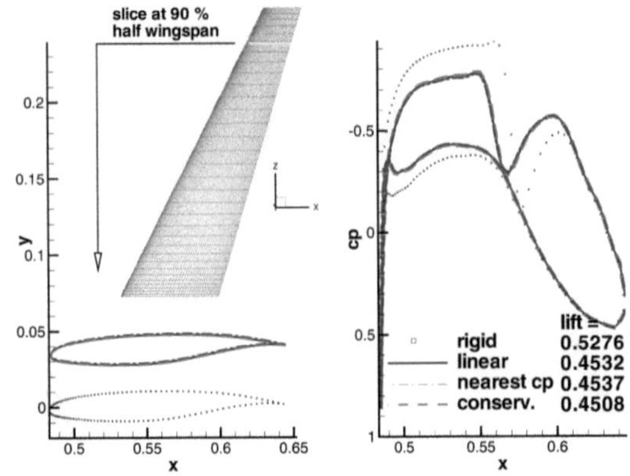

Fig. 4. Comparison of resulting c_p-distribution close to the wing tip in aeroelastic equilibrium in dependency of the interpolation technique used (generic LANN wing [14], Ma= 0.82, $\alpha = 0.6°$, inviscid)

differences observed between the coupled computations. The agreement in the global lift in equilibrium is also very good.

Interpolation of Deflections

Of course a nearest neighbor search is not an appropriate technique for the interpolation of the deformations. A linear interpolation of the deflections can also become problematic if the matching of the coupling surfaces is not good, the resolution is very different or if no connectivity is given on the CSM side. As is known from literature, see for example [1], interpolation schemes based on radial basis functions are appropriate for this task, even if only scattered data is present. This approach is also followed here. These techniques are very well suited for smooth functions, and usually the deformation of aerodynamic components are smooth. Figure 5 shows for the previous example the deformations in y-direction (normal to the wing plane). The solid line corresponds to the deformations computed by ANSYS, the dashed lines to the interpolated values.

Problems can arise if different components of an aircraft have to be taken into account, like for example an aircraft in start- or landing configuration. Figure 6 top shows such a configuration. In the lower half of the figure the deflection in z-direction (normal to the wing plane) is shown in a slice at $y = 1$ m (71% of half wing span). The symbols correspond to the computed deformations. The different behaviour of slat, wing and flap becomes clear. For the same x-coordinate the deformation of the slat and flap is slightly higher. In other words: The gaps between wing and the two components have changed. This cannot be handled properly by a single interpolation function properly, as becomes clear from the interpolated y-displacement (solid line, left). This situation can be solved by using local interpolation functions, which are calculated for each

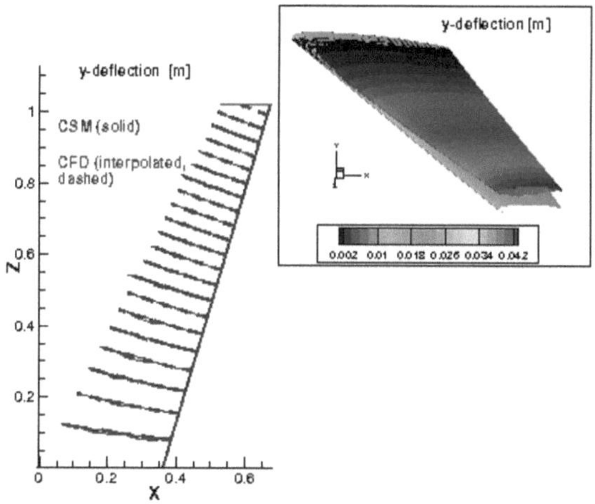

Fig. 5. Computed and interpolated deflections normal to the wing plane (generic LANN wing, lower and upper surface)

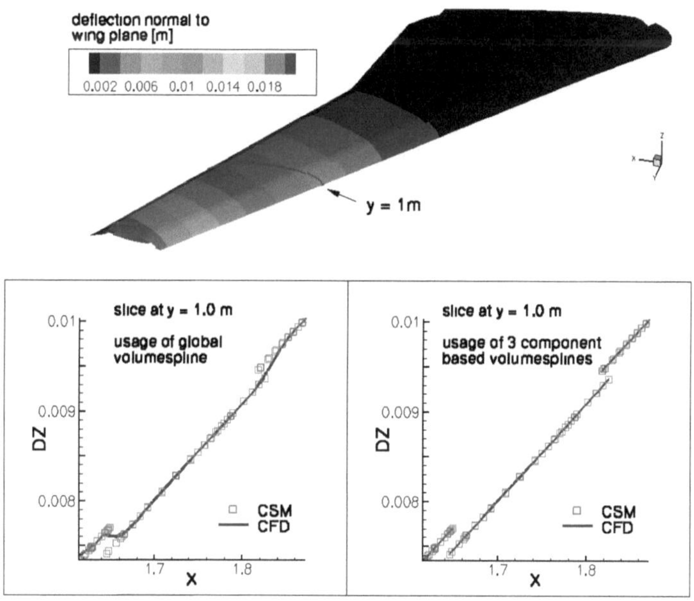

Fig. 6. Computed (symbols) and interpolated deflections (red line) normal to wing plane for a high lift application DLR F11 wing in starting configuration)

component. Now the agreement of calculated and interpolated displacement is excellent as shown in the lower right part of Fig. 6.

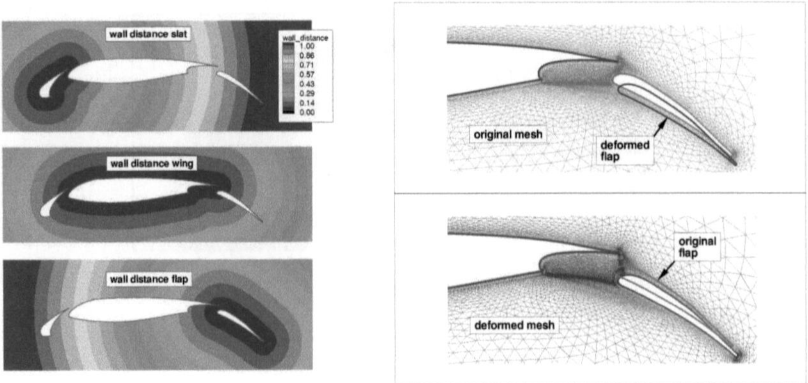

Fig. 7. Left: Wall distance distribution for each component of 2D high-lift application (TC11 from EUROLIFT I) Right: Comparison of original and deformed mesh of TC11 configuration (flap surface has been translated)

2.4 Volume Mesh Deformation

If the CFD surface mesh is deformed, the volume mesh has to be adapted as well. For TAU a powerful mesh deformation procedure has already been developed [4], but for the block-structured FLOWer code no robust tool was available. Therefore a new tool has been developed, which can be used for block-structured meshes as well as for hybrid unstructured meshes. This tool has been used for all applications shown in this paper. The basic idea is to apply the interpolation functions calculated for the surface mesh deformation to the nodes of the volume mesh as well. Additionally, the resulting deflections can be superimposed with a blending function based on the wall distance in order to achieve zero deflections for a specified distance from the wall, for example at the farfield. This idea can also be used if more than one interpolation function is used like for a high-lift application. Three interpolation functions can be created, one for each component. A global interpolation function is created by weighting of each component function based on the wall distances. Therefore wall distances are calculated for each relevant component, as sketched in Fig. 7 left.

The deformation of a mesh node is calculated as

$$\Delta\mathbf{r}(x,y,z) = \sum_{i=1}^{n} w_i \Delta\mathbf{r}_i(x,y,z) \,, \tag{1}$$

where $\Delta\mathbf{r}_i(x,y,z)$ is the resulting deformation based on interpolation function i, and is the corresponding weight. The sum of all weights is equal to one:

$$\sum_{i=1}^{n} w_i = 1 \,. \tag{2}$$

The weights are calculated as

$$w_i = \frac{\left(\frac{1}{\text{walldist}_i}\right)^p}{\left(\frac{1}{\text{walldist}_j}\right)^p} . \quad (3)$$

In equation (3) $\text{walldist}_j(x, y, z)$ is the wall distance of the mesh node (x, y, z) relative to component j and p is a user specified exponent. If the wall distance for component j is smaller than a specified epsilon value, then w_j is set to 1 and the other weights are set to zero. Figure 7 right shows as an example the movement of a flap for a 2D high lift application. For the components slat and wing, zero deflection is prescribed, whereas a translation downward and in flow direction is used. The top half shows the original undeformed mesh. The bottom half shows the result of the deformation using the global interpolation function based on the blended component interpolation functions. Several tests have been made to check the resulting quality of the deformed meshes. For details the reader is referred to [5].

2.5 Coupling Management

Currently a coupled computation is controlled by a shell-script. All components can run locally on the computer where the script is launched. Usually, however, only a part of the components runs locally. The CSM code usually runs on a separate computer (question of license model used). Therefore all input files that are needed are copied automatically to a temporary directory on the remote host (using secure copy). The ANSYS job is then started via secure shell. After successful computation the file containing the resulting deformations is copied back as input for the interpolation module. For large scale applications the CFD codes usually run on batch computers in parallel. Depending on the target host, a job script is created automatically. All input files needed for FLOWer/TAU are transferred to a temporary directory of the remote host via secure copy, and the resulting surface solution is transferred back to the local (controlling) host. After finishing the CFD computation control is given back to the local shell script. For each coupling cycle the user has the possibility to adjust parameters of the CFD computation. For example the angle of attack can be increased after maybe 10 coupling cycles to perform a coupled computation for the next point of a polar.

2.6 Validation

Rectangular Swept Wing

For a first validation of the process chain the rectangular swept reference wing of the colaborative research area 401 of the Technical University of Aachen, dealing with aeroelastic research, is selected [15]. At the "Lehr- und Forschungsgebiet für Mechanik" in Aachen the DLR FLOWer code has been coupled to a Timoshenko beam model [2]. The approach has been successfully validated during the last

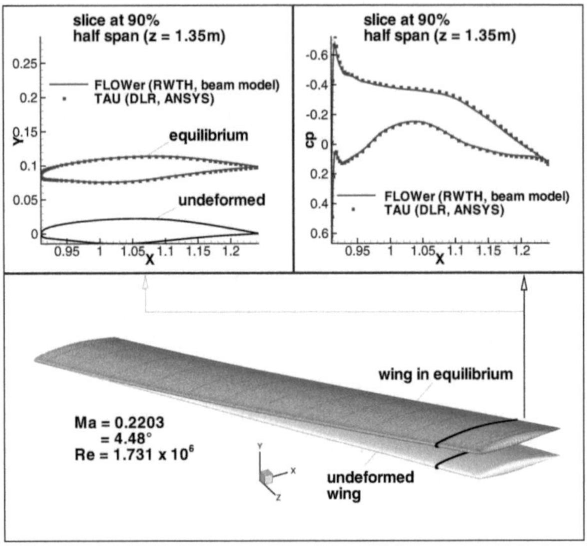

Fig. 8. Comparison of results of solutions of TAU coupled with ANSYS and FLOWer counped with a Timoshenko beam model

years, see for example [3, 15]. The wing, shown in Fig. 8, has a span of 1.5 m and a sweep angle of 34°. A structured CFD mesh, computational results and a beam model have been provided by the colleagues in Aachen. Based on the characteristics of the beam model a FEM model has been created for use with ANSYS. The block-structured FLOWer mesh has been converted to a hybrid mesh, which can be used with the TAU code. The coupled computation has been made with TAU coupled to ANSYS using the DLR process chain. These results have been crosschecked with validated results of Ballmann et al., see Fig. 8. In the lower half of the figure the undeformed wing and the wing in static equilibrium is shown for a Mach number of 0.22, a Reynolds number of 1.7×10^6 and an angle of attack of 4.48°. Slices at 90% half span are shown on the top half. The reference results are plotted as solid line, the results of TAU coupled with ANSYS with symbols. The equilibrium shape as well as the pressure coefficient distribution is in excellent agreement with the reference solution.

Landing Aircraft Configuration

Within the DLR internal project FORMEX measurements of the deformations of the F11 model in landing configuration have been made during the wind tunnel tests in the DNW-NWB in Braunschweig. The on flow Mach number is of course the same like for landing configuration in EUROLIFT (0.2), but the Reynolds number is with 1.57×10^6 one order of magnitude lower compared to the EUROLIFT ETW experiment. The static pressure is 1 bar compared to 2.59 bar in ETW (see also Sect. 3). Although the static pressure is relatively small, visible

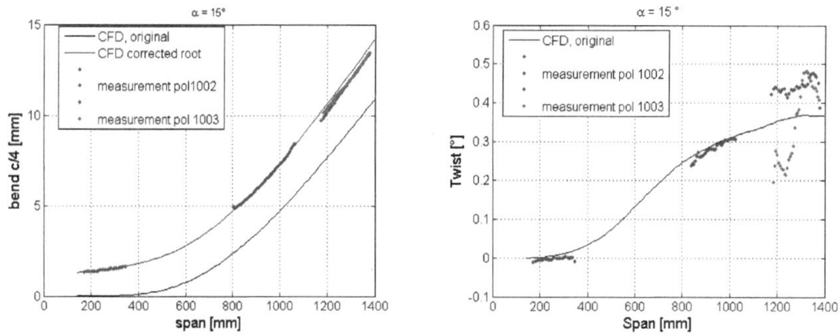

Fig. 9. Comparison of measured and computed bending at $c/4$ (left) and twist (right) for $\alpha = 15°$

deformations up to 15 mm have been found. Figure 9 left shows the comparison of measured and computed deformations normal to the wing plane at the local quarter point for an angle of attack of 15° and Fig. 9 right shows the corresponding twist distribution. The structure model used in the coupled computations is the same like described in Sect. 3.1. The agreement of the bending as well as the twist is very well. For the comparison of the bending distribution the numerical results have been translated, because there is a constant shift between numerical and experimental data due to the mounting of the wind tunnel model. For more details the reader is referred to [5].

3 Study on Model Deformation under ETW Conditions

Within the European project EUROLIFT I experiments have been made for high lift applications under high Reynolds number conditions in the ETW wind tunnel in Cologne. During the experiments visible deformations of the wing have been noticed. Of course the question arose whether the deformations under high loads have a large or only a small, negligible influence on the aerodynamic behavior. This question has been addressed in task 1.1.1 in the follow-on project EUROLIFT II. In the following, results achieved with the new process chain are shown. The DLR-F11 (KH3Y) configuration in start and landing configuration has been investigated. For a coupled computation several difficulties have to be solved, which are partly caused by the high complexity of the geometry and the resulting computational meshes. Main aspects have already been addressed in previous chapters. Figure 11 shows some details of the complexity of the landing configuration (TC462) including. fuselage, belly fairing, wing, slat, flap, flap-track fairings, pylon, engine, gondel strake, slat tracks and pressure tube bundles. The mesh contains in total 14 million nodes. The starting configuration used (TC322) is slightly simpler (fuselage, belly fairing, wing, slat, flap, flap-track fairings). The mesh contains 7 million nodes. All meshes used in this investigation have been generated by NLR with the FASTFLO grid generation system.

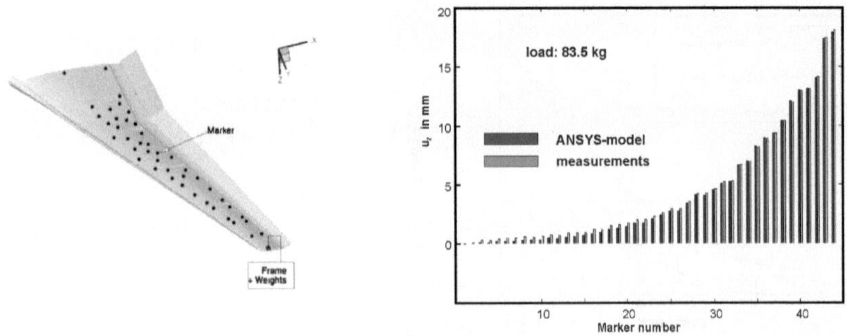

Fig. 10. Left: Position of the markers and the point where the force is applied within the static deformation test. Right: Measured and computed deflections for the marker positions 1–45 for a load of 83.5 kg.

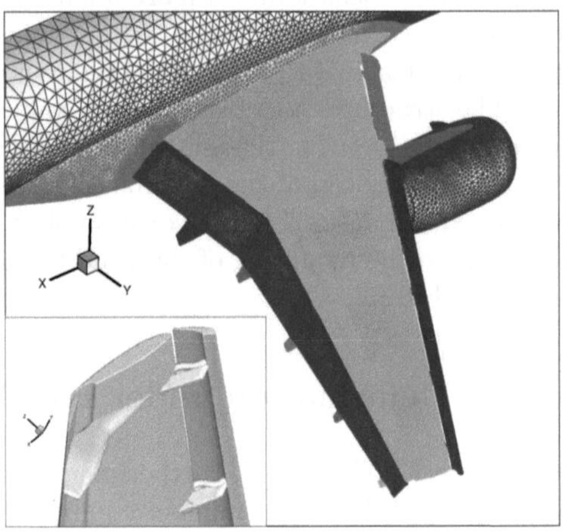

Fig. 11. Details of 3D DLR-F11 landing configuration including fuselage, belly fairing, wing, slat, flap, flap-track fairings, pylon, engine, gondel strake, slat tracks and pressure tube bundels (mesh generated by NLR)

3.1 Structure Model

Structural FEM models have been created by the company Leichtwerk [17] with the commercial FEM package ANSYS. 29000 elements have been used to mesh the starting configuration and 21000 for the landing configuration. Mainly quadratic volume elements have been used (SOLID95) for meshing. Only for the tracks another element type has been selected (SHELL91).

For validation and calibration of the FEM models designed within task 1 activities of EUROLIFT II DLR has performed a static deformation test of the KH3Y (F11) configuration [9]. The slat and flaps were deflected according to the landing settings. The lower side of the wing was equipped with 45 markers suited for deformation detection via a stereo camera system. The wing was loaded with different weights imposed at one single point close to the wing tip, see Fig. 10 left. The loads range from 3.5 kg up to 83.5 kg. Figure 10 right shows a comparison of the measured deflections normal to the wing plane for all 45 markers and the deflections computed with ANSYS for a loading of 83.5 kg). The agreement of measured and computed data is good.

3.2 Results of Numerical Study

The following table gives an overview of the computations done. To check the influence of the deformations, reference computations without coupling have been done first. They were used as initial input for the coupled solutions.

Table 1. Overview of computations with and without coupling for starting andlanding configuration

Configuration	Angle of attack (no coupling)	Angle of attack (with coupling)
Start conf. TC322 (Ma = 0.176, Re = 15×10^6, $p_\infty = 2.95$ bar)	$7°, 11.9°, 18.5°, 20°$	$7°, 11.9°, 18.5°$
Landing conf. TC462 (Ma = 0.203, Re = 15×10^6, $p_\infty = 2.59$ bar)	16.5°	16.5°

Figure 12 left shows the undeformed shape of the wing close to the tip region in comparison with the shape in aeroelastic equilibrium for three angles of attack (starting configuration). The maximum deflection in z-direction ranges from about 22 mm for $\alpha = 7°$ up to 32 mm for $\alpha = 18.5°$ which is indeed a visible deformation. Of course these values have to be put into relation to the half span of the wing which is 1400 mm. Then we end up with 1.6% for $\alpha = 7°$ and 2.3% for $\alpha = 18.5°$. For the landing configuration, see Fig. 12 right ($\alpha = 16.5°$), the maximum computed deflection has been computed as 34.6 mm (2.5%). Figure 13

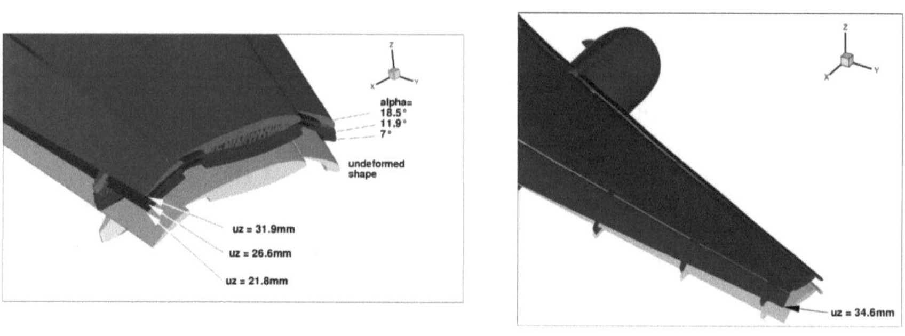

Fig. 12. Surface of configuration in aeroelastic equilibrium for TC322 (starting, left) and TC462 (landing, right) in comparison to undeformed shape

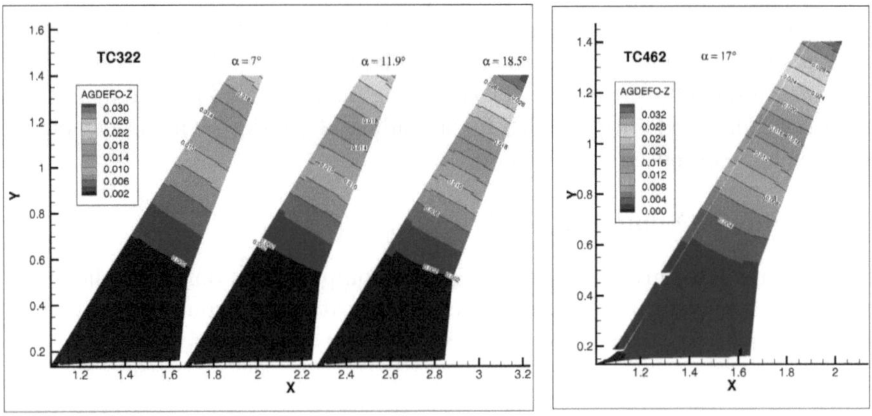

Fig. 13. Left: Computed deformations for starting configuration for three angles of attack. Right: Computed deformations for landing configuration for $\alpha = 17°$.

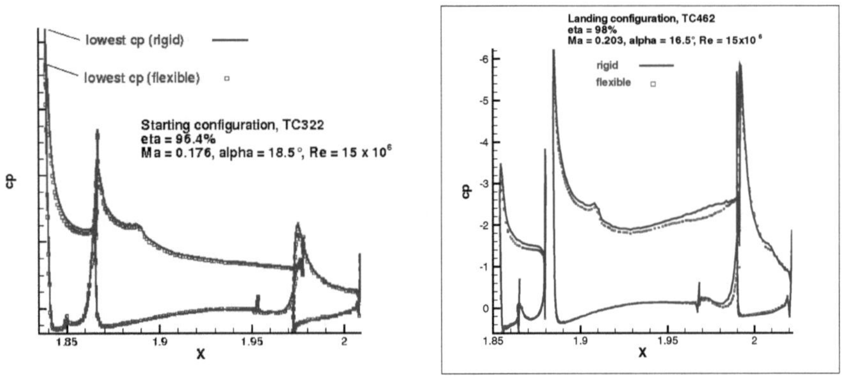

Fig. 14. Comparison of resulting c_p distribution with (symbols) and without coupling (solid line) close to the tip for the starting configuration (left) and landing configuration (right)

shows additionally isolines of the deformations normal to the wing plane for the starting and landing configuration. For both configurations a small gap change for slat and flap is visible. The relatively small maximum deformation of the wing explains why even close to the wing tip the change of the pressure coefficient distribution is visible, but relatively small, see Fig. 14 left for starting and Fig. 14 right for landing configuration. In Fig. 15 the lift versus α curve is plotted with and without coupling in comparison to experimental data available from EUROLIFT I. As could be expected, with coupling the lift is decreased slightly, because of the reduction of the wing's twist caused by bending. All in all it can be summarized that aeroelastic effects are visible for current high lift applications, but they seem to play only a secondary role. Within EUROLIFT II experiments

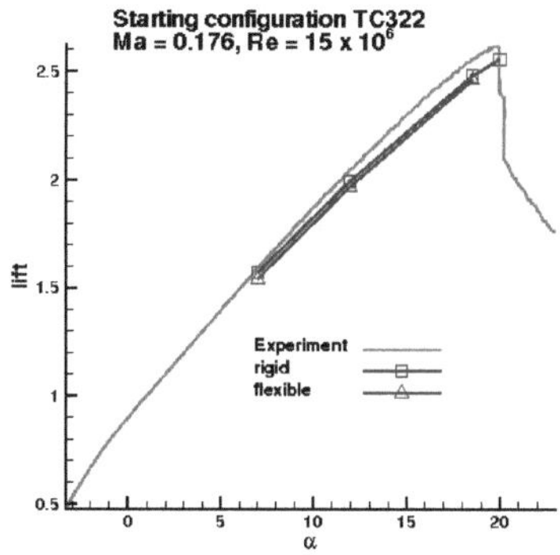

Fig. 15. Lift versus angle of attack for starting configuration

have been made for current geometry using a 70% higher Reynolds-number. Here a larger impact of aeroelastic effects is expected. Computations using the process chain developed are planned.

4 Summary and Conclusions

Within EUROLIFT II a process chain for the computation of the aeroelastic equilibrium, including FLOWer/TAU on CFD side and ANSYS on CSM side has been developed. Two new components have been created. One for the component based data interpolation between CFD and CSM coupling surfaces in both directions and a volume mesh deformation tool, which can be used for either block-structured or hybrid, unstructured meshes. The process chain can be run automatically without user interaction on a single workstation as well as on a heterogeneous network including multiprocessor batch computers on the CFD side. It has been proven, that the process chain can also be used for very complex high-lift configurations. The mapping of the deformations from the CSM surface to the CFD surface is realized with interpolation functions based on radial basis functions. It has been shown, that it is of great importance, to use component based interpolation functions to transfer e.g. the gap-change between different components like wing and flap correctly. A first successful validation for an isolated rectangular swept wing has been presented, showing excellent agreement with results of a validated process chain developed at the Technical University of Aachen. But also for very complex high lift applications a good agreement has been found for bending and twist distributions measured in the DNW-NWB in Braunschweig (DLR project FORMEX). Within EUROLIFT II coupled computations have been performed for the F11 starting and landing configuration under ETW conditions to check, if aeroelastic effects play an important role compared to other geometrically effects investigated within the other subtasks of task 1.1. Therefore a FEM ANSYS model has been created and validated with experimental data from a static deformation test. It has been found, that aeroelastic effects are visible, but they seem to be only of secondary interest for the static pressures investigated. As could be expected, with coupling the lift is decreased slightly, because of the reduction of the wing's twist caused by bending. Within EUROLIFT II also experiments have been made for higher static pressures. Here a larger impact of deformation effects is expected. It is planned to apply the process chain to one of these test cases in the near future.

Acknowledgement

The author likes to thank G. Britten, C. Braun, L. Reimer and J. Ballmann of the Lehr- und Forschungsgebiet für Mechanik, RWTH Aachen, for making available the beam model and the reference data of the wing test case described in Sect. 2.6.

References

1. Beckert, A., Wendland, H.: Multivariate interpolation for fluid-structure-interaction problems using radial basis functions. Aerospace, Science and Technology (AST) 5, 125–134 (2001)
2. Britten, G., Braun, C., Hesse, M., Ballmann, J.: Computational Aeroelasticity with Reduced Structural models. In: Ballmann, J. (ed.) Flow Modulation and Fluid-Structure Interaction at Airplane Wings. Notes on Numerical Fluid Mechanics and Multidisciplinary Design, vol. 84, pp. 275–299 (2003)
3. Braun, C., Boucke, A., Ballmann, J.: Numerical Study of the Influence of Dynamic Pressure and Deflected Ailerons on the Deformation of a High Speed Wing Model. In: The 7th Results and Review Workshop of the HLRS, at the High Performance Computing Center Stuttgart, October 4-6 (2004)
4. Gerhold, T.: Efficient Algorithms for Mesh Deformation. In: ODAS symposium Toulouse (2006)
5. Heinrich, R.: CFD study on model deformation using a TAU-ANSYS coupling procedure, EUROLIFT II technical report, D1.1.1-9 (2007)
6. Himisch, J., Horstmann, K.H., Streit, T., Nagel, B.: Design of a retrofit winglet for a transport aircraft. In: ODAS symposium Toulouse (2006)
7. Kroll, N., Fassbender, J.K. (Hrsg.): MEGAFLOW – Numerical Flow Simulation for Aircraft Design, 10.–11.12.2002. Notes on Numerical Fluid Mechanics and Multidisciplinary Design (NNFM), vol. 89. Springer, Closing Presentation DLR Project MEGAFLOW, Braunschweig (de), Heidelberg (2002)
8. Kroll, N., Gauger, N., Brezillon, J., Becker, K., Schulz, V.: Ongoing Activities in Shape Optimization Within The German Project MEGADESIGN. In: CD of ECCOMAS 2004, Jyväskylä (FI), July 24-28 (2004)
9. Puffert-Meißner: Static Deformation Test, EUROLIFT II report (1995)
10. Rudnik, R.: Towards CFD Validation for 3D High Lift Flows – EUROLIFT. In: Book of Abstracts – ECCOMAS CFD 2001, S. 70, ECCOMAS Computational Fluid Dynamics Conference 2001, Swansea (gb), 04.–07.09 (2001) ISBN 0905091116
11. Rudnik, R., Geyr, H.v.: The European High-Lift Program EUROLIFT II. In: Aeronautics Days 2006, Vienna, June 19-21 (2006)
12. Rolston, S., Elsholz, E.: Initial Achievements of the European High Reynolds Number Aerodynamic Research Project HiReTT, AIAA paper 2002-424 (2002)

13. Schwamborn, D., Gerhold, T., Heinrich, R.: The DLR TAU-Code: Recent Applications in Research and Industry. In: Proceedings of European Conference on Computational Fluid Dynamics, ECCOMAS CDF 2006, Delft The Netherland (2006)
14. Zwaan, I.R.J.: LANN Wing Pitching Oscillations, Compendium of Unsteady Aerodynamic Measurements, AGARD-R-702 (August 1982)
15. SFB 401: Strömungsbeeinflussung und Strömungs-Struktur-Wechselwirkung an Tragflügeln, Arbeits- und Ergebnisbericht (2002)
16. `http://www.ansys.com`
17. `http://www.leichtwerk.de`
18. `http://www.mscsoftware.com`

Fluid-Structure Coupling: Simplified Structural Model on Complex Configurations

Eberhard Elsholz

Airbus, Airbusallee 1, 28199 Bremen, Germany

Summary. In order to provide a more realistic aerodynamic simulation of the flying aircraft, a simplified structural method "WingDACC" was coupled with the DLR TAU code and is applied on simple wing/body as well as complex high-lift cases. The results obtained for the flexible wing are considerably improved when compared with rigid wing computations. Details of the technique and the testcases are addressed and the results are discussed.

1 Introduction

It is well known that wing deformation might affect the flow over the wing to a certain degree, especially in the transonic flow regime. This not only holds for flexible real-wing constructions but also for windtunnel models which nearly behave like made from solid material and therefore are of considerable stiffness. Due to this, the pressure distributions over the wing might be affected and therefore the aerodynamic forces are changed, compared to rigid wing assumptions. In order to deal with these effects, a simplified structural model was developed and coupled with the flow solution obtained by the DLR Tau code [1] which then was embedded in a highly automated deformation tool "WingDACC".

2 Short Description of WingDACC Method

The WingDACC tool ("Wing Deformation — Analysis, Correction, Calibration") was developed at Airbus and is a wing deformation method based on

- Simplified deformation theory (cantilever beam bending).
- Simplified structural model (solid material), immediately derived from wing/wing solution cuts in streamwise direction, evaluated by BSURF.
- Specifically designed for TAU-code applications.

Main features

- WingDACC provides the 3D offset of all wing surface points of the CFD mesh.
- Then WingDACC applies volume mesh deformation, such as [1, 2, 3].

N. Kroll et al. (Eds.): MEGADESIGN and MegaOpt, NNFM 107, pp. 169–178.
springerlink.com

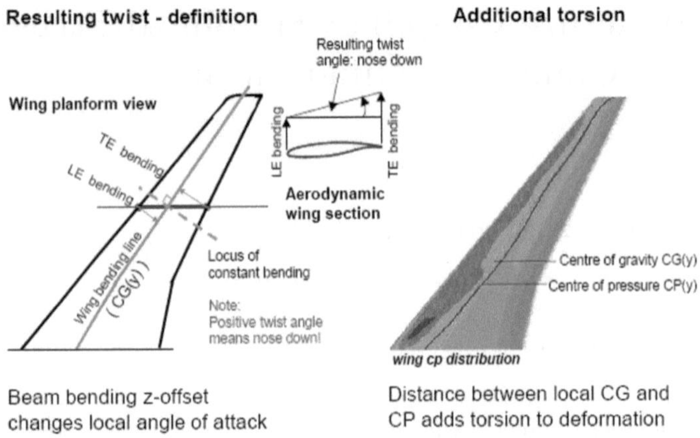

Fig. 1. a) Twist due to bending of a swept wing and b) additional torsion

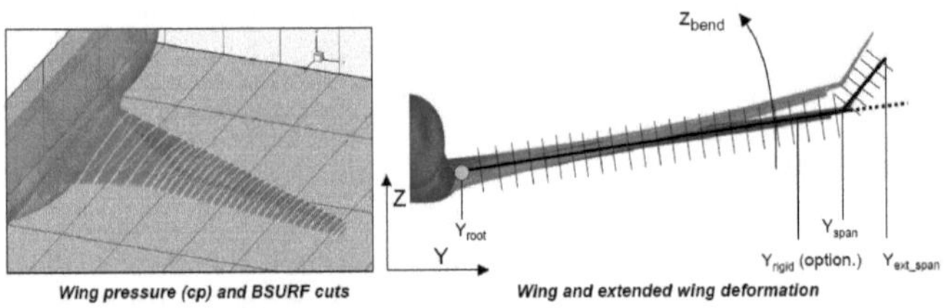

Fig. 2. Wing bending along circular path

This can be done in a single stand-alone step or as a fully integrated CFD-CSM-coupled loop established by the python script Setup_taujob.py The main advantages of the WingDACC tool are seen as follows:

- It is easy to use, with a minimum of input required
- Within a wide range of applications, i.e. from clean wing/body to complex high-lift configurations
- Validated for good accuracy,

where the validation is ongoing for the more complex high-lift cases.

The method takes advantage of the streamwise cuts produced by BSURF to evaluate the spanwise neutral line (center of gravity) for a solid material wing, made from steel or aluminum. Subject to pressure loading, the wing bends up, assuming constant bending normal to the neutral line. This results in direct bending and associated nose-down twist of the aerodynamic sections, see Fig. 1a. In addition, the torsion moment resulting from the center-of-pressure line mismatching the neutral line is taken into account, Fig. 1b.

Then the wing surface points of the CFD mesh are bent-up and twisted where the amount of bending follows a circular path around the wing root axis, Fig. 2.

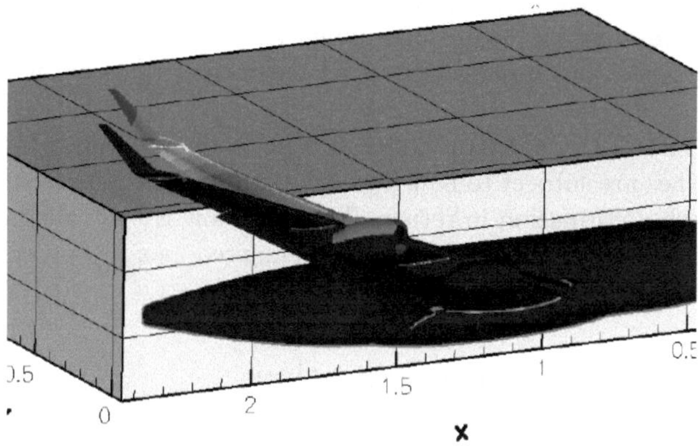

Fig. 3. Sample of surface mesh deformation

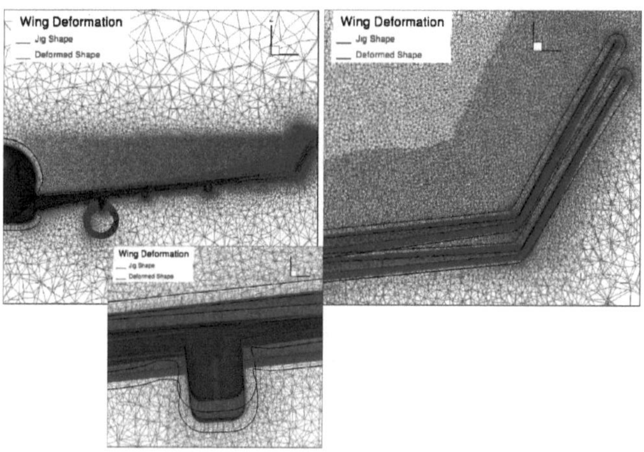

Fig. 4. Sample of volume mesh deformation

Note that the cuts produced by BSURF should be approximately normal to the bending line for increased accuracy.

It is worthwhile to note that the model jig shape of the wing — or vice versa, the offset of the CFD mesh in 1g flight design wing shape from (unloaded) jig shape has to be known. In case of windtunnel models for transonic testing, usually an offset in twist holds but no offset in bending.

In Fig. 3, a sample of successful surface mesh deformation is given, where the colours indicate the amount of deformation from the wing's original shape (blue colour). The next step is the final volume mesh deformation, where different tools [1, 2] may be involved. Fig. 4 gives examples of successful volume mesh deformation. In case of wing add-ons such as engine/pylons, FTFs etc., appropriate section handling has to be applied within the WingDACC procedure, i.e. all spanwise stations intersecting other parts than the wing itself should be dropped.

In case of full CFD-CSM loop between TAU code and WingDACC tool, the deformed meshes are subject to new flow solution, which then are subject to new wing- and mesh deformation in return. The data flow is controlled by the script Setup_Taujob.py, where WingDACC is implemented as a pre-processing step to the TAU code.

3 Results on Cruise Configuration

3.1 Near Design Point

In case of a clean wing body configuration in transonic flow regime, a complete polar was computed for

- Rigid wing at 1g flight design shape and
- Flexible wing assumed as made from solid steel using WingDACC.

Fig. 5 summarizes the main results in comparison to available experiments in ETW (Eurpean Transonic Windtunnel). The improvements of the polar itself as well as the outboard pressure distributions becomes obvious in case of the flexible wing. Residual convergence and accuracy of the coupled CFD-CSM process is shown in Fig. 6. The process was stopped after 5 coupling steps, when the resulting twist did not change any more, see the blow-up view in Fig. 6.

3.2 Off-Design

Flow at low Mach number was also tested in ETW, using the same model as for the high-speed tests before. Computations of rigid and flexible wing in this far

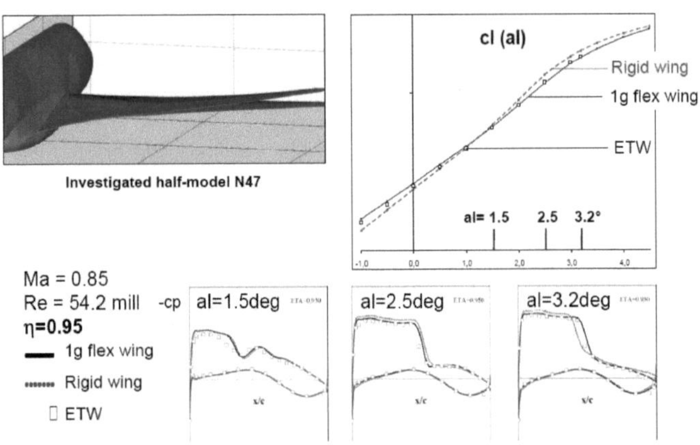

Fig. 5. Results of coupled CFD-CSM on clean wing/body: Lift polar and pressure distributions at various incidences

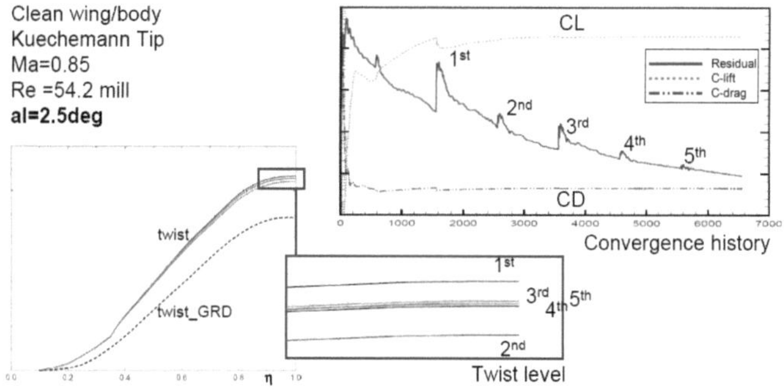

Fig. 6. Results of coupled CFD-CSM on clean wing/body: Convergence

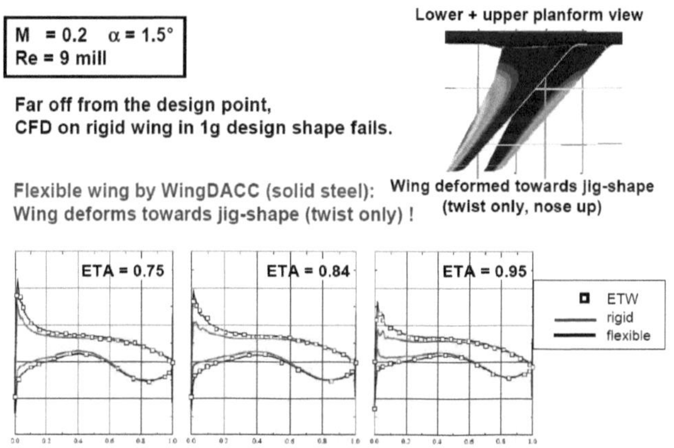

Fig. 7. Wing deformation at off-design conditions: Low Mach-number

off-design regime again show remarkable improvements in pressure distributions, Fig. 7. During flexible wing computations, the CFD mesh deforms towards the jig shape of the model (i.e. changing twist, see Fig. 7), since the loads at low Mach number and low windtunnel pressure become small compared to the transonic tests. In consequence, the computed solutions on the flexible wing approximate the measurements within much better accuracy than the rigid computations within the 1g flight design-shaped CFD mesh.

4 Results on High-Lift Configuration

In case of high-lift configurations, the WingDACC tool automatically recognizes the additional elements of the wing and computes their additional aero-forces from the BSURF sections. Again, it has to be taken care that the generated cuts do not intersect engines or pylons, since this would affect the computed stiffnesses of the wing at these sections, see above.

4.1 Integrated Approach

The simplified "integrated approach" is defined as follows:

- Lift components of the additional elements apply on the main wing (bending and torsion
- Wing and high-lift elements are assumed to be rigid sections
- Additional elements do not contribute stiffness to the whole configuration.

In terms of integrated approach, all mesh surface points are subject to deformation (twist and bend) as described before.

In Fig. 8, a typical low-speed surface deformation is presented. Again, the colouring refers to the amount of deformation from the original grid.

Note that in these high-lift cases, the manufactured windtunnel models do *not* offset from the designed wing shape. Here, a model made from solid aluminum is assumed.

After successful choice and application of a suitable volume mesh deformation method [2], different incidences were computed and compared to their rigid solutions. The flexible solutions typically converged within 4 CFD-CSM loops. However, the solutions do not show major differences between rigid and flexible calculations $\alpha = 15.18°$, see Figs. 10a,b.

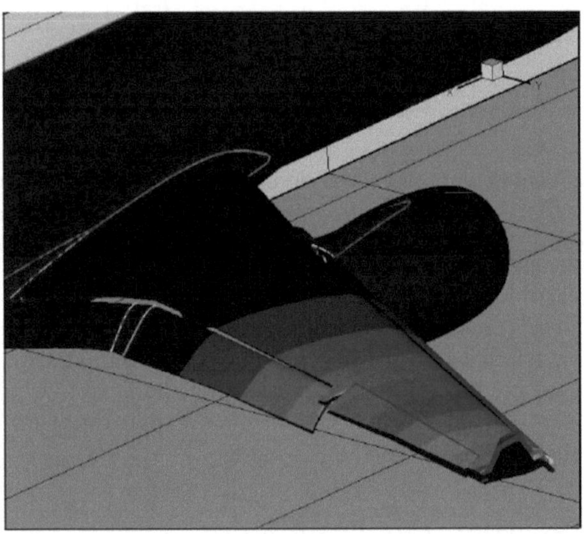

Fig. 8. High-lift configuration: Rigid and deformed wing (integrated approach)

4.2 Overlay Approach

In extension of the integrated approach to high-lift configurations, an overlay-approach may apply where the high-lift elements undergo additional deformation between their supporting stations. The deformation may apply normal to each element after the complete structure has been deformed in the "integrated" definition.

This approach will affect the sizes of gaps between elements and main wing in a way that gaps between slat and wing are widened while gaps between flaps and wing appear narrowed, see Fig. 9 showing the overlay part of deformation identified by colour.

In order to get a first impression on the resulting effects on the flow, additional assumptions were made — partly in contrast to the experiment in BLSWT (Bremen Low-speed Windtunnel)

- High-lift elements were treated for reduced stiffness
- Additional slat-brackets of the experiment were omitted for the computations. Here, the slat is assumed to be supported only at its edges (Fig. 9).

Within Megadesign project, a very first solution is available. This solution just includes a single step in CFD-CSM coupling, not yet a converged loop. However, the flow reaction becomes obvious already:

1. From the skin-friction pattern in Fig. 10c, a more downstream separation occurs on the outer flap. This may be caused by the narrowed gap which stabilizes the boundary layer flow by increased tangential blowing.
2. From the pressure distributions in Fig. 11, comparing to the rigid solution and the experiment,
   - a decrease in accuracy is observed on the lower side of the slat,
   - while increased accuracy is observed on the upper side of the flap.

Recalling the deviations between CFD and experiment, one might conclude that

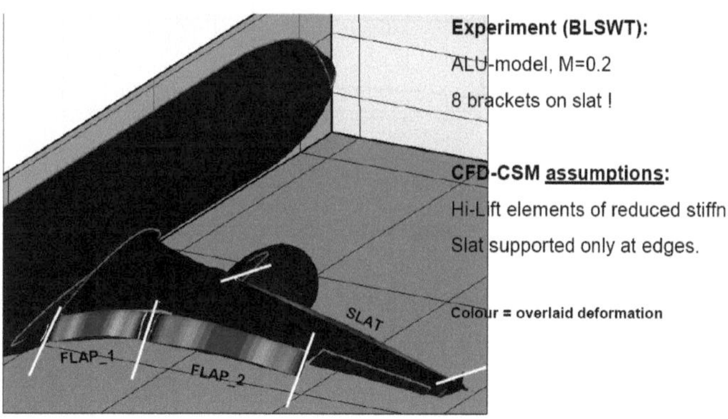

Fig. 9. High-lift configuration: Deformed elements and supports (overlay approach)

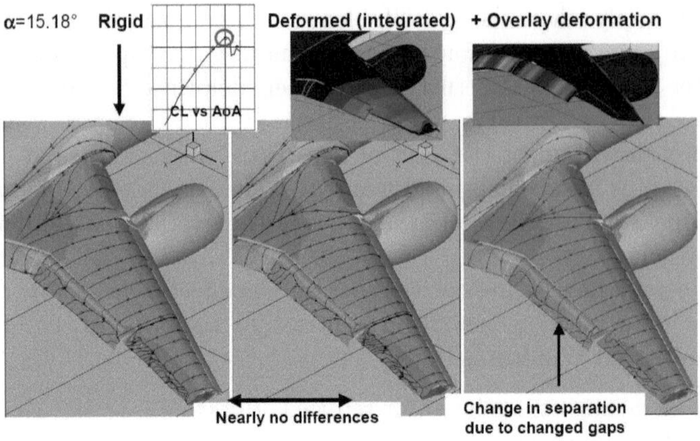

Fig. 10. Skin-friction pattern over high-lift configuration. Left: Rigid wing, Middle: integrated approach, Right: Overlay approach.

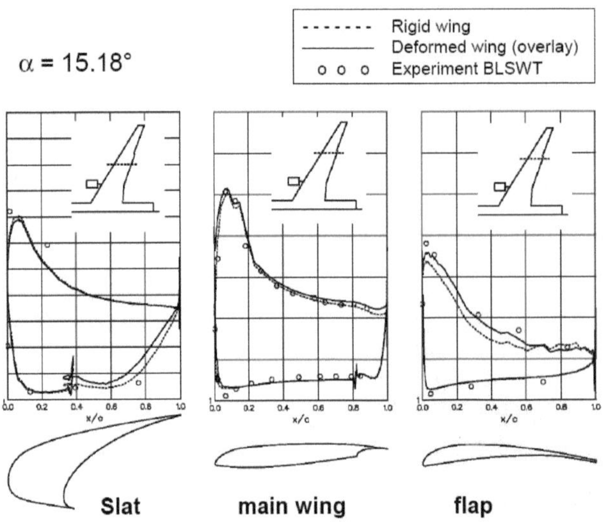

Fig. 11. Pressure distributions over high-lift configuration. Dashed line: rigid wing, solid line: fully flexible (overlay approach).

- The slat in the experiment was kept much more rigid than assumed in CFD — remember it was fixed by several brackets in between
- The flap might have been deformed in a similar way as had been assumed here.

5 Future Extensions

Current and future extensions of the WingDACC tool mainly address the simulation of a real wing, not a model wing anymore behaving like made from solid material. Approaching this goal, two major extensions have to be introduced:

- Changes in stiffness computations by introducing the stiffness of the wing box, formed by front- and rear-spar of predefined strength and an equivalent skin thickness of the wing.
- The effect of gravity, acting on wing structure and fuel mass as well as on engines — plus the effect of thrust as an additional torsion moment

At the time of writing, the introduction of gravity and thrust forces approaches its final testing stage while the aforementioned wing box stiffness approximations are planned for the next future.

6 Summary/Conclusions

It had been shown that introduction of model wing flexibility into aerodynamic CFD prediction can significantly improve the results.

A simplified method of wing deformation had been described, which is based on cantilever beam bending theory and solid material assumption (steel, aluminium). The method applies on a wide variety of configurations, i.e. from clean wing/bodies up to complex high-lift configurations including engines etc. This method was coupled to the DLR TAU code for full CFD-CSM interaction.

Applications of the coupled procedure addressed near- and off-design flow regime of a wing/body model in ETW, where the results were improved significantly compared with the rigid wing computations.

For a high-lift configuration, encouraging results were obtained in comparison with experiments in BLSWT, where the overlay bending of the high-lift elements was included.

The method used here, is able to produce very good results very quickly and from a minimum amount of input — even for complex high-lift cases. This is due to the simplified theory and the solid material assumption applied.

Acknowledgement

The results and techniques discussed here were strongly supported by H. Jakob*, O. Knobloch*, H. Schuemann and R. Heinrich+

* Airbus Bremen, + DLR Braunschweig

References

1. Gerhold, T.: Overview of the Hybrid RANS Code TAU. In: Kroll, N., Fassbender, J.K. (eds.). Notes on Numerical Fluid Mechanics and Multi-disciplinary Design, vol. 89, pp. 81–92. Springer, Heidelberg (2005) ISBN 3-540-24383-6, ISSN 1612-2909
2. Heinrich, R.: Mesh deformation based on volume splines. 2006–2007, private communication
3. Jakob, H.: Method for Determination of Wing Deformation from Measured Pressure Distributions, EUROLIFT II TR D1.1.1-7 (December 2005)

Part IV

Improvement of Shape Optimization Strategies

Development of an Automated Artificial Neural Network for Numerical Optimization

Olaf Frommann

Synaps Ingenieur-Gesellschaft mbH
Hermann-Köhl-Str. 7, 28199 Bremen, Germany
o.frommann@synaps-ing.de

Numerical Optimization requires a large amount of intermediate computations for the design data sets suggested by any optimization strategy. The results of these computations are necessary in order to find directions to the optimum. Nevertheless, most results are useless from the quality standpoint of view. Thus, it is desirable to avoid these, which would save a lot of time and money. It will be shown that the application of Artificial Neural Networks can serve in this sense and result in computational savings of about 77%. The problem in this context, i.e. the choice of an appropriate network topology, is discussed and solutions, resulting from extensive numerical investigations, are presented. Finally, the application to a challenging multimodal optimization problem, which serves as a surrogate for multidisciplinary optimization with comparable multimodal solution spaces, demonstrates the power of this approach.

1 Introduction

Evolutionary Strategies and Genetic Algorithms are powerful tools for optimizing complex designs with highly multimodal solution spaces. However, there exist two drawbacks limiting the practical application. The first one is the high number of necessary objective function evaluations, which in practise means a very high number of time consuming and costly simulations. The other one is inherent to optimization. In order to find the way towards an optimal solution, any optimization algorithm produces to some extend sets of design parameters which yield most of the time inferior qualities. Nevertheless, these qualities need to be assessed resp. evaluated, although it is a waste of time in some respect. Therefore, it is in general desirable to avoid these calculations. The only way to accomplish this is to have some kind of mechanism to forecast the quality of solutions with less computational effort.

In any case, this requires an approximation of the solution space topography. There exist three approaches for the approximation of the resulting solution space topography:

N. Kroll et al. (Eds.): MEGADESIGN and MegaOpt, NNFM 107, pp. 181–190.
springerlink.com

1. Use of fast running, lower order simulation codes for the prediction
2. Interpolating algorithms (Response Surfaces)
3. Extrapolating algorithms (Artificial Neural Networks)

The first one suffers from the disadvantage that lower order codes may be unable to predict effects, which might be of interest. This limits the success of the forecast significantly. On the other hand, interpolating approaches, like usual response surface technologies based upon polynomials or splines, are quite weak in case new regions of the solution space need to be investigated. Therefore, extrapolating algorithms like Artificial Neural Networks (Artificial Neural Network) offer in principle the possibility to explore these new regions more reliably and are most promising.

2 Assessing the Optimization Problem

The application of any forecasting method means additional computational effort. Therefore, it is necessary to decide, whether it is worthwhile to do so. There exist two things which need to be considered: the number of design parameters and the time for a single analysis. If both are high, the optimization problem can be regarded as a difficult one, because it will take a lot of time. The question left is, how to assess optimization tasks. For this purpose, the so called "Toughness Factor" has proven to be a good indicator. It is calculated by

$$T = 1 + \ln(n \cdot t) , \tag{1}$$

where n denotes the number of design parameters and t the time for a single evaluation in minutes. From experience one knows that for any design parameter at least 20 variations are necessary to reach an approximate optimal solution. For evolution strategies this is about three to four times more. Assuming 10 design parameters and a simulation time of one hour, this yields eight days and eight hours of calculations. Even with this quite short calculation times and small number of design parameters the effort is significant. For the toughness factor one calculates a value of $T = 7.4$. Also from experience one knows that optimization problems with a value of $T \leq 5$ can be considered as easily manageable. Thus, even this quite simple problem falls into the region of difficult optimization tasks. In fact, if one plots the toughness factor in dependence on the number of design parameters and the computation time, one notices that most of the tasks lie in the difficult area, Fig. 1. It is obvious that there exists the necessity to avoid the calculation of designs with inferior quality, at least in cases with "Toughness"-values above five.

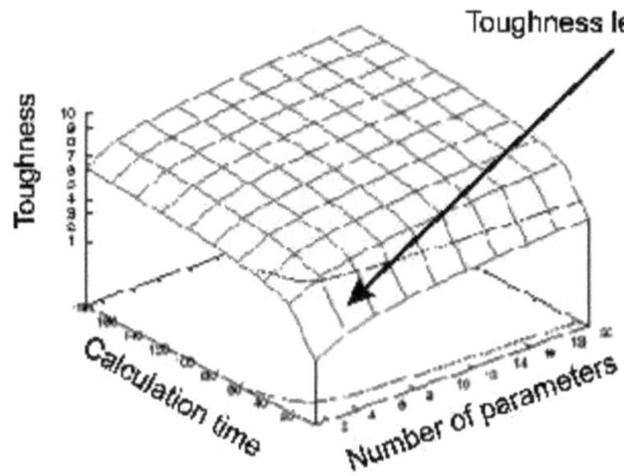

Fig. 1. "Toughness" in dependence on the number of design parameters and calculation time

3 Principles of Artifial Neural Networks

Artificial Neural Networks serve a wide range of applications. The most interesting property is the capability of extrapolating dependencies in contrast to interpolating methods like response surfaces. A Network of the so called feedforward type consists of layers, which themself consist of neurons. There exist three types of layers: an input layer, an output layer and hidden layers in between. The neurons of two adjacent layers are connected to each other, Fig. 2. This way, the neurons of a previous layer serve as input for the next layer, which produces an output for each neuron based upon the input of previous neurons. The connections are represented by weights, i.e. numbers describing the strength of the connection. The neurons are characterized by a sigmoidal function, e.g. $g(x) = \tanh(x)$, which is called the activation function. Assuming

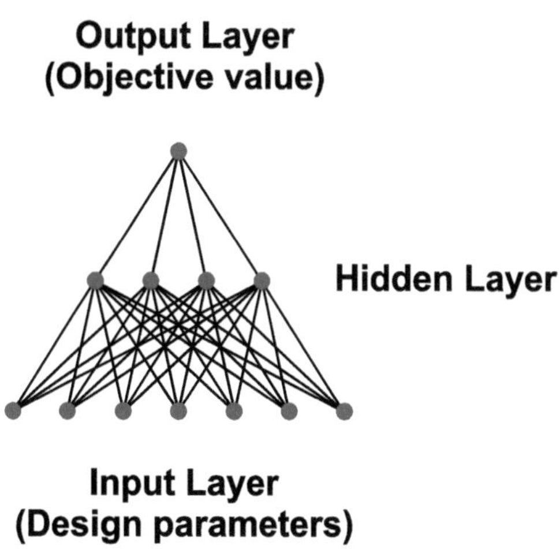

Fig. 2. Topology of feedforward Artifial Neural Networks

a connection J_{ij} from the jth input neuron to the ith output neuron, the weighted sum of all input signals is

$$s_i = \sum_{j=1}^{N} J_{ij}\xi_j \ , \tag{2}$$

where ξ_{ij} denotes the weight. This value is used to calculate an output O_i for the neuron i by use of the activation function as follows:

$$O_i = g(s_i) = g\left(\sum_{j=1}^{N} J_{ij}\xi_j\right) \tag{3}$$

This output value is then again an input for neurons in subsequent layers. This way, an Artificial Neural Network is feeded by the input layer, transforms these inputs through the hidden layers and produces results at the output layer.

The determination of the interconnecting weights is called learning. During the learning phase a Network is fed with a set of input values and the output values are calculated. The difference of the calculated output and the known output serves as an indicator for the quality or standard of knowledge. There exist several possibilities of learning, while the so called backpropagation is the standard one. In this case, the error, i.e. the difference of the outputs, is propagated backwards through the Network and the weights are adjusted accordingly. However, due to the fact that this is generally spoken more or less a gradient approach for the optimization of weights, it often gets stuck in local minima for the weights. This becomes worse in cases of highly nonlinear practical problems [1,2]. Therefore, in this case an evolution strategy was chosen for the training phase of the network to overcome these difficulties. This means an additional effort in comparison to backpropagation, but has proven to be much more successful.

Although the underlying mathematics for Artificial Neural Networks are quite simple, the possibilities for the topology in terms of the number of hidden layers and the number of neurons per layer is infinite. On the other hand, the prediction capabilities strongly depend on the topology of the Artificial Neural Network. If the topology was chosen to be too simple, the network will be unable to predict anything reasonable. If it has too many neurons, it will degenerate to an interpolating method which only can predict coherences in between known ranges. This awkward situation was the motivation for intensive investigations within the MEGADESIGN project funded by the German Federal Ministry of Economics and Technology.

4 Application to Optimization

With respect to optimization, the inputs are represented by the design variables and the output is the objective function value. Initially, one needs some sets of design variables with their corresponding objective function value. These are used for the training of the Artificial Neural Network. Afterwards, the Artificial Neural Network will be queried instead of doing real simulations. From time

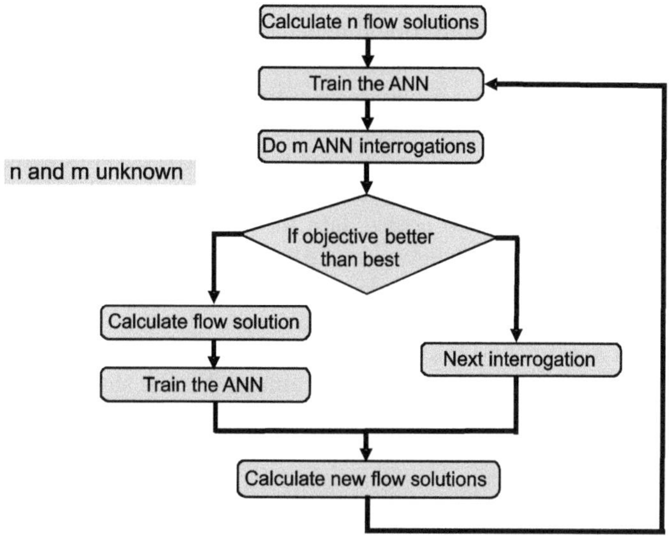

Fig. 3. Integration of Artificial Neural Network into the optimization loop

to time additional simulations resp. calculations of the solution space will be carried out in order to intensify the shape of the approximated topography and avoiding the Artificial Neural Network to get stuck in some region. The overall process is shown in Fig. 3.

Additionally, to ensure always correct improvements in the objective function value, everytime when an improvement is forecasted by the Artificial Neural Network, this set of design variables is scheduled for calculation. Afterwards the result will be added to the already available learning data sets and the Artificial Neural Network is again trained with this enlarged amount of data. This also yields an improvement of the topography approximation and therefore in reliability of forecasts.

However, it is unknown how many interrogations and how many real calculations will produce the best result. It is only known that it strongly depends on the number of design variables and the number of data sets per iteration of an optimization method. For instance, evolution strategies allow for a larger number of Artificial Neural Network queries, because they produce a lot more data sets compared to e.g. a downhill simplex algorithm. To solve this problem the only way is to do extensive numerical experiments.

5 Investigation of Suitable Artificial Neural Network Topologies

In order to answer the main question of how an Artificial Neural Network should be constructed, i.e. the number of hidden layers and the number of neurons per layer, one needs some strategy for the investigation. One may asume that it would be advantageous to use as many layers and as many neurons as possible. This, however, increases the computational effort for the training phase excessively, because the interconnects between the neurons and thus the number of weights increase exponentially. Also, if one decides to do so and limits the number of neurons to the available computing power, it is not a good idea. Investigations of Artificial Neurol Networks for decades point out that the network will over-adapt to the solution space topography, leading to decreased extrapolation capabilities.

It may look self-evident to create data for some solution space topography and to train several different networks with a subset of these data sets. Afterwards, the Artificial Neural Networks will be queried with the remaining data sets and the output compared to the known result. This information can be used to calculate an average and maximum error to decide, which topology performs best. For example, assuming the three parameter NACA shape function for an airfoil, it is possible to calculate a fully factorial exploration of the design space and randomly extract 1% of data for the training. In this case, 907925 flow calculations using a full potential code have been carried out. An overall number of 625 Artificial Neural Networks, having $m = 1$ to 25 neurons in one or two layers, have been trained and investigated. The result was that, except for very simple topologies with very few neurons, the approximation capability was almost the same for each network. The best Artificial Neural Network had an average deviation of 5% and had 14 neurons in the first and 25 neurons in the second hidden layer. It is obvious that this topology seems to be best suited for this case. However, applying this to a real optimization case, i.e. letting it start untrained from scratch and letting it learn during the course of optimization, yielded no good results. The reason for this is the above mentioned over-adaption as well as the disregard of the talent to adapt locally on the fly.

Therefore, the only way to gain some insight into the dependencies is to integrate all Artificial Neural Networks into a real optimization loop with training on the fly and see, how they perform in this circumstance. A lot of optimizations are needed and have been carried out using the deterministic downhill simplex algorithm with a lot of conceivable combinations of hidden layers and number of neurons per layer for different numbers of input neurons, which represent the design paramters. It was not assured, whether such an interdependence actually exists. Fortunately, the result was a clear correlation between the number of design parameters and the necessary topology of the Artificial Neural Network. Figure 4 shows contour lines of the best objective value for the three parameter NACA shape function found for various Artificial Neural Network topologies in dependence on the number of neurons within the hidden layers. It becomes obvious, that there actually exists a region or corridor for combinations of the number of neurons which perform best. These Artificial Neural Networks yielded approximately the same result as compared to an optimization without Artificial Neural Networks.

It has been found that a certain ratio of neurons in dependence on the number of design parameters in the first and second layer yields always best approximation and extrapolation results. Depending on these results a rule for the setup

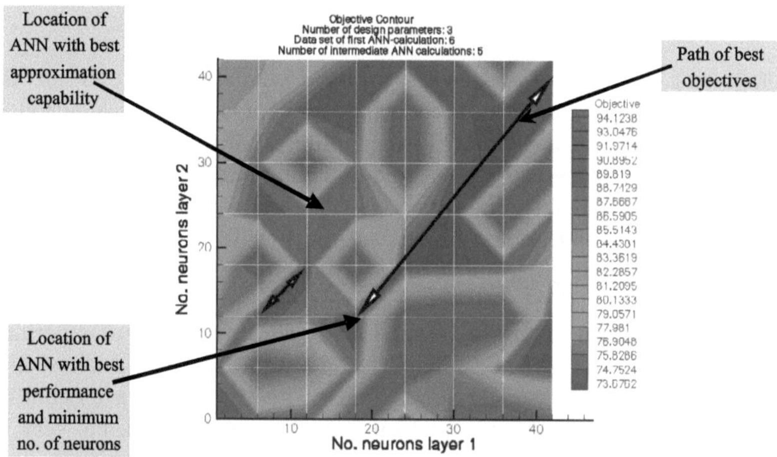

Fig. 4. Best objective function value found during optimization

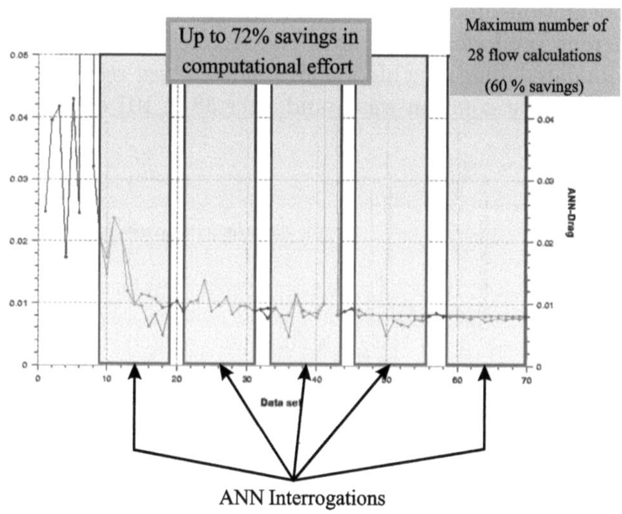

Fig. 5. Comparison of forecasted and real objective function values

of appropriate Artificial Neural Networks for application to optimization has been derived. This rule has been implemented into the SynapsPointer® Pro optimization framework, which now offers an easy to use and powerful tool. Test were carried out on several cases and the result is a significant reduction of the necessary real evaluations of the objective function, i.e. simulations, of about two third, Fig. 5. This heaves the application of Numerical Optimization and particularly Evolution Strategies to real world multidisciplinary problems to a practical level, avoiding the waste of computational effort.

6 Application to Real World Problems

In the following the result of a real world problem will demonstrate the advantage of the proposed approach. It is the optimization of a 35% thick airfoil for a wind energy turbine rotor. The goal was to find an airfoil that does not suffer from excessive flow separartion while providing maximum lift-over-drag ratio and maximum lift at this point. The objective function was created by avoiding a weighted sum and using the Fuzzy Logic approach, which has been proven to be very efficient and reliable [3,4,5,6]. The flow calculations for the airfoil polar were carried out using the XFOIL code from Mark Drela [7], which needs only a few minutes of computation time. The reason for this choice was only of experimental character in order to have the opportunity to do a lot of calculations within the available time. Of course, it could also be substituted by higher order codes. The solution space topography of this problem is highly multimodal, i.e. very many local minima, which makes it quite difficult to find the global optimum.

The applied optimization algorithm was a simulated annealing code, while the airfoil shape was generated by the SynapsAirfoil code, which is based upon conformal mapping with five design parameters. This normally requires about 100 flow calculations in order to find a near optimal solution. Several runs, due to the inherent randomness, without the Artificial Neural Network using the standard simulated annealing algorithm yielded a best result of about 119 for the L/D-ratio. The best solution was found after 38 of 101 calculations, while after

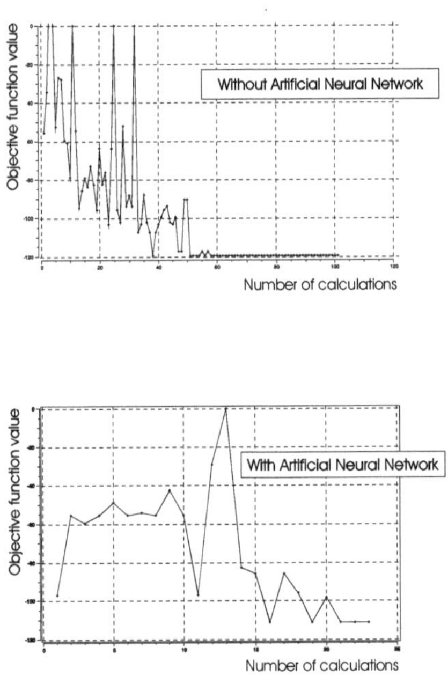

Fig. 6. Comparison of necessary calculations with and without Artificial Neural Network

about 50 flow simulations no improvement occurs, Fig. 6. Other runs resulted in inferior performance due to the mentioned randomness. In contrast, applying the automated Artificial Neural Network to this problem resulted in a best solution of 111 for the L/D-ratio after 16 of 23 real simulations, which is quite similar with respect to the quality. Keeping in mind that in this case randomness also matters in several ways, i.e. within the simulated annealing code and for the training using evolution strategies, the difference is of minor importance. The resulting airfoils are quite similar, Fig. 7.

Depending on the basis of comparison, i.e. the number of calculations, the savings in computational time are at least about 58% if one takes the number of the best iteration (16 compared to 38). However, in practice one does not know when the best calculation will happen, particularly in the case of stochastic methods. In these cases the above mentioned 101 calculations need to be carried out to be more or less sure that it could have been found. In that case the savings are 77% (23 compared to 101 calculations), which means a great advantage. The calculation time for the training of the Artificial Neural Network lays in the range of minutes while the evaluation of the Artificial Neural Network for a specific data set is negligible, which means the real savings in time for a usual simulation code are in the same range as the savings of data sets.

Altogether it can be stated that the application of Artificial Neural Networks for the forecast of potentially good solutions seems to be efficient. The fact that in this highly multimodal and thus complicated case the result of the run without Artificial Neural Network has not completely been reached does not imply a

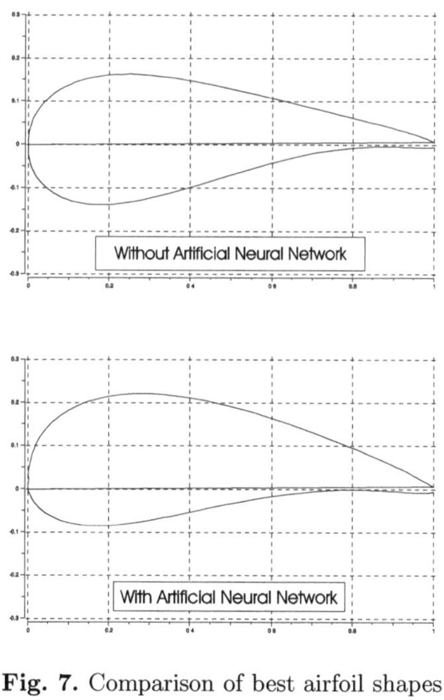

Fig. 7. Comparison of best airfoil shapes

weakness. As mentioned, stochastic effects play a major role compared to the forecasting capability.

7 Summary

It has been shown that the application of Artificial Neural Networks for the forecast of potentially good solutions is promising and successful. Associated problems like the choice of the Artificial Neural Network topology were solved by extensive investigations of a large number of possible Artificial Neural Networks within an optimization loop. The results were used to implement this approach into the well known SynapsPointer® Pro optimization framework, which provides users easy-to-use Artificial Neural Networks for time savings. The application to a real world aerodynamic optimization problem served as an example for the reliability of this solution, yielding savings in computational effort between 58% to 77%.

Acknowledgements

The author is grateful for the funding by the German Federal Ministry of Economics and Technology within the MEGADESIGN project. Without this funding the enormous amount of calculations and effort in implementation, testing and improvements would not have been possible.

References

1. Sontag, E.D.: Some Remarks on the Backpropagation Algorithm for Neural Network Learning. Technical report SYCON-88-02, Rutgers Center for Systems and Control (1988)
2. DasGupta, B., Siegelmann, H.T., Sontag, E.D.: On the Complexity of Training Neural Networks with Continuous Activation Functions. IEEE Transactions on Neural Networks (1995)
3. Frommann, O.: Objective Function Construction for Multipoint Optimization Using Fuzzy Logic, Eurogen 97, Triest (1997)
4. Frommann, O.: Conflicting Criteria Handling in Multiobjective Optimization Using the Principles of Fuzzy Logic. In: Applied Aerodynamics Conference, Albuquerque, NM, AIAA Paper 98–2730 (1998)
5. Frommann, O.: Objective Function Construction for Multipoint Optimization Using Fuzzy Logic. Surv. Math. Ind. 167/177 (2000)
6. Frommann, O.: Bewertung multipler und gegensätzlicher Qualitätskriterien in Multidisziplinärer Optimierung, Deutscher Luft- und Raumfahrtkongress, Leipzig (2000)
7. Drela, M.: XFOIL Subsonic Airfoil Development System, `http://web.mit.edu/drela/Public/web/xfoil`

modeFRONTIER©, a Framework for the Optimization of Military Aircraft Configurations

L. Nardin, K. Sørensen, S. Hitzel, and U. Tremel

Aerodynamics and Methods, EADS Military Air Systems, Germany

Summary. Design optimization is a process aimed at selecting the best design, referred as objective function, satisfying certain requirements, called constraints, by modifying the input variables or design parameters. modeFRONTIER is a software that allows to use data coming from many different sources to evaluate several designs and, through the use of an optimization algorithm, can change the input parameters in order to find a configuration that performs better than the others. The preferred methodology to carry out the optimization is usually the use of numerical analysis, instead of experimental or analytical analysis, because it combines good accuracy of the predicted solution with fewer time and resources needed for the investigation.

1 Description of modeFRONTIER Capabilities

The design and simulation process can be a rather complex flow of actions to be controlled in an orderly way. modeFRONTIER provides an environment which allows the proper set-up and control of the complete process including the provisions necessary to call appropriate optimization tools. Figure 1 gives an idea of the systems principles. On the horizontal path the data flow of an optimization is sketched. It corresponds to the complete way of a full simulation including pre and post-processing. Input parameters can be of various types, like geometrical entities, design point parameters, etc. After the simulation (where the horizontal path meets the vertical one), the outputs serve to compute objectives and constraints of the optimization.

The vertical path shows the optimizer logic. The Design Of Experiments (DOE) presents a first layout of designs, for which the input parameters of the data path are already specified. Those parameters start, in the examples later explained, some configuration shape e.g. via CAD, which then is subject to meshing, subsequent simulation and analyses.

The objectives and constraints evaluated from the simulation outputs together with the method of optimization chosen (scheduler) determines the search of the next design or family of designs by monitoring the development of the design objective(s).

modeFRONTIER comes with a variety of built-in tools and it can be used, apart from the optimization itself, also for other purposes. For example it is

N. Kroll et al. (Eds.): MEGADESIGN and MegaOpt, NNFM 107, pp. 191–205.
springerlink.com

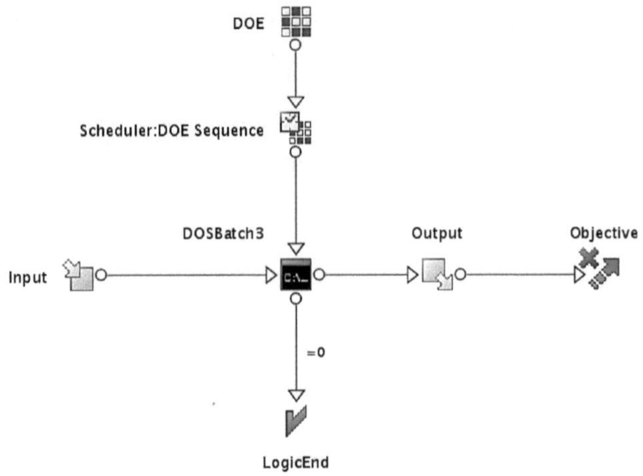

Fig. 1. Data and logic path in the modeFRONTIER control and optimization environment

simply and straightforward to set up a DOE analysis, which is an evaluation of predefined designs, without the goal of optimizing them, but only trying to get more information of the space of variables, e.g. like the polar curve of an airfoil.

Another feature of modeFRONTIER is the availability of a set of Response Surface Methodologies (RSM). They range from K-nearest to Artificial Gaussian Process, from Kriging to Gaussian Process and are needed to extrapolate the distribution of the variables in the whole space, providing just a limited set of designs. They are particularly important when the simulation analysis is time-consuming, because it is possible to obtain an extrapolated value for some design points; therefore it is possible to skip the simulation of those points.

For complex optimization with several objectives, it is also providing a Multi-Criteria Decision Making tool (MCDM) which, on the base of some preferences expressed by the user, is able to assign a ranking to the different solutions found during the optimization process, helping the designer to choose among them.

modeFRONTIER is also suited for robust optimization, i.e. it is able to provide an optimized solution which is poorly affected by the perturbation of input parameters, which can be the case if e.g. the velocity or the angle of attack slightly varies.

Figure 2 illustrates the conceptual optimization process of a modern aircraft, including several disciplines. Through the use of modeFRONTIER it is possible to control each of the different tools needed for such an optimization.

An optimization environment such as modeFRONTIER provides the link for all of the steps of the optimization process: design, simulation and analysis of the results, as described in Figure 3. It controls the overall strategies, the selection of CAD-models, the demands on the meshing resolution, starts the flow simulations,

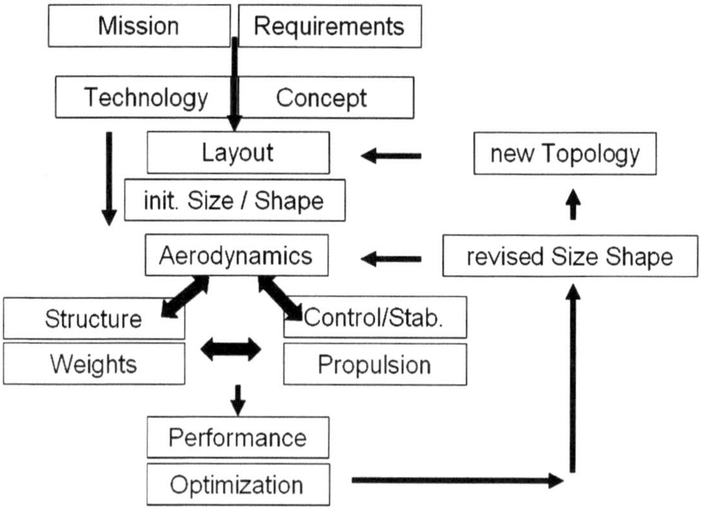

Fig. 2. Design cycle process of a modern aircraft

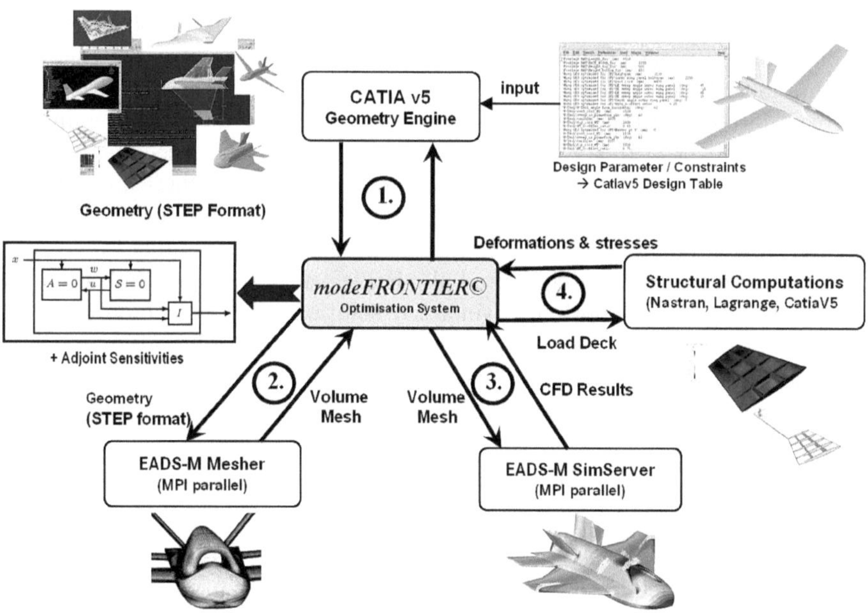

Fig. 3. modeFRONTIER linking with the simulation tools

analyses the results and determines the objectives, possibly with a multidisciplinary connection by weight estimations or by structural sizing through FEM. Depending on the optimization procedure selected, modeFRONTIER then determines the next design which have to be analysed by redefining its parameters. Simultaneously it includes design constraints and limits.

There exist several optimization algorithms with different characteristics. They range from deterministic methods, based on gradients, quadratic approximation or logical search moves to evolutionary algorithms, which are mainly divided in genetic algorithms and evolution strategies. The main parameters which drive the designer's choice about the methodology to use are the presence of one or more objective functions, the typology of variables (continuous, discrete or category variables) and the time available for the whole optimization design process. This last parameter is a function of number of variables and objectives and of simulation time of a single configuration.

2 Single and Multi-objective Strategies

Generally speaking, optimization can be either single-objective or multiobjective. An attempt to optimize a design or system where there is only one objective usually entails the use of gradient methods where the algorithms search for either the minimum or maximum of an objective function, depending on the goal. One way of handling multiobjective optimization is to incorporate all the objectives (suitably weighted) in a single function, thereby reducing the problem again to a single objective one. This technique has the disadvantage, however, that these weights must be provided a priori, which can influence the solution to a large degree. Moreover, if the goals are very different in substance it can be difficult, or even meaningless, to try to produce a single all-inclusive objective function.

True multi-objective optimization techniques overcome these problems by keeping the objectives separate during the optimization process. It should be kept in mind that in cases with opposing objectives (an example would be to try to minimize a beam's weight, and also its deformation under load) there will frequently be no single optimum, since any solution will be a compromise. The role of the optimization algorithm is then to identify the solutions which lie on the trade-off curve, known as the Pareto Frontier.

These solutions all have the characteristic that none of the objectives can be improved without prejudicing another.

3 Design of Experiments

Design Of Experiments (DOE) techniques are needed to explore the design space and they can give information to be used in two different ways:

- they can be directly used to sample the variables space and obtain information about the variables themselves, i.e. if there are any useful correlations among them or if there are any variable which can be discarded, etc.

- they can be used to provide the optimization algorithm with a suitable initial set of designs.

Among the large set of different DOE techniques, Sobol (pseudo-random) methodology was used in order to provide a good set of initial designs for the genetic algorithm, because other techniques, such as Factorial DOEs, Latin square, etc. need usually several designs to correctly define a problem with a great number of variables. Mostly that number of designs is too big to be used as an initial population. These statistical DOEs were therefore only used during an initial phase of the design process, in which only the designs in the DOE are evaluated and there is no further optimization. Therefore the total number of DOE designs in this phase has not to be limited.

4 Multi Objective Genetic Algorithm

The Multi Objective Genetic Algorithm (MOGA) carries out the optimization as the evolutionary process of the living beings. Each individual is represented by a binary string and a fitness value is assigned to it. The fitness is a score that indicates the ranking position of the design compared to the other solutions with respect to the objectives [1].

Some operators are used to define the genetic process by modifying the variables values of the individuals and therefore by creating a new generation:

- The *cross-over* operator is the genetic recombination of two individuals to create two new solutions that will obtain some bits from one parent and the rest from the other one. The software implements a ***directional cross-over*** an improvement of the classical method that creates new designs assuming that a direction of improvement can be detected by comparing the fitness values of two reference individuals and it gives more efficiency to the algorithm.
- The *selection* operator allows some designs to survive to the next generation; the possibility for each design of being selected is correlated with its fitness value, i.e. a good solution has more probability to survive than a bad one. To avoid premature convergence of the method a *sharing methodology* is applied to the selection operator: if a design has a great possibility of being selected, because it has a good fitness value, the solutions close to it will have a poor possibility of being selected.
- The *mutation* operator randomly changes some bits of some individuals in order to increase the possibility of covering all the designs space.
- The *elitism* operator preserves the best design for each objective to the next generation. The proprietary version of the elitism operator used in this MOGA algorithm enables a smart multisearch elitism. This new elitism operator is able to preserve some excellent solutions without bringing premature convergence to local optimal fronts, a drawback of former versions of the elitism operator.

The main characteristic of the genetic algorithm is its robustness, intended as a not premature convergence, if compared to other optimization methodologies. On the other hand the accuracy and the converge speed are worse than what other schedulers are able to do.

5 Simplex

As explained in [2], if $\boldsymbol{n}$ is the number of degrees of freedom of the model, its input variables, the simplex is a polyhedron with $\boldsymbol{n+1}$ vertices in a $\boldsymbol{n}$-dimensional space, thus in two dimensions it is a triangle, in three dimension is a tetrahedron, and so forth. The aim of this methodology is to move the simplex geometrical figure into the neighbourhood of an optimum by replacing its vertices. The simplex methodology compares the values of the objective function at the $\boldsymbol{n+1}$ vertices and then it moves this polyhedron gradually toward the optimum point during the iterative process. For each iteration of the algorithm, it attempts to replace simplex vertices with poor fitness values with vertices whose fitness values are better. The movement of the simplex is given by three operations: *reflection*, *expansion* and *contraction* as illustrated in Fig. 4.

In the ***reflection*** movement the point presenting the worst objective function is reflected with respect to the other points. The idea is that if the function is sufficiently smooth and it does not present steep valleys, repetitive reflections lead to the optimum. In the sketch besides the worst vertex A of the simplex ABC is reflected obtaining a new simplex CBD.

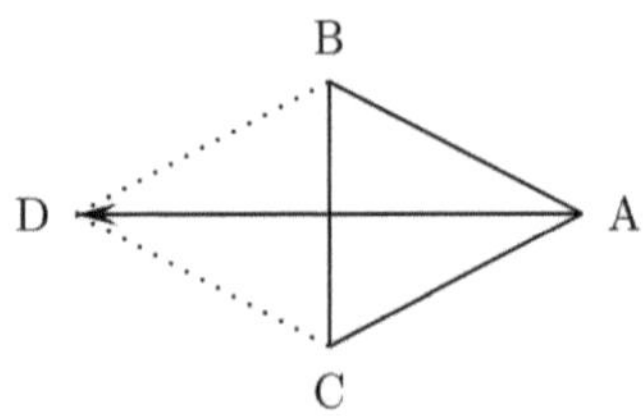

Original simplex in a 2 dimensional space (ABC) with reflected point (D).

If the reflection movement produces a point which fitness value is better than all the other points, it can be expected that moving toward that direction even better values can arise. This operation, exemplified in the sketch besides is called the ***expansion*** of the simplex.

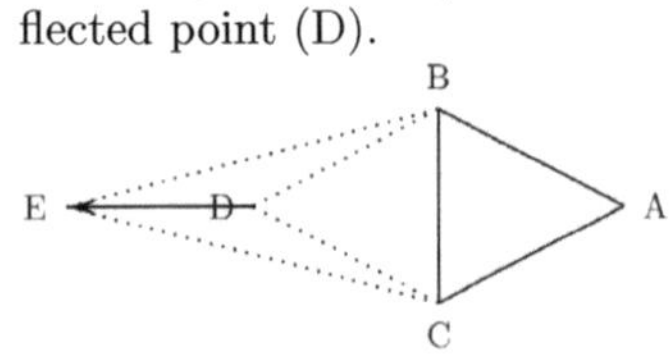

Expansion of the reflected vertex D to E.

If the reflection process results in a point with a fitness value that is worse than all the other vertices except the starting point (in the example Figure, this happens when D is better than A but worse than B and C), a ***contraction*** of the simplex is carried out by changing the new point with a point closer to the old one.

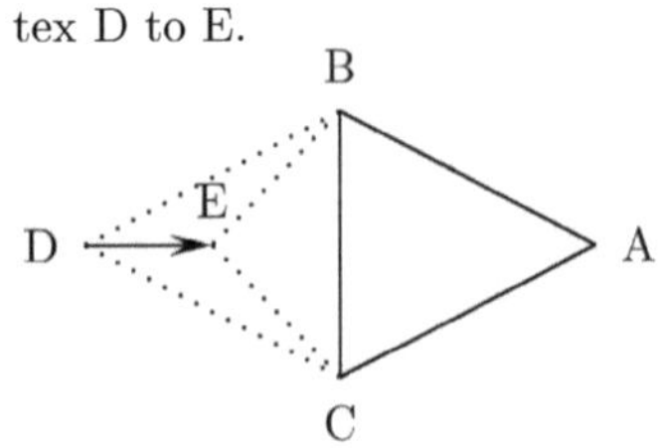

Contraction of reflected vertex D to E.

Fig. 4. Simplex methodology search strategies

Therefore the methodology is to find a new point of the simplex and to replace the old point with the new one (reflection) and, depending on the quality of the new reflected point, another move can be done by expansion or contraction. The procedure is repeated until the planned total number of iterations is reached or the convergence is found. The latter case is obtained when the geometric distance of the moved points found is less than a fixed tolerance.

The simplex methodology is a mono-objective optimization scheduler, so it is not able to tackle complex problems with more than one objective. Also the robustness is not at the same level of the genetic algorithm but it is fairly better than that of a gradient based algorithm.

The advantage of the simplex methodology compared to the genetic algorithm is the convergence speed of the solution, therefore it is possible to think about it as a good compromise between classical gradient based methods and evolutionary strategies.

6 Gradient Based Algorithm, SQP

Sequential Quadratic Programming methods are, among gradient based methodologies, one of the most powerful nonlinear programming algorithms we know today for solving differentiable nonlinear programming problems [3]. The search direction is found by solving a quadratic programming sub-problem given by a quadratic objective function and linear constraints. The objective function is augmented using Lagrange multipliers and an exterior penalty so that the resulting one dimensional search is unconstrained.

The optimization software used is able to accept gradients calculated from external programs, like simulation software which uses the adjoint methods to calculate the sensitivities. Of course the software can as well schedule the simulations in order to calculate the gradients by finite differences, by simply perturbing the variables for which no sensitivity information is provided and simulating the new configurations. With this methodology it is possible to calculate the gradient through central or forward differences. Gradient based algorithms are among the fastest optimization methodologies, but the optimum found by them is always a local one. Therefore, in order to have chances of finding a non-local optimum, either the variables space is smooth or it is advisable to start several optimizations from different initial points.

7 Multi Objective Game Theory

The Multi Objective Game Theory (MOGT) methodology is based on the competitive game theory proposed by John Nash [4]. The objectives and the variables of the optimization problem are decomposed among the players that, through the application of a defined number of Simplex iterations, try to optimize their own objective function, influencing each other by the sharing of the best variables obtained after each step of the game as explained in [5]. The number of

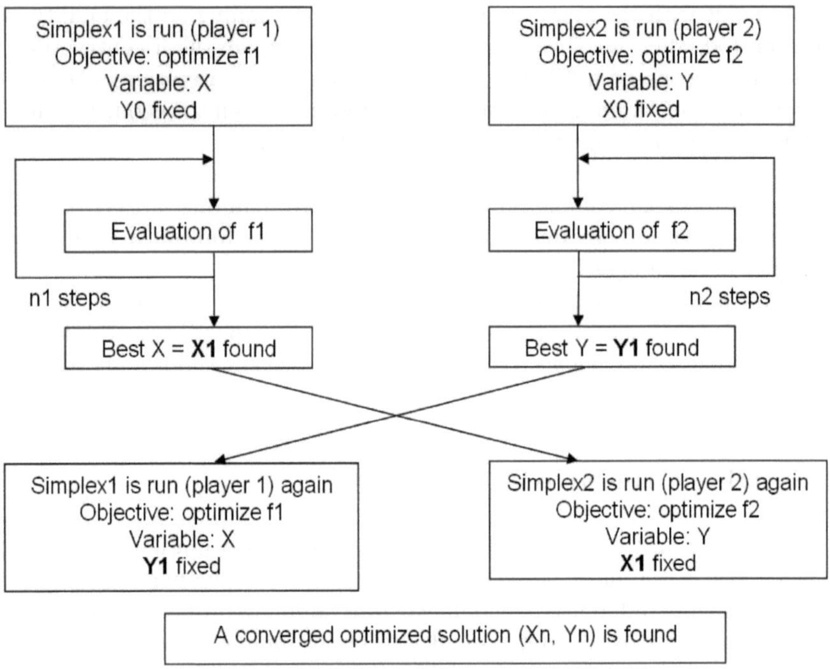

Fig. 5. Logic of the MOGT algorithm

players equals the number of objectives. It is to be underlined that this optimization scheduler combines the fast convergence speed of Simplex with the multi-objective strategy, introduced by Nash's competitive game theory.

In a competitive game, the players act following different objectives. For example, in a two-objective optimization, player A have to choose his strategies in order to minimize his function f_A, while player B have to minimize the function f_B, see Figure 5, taken from [6]. In other words, each player is forced to optimize his variables following his objective, but it is constrained by the value of the variables that have been found at the end of each step by the other players, and that become fixed during his search. The aim of a competitive game is to find a Nash equilibrium point which in [7] is defined in mathematical terms as $(x^{ast}, y^{ast}) \in A \times B$ if and only if:

$$\begin{cases} f_A(x^{ast}, y^{ast}) = \inf_{x \in A} f_A(x, y^{ast}) \\ f_B(x^{ast}, y^{ast}) = \inf_{y \in B} f_B(x^*, y). \end{cases}$$

Of course, as generally both the functions depend on the two domains, the strategies of one player influences the choices of the other one: the two players act simultaneously until an equilibrium is found: in that case, each player has minimized his own function with a common pair of strategies. The initial decomposition of variables and objectives is random but, in the following steps, it is changed accordingly to statistical analysis: using it and in particular the t-Student coefficient it is possible to decide, at the end of each player step (that is after a certain number of Simplex iterations), if a variable is statistically significant for the player to which it was assigned or not and, in the latter case, i.e. if the significance percentage is lower than an assigned threshold, the variable is given to another player in the following step. In other words, the significance percentage expresses the probability that a variation of the objective function is really produced by a variation of the variables belonging to the same player instead of an effect of the statistical variance around the mean value of the objective function.

8 Example: modeFRONTIER Optimization with Genetic Algorithm

To check on the set up time of an optimization process which was defined as the major milestone M1 in the Megadesign project, a rather simple wing was optimized for maximum lift / drag ratio. Flight conditions were chosen to be at Mach = 0.85, $\alpha = 1°$. The wing area was to be kept constant and the pitching moment was forced to stay negative. The sweep was allowed to vary in between +- 60°, while the half span was to be found in between 0.75 and 1.127 m with a taper ratio to be among 0.2 and 0.8. The aspect ratio was unlimited.

The results are subject to constraints. Here the MOGA evolutionary optimization tool was chosen which after analysis of the results provides the variables for a new design. The set up of a first generation is assisted by an automatic tool (DOE) to provide a set of initial designs. All constraints on objectives and/or parameters can be handled in a GUI supported interactive way, which makes the

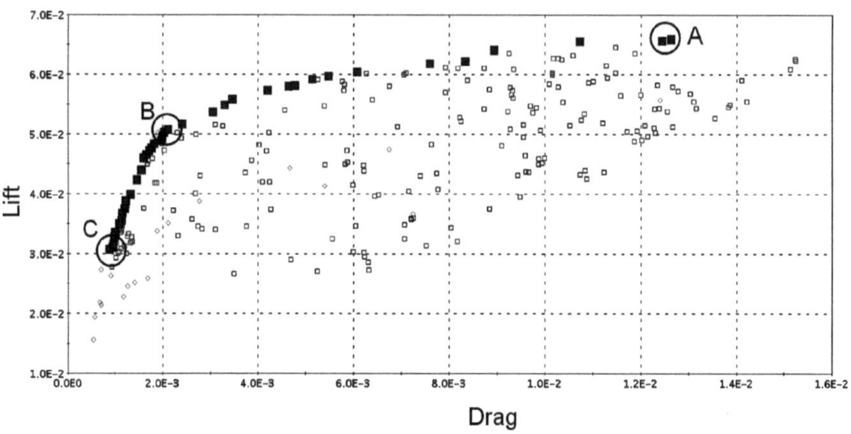

Fig. 6. Pareto front of lift and drag for the wing design example

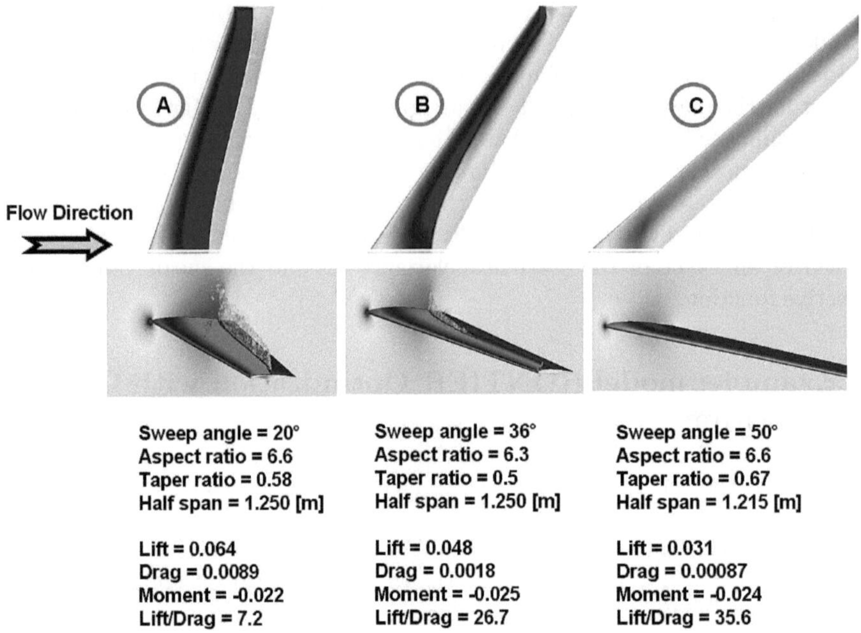

Fig. 7. Wing design examples coming from the optimization

system rather user-friendly. Assuming the CAD-model and the solution process are well tested the complete cycle can be set up in short time depending on the complexity of the task. The case given was set up in 25 minutes and achieves the targeted aim for the Megadesign milestone M1.

The overall result of all wing designs can be seen in Figure 6 while Figure 7 shows some typical results from different locations of the pareto-front of the optimization. By and large the results show the typical effects of wing sweep. Example A is a rather effective lift-producer, while its drag is high due to the transonic shock configuration also depicted in the lower pictures in Figure 7. Result B shows very much reduced shock systems due to its much higher sweep which is reflected in a much lower drag. The excessive sweep of C causes both the lift and drag drop to lower values. This purely academic result would be penalized if weight constraints would be applied.

9 Example: modeFRONTIER Optimization through Adjoint RANS-Methodology

The implementation of the continuous adjoint gradient based optimization method (Figure 8) was first tested with 2-D airfoil examples for which the

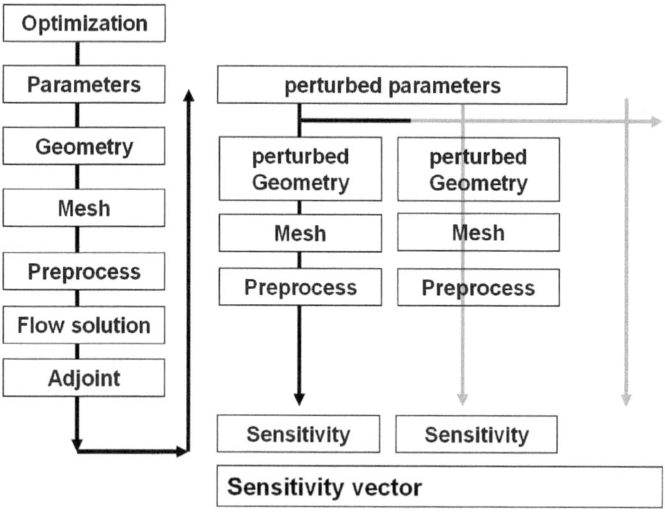

Fig. 8. Scheme to evaluate sensitivities via the adjoint method

optimum solution is known. At Mach = 2.0 the relative mix of three profiles was selected with 2 independent and one dependent parameter.

The local shape was determined via a weighted function as $y(x) = \sum_{j=1}^{3} c_j y_j(x)$ with $\sum_{j=1}^{3} c_j = 1$ where y_j describes a basic airfoil – here out of NACA0006, NACA0010-64, NACA0010-66, while c_j is the weighting factor defining their relative contribution to the final profile.

At supersonic speeds the airfoil thickness is the primary factor for the drag. Therefore at the end of the optimization the weighing factor of profile NACA0006 must be equal to one (and the other to zero), as this the minimum drag design. Since the problem is dominated by compressibility effects the optimization was performed on the basis of the Euler-equations.

Different optimization methods were tested in which the NLPQLP-method (version of the sequential quadratic programming method) proved to be the most efficient one. As to be expected from a gradient based method the convergence was very dependent on the starting conditions. In Figure 9 two different optimizations convergence are plotted, one of which was started at a weighing parameter for profile NACA0006 of 0.9, the other with 0.3 only.

To check on the effect of constraints and the efficiency on typical airfoil parameters an airfoil was optimized for a Mach = 1.2, $\alpha = 1°$ (Figure 10). As before a 2-D structured mesh was used to ensure the same mesh-topology for the baseline and perturbed geometries.

Figure 11 shows the drag development and the simultaneous history of the constraint lift and pitching moment. The pitching moment C_M had an upper and lower bound to enforce an airfoil with a small pitching moment. Examples

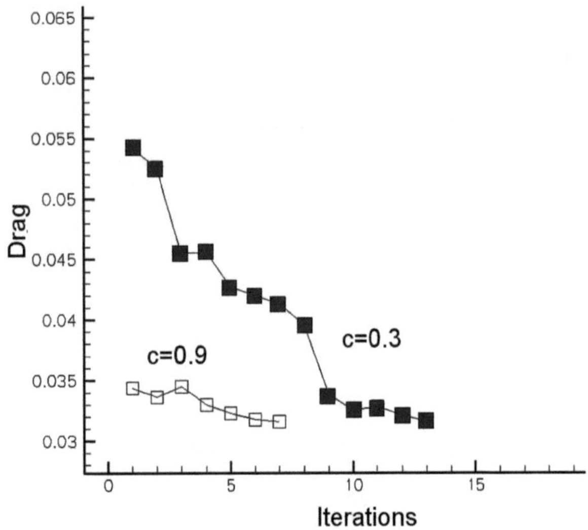

Fig. 9. Convergence of adjoint optimizations

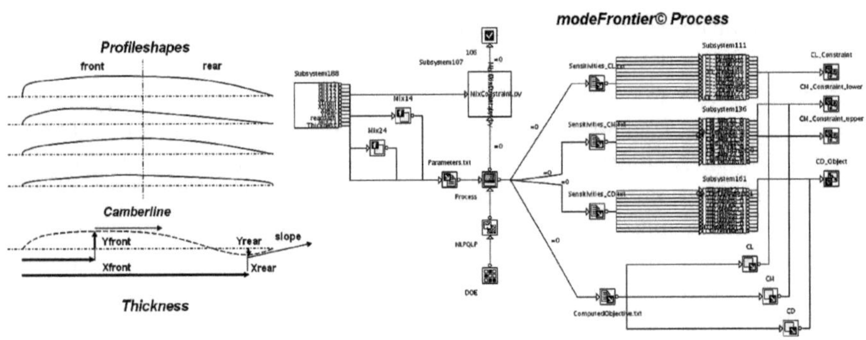

Fig. 10. Adjoint airfoil optimization for minimum drag at Mach = 1.2, $\alpha = 1°$, airfoil parameters and modeFrontier process

of pressure distributions show the rearward cambering effects to control the pitching moment.

The application of the adjoint approach to RANS-solutions requires the so called discontinuous adjoint method as implemented into the TAU code [8]. The evaluation of the aspect ratio sensitivities of a simple wing (Figure 12) was tested and compared with the results of a discrete evaluation. For the basic wing with an aspect ratio of 2.0 and the RAE-2822 airfoil the flow condition Mach = 0.5, $\alpha = 2°$ was chosen. At constant lift the drag should be reduced by an increase in aspect ratio to 2.275.

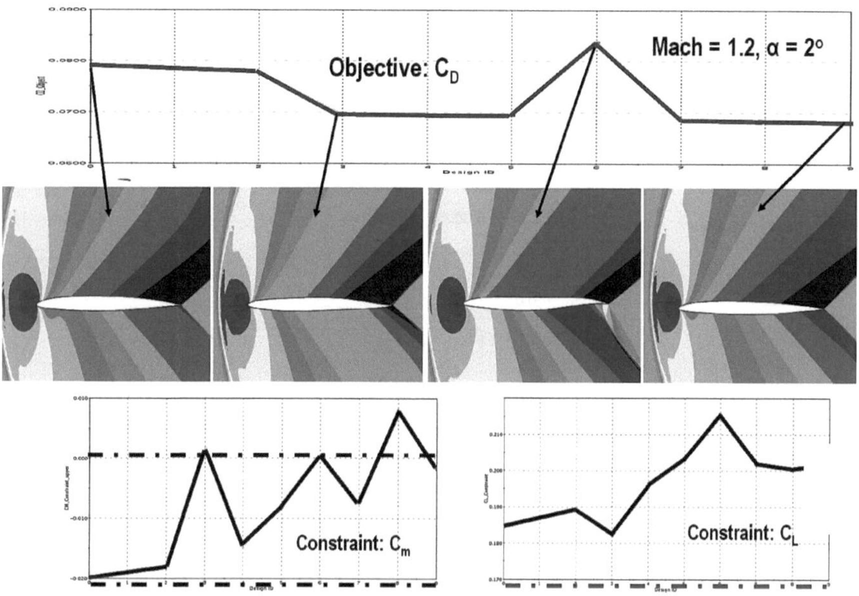

Fig. 11. Adjoint airfoil optimization for minimum drag at Mach = 1.2, $\alpha = 1°$, drag objective, lift and pitching moment constraints and pressure distributions for selected designs

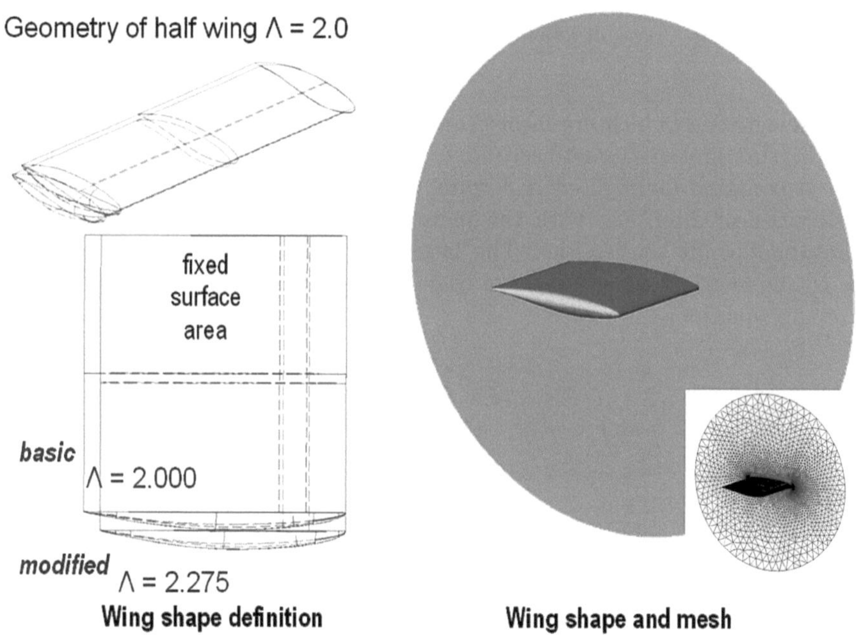

Fig. 12. Wings used to test the evaluation of aspect ratio sensitivities via an adjoint method

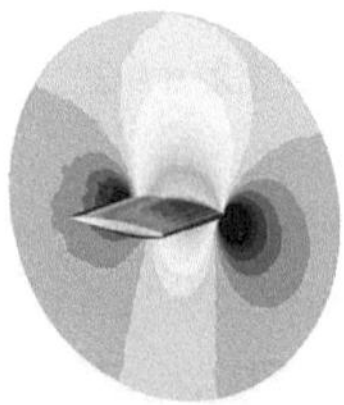

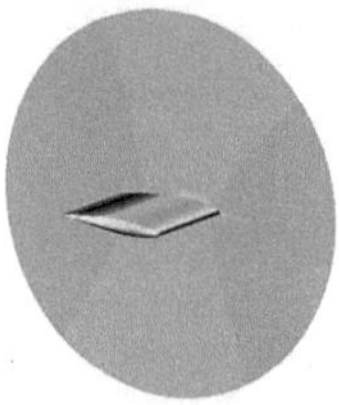

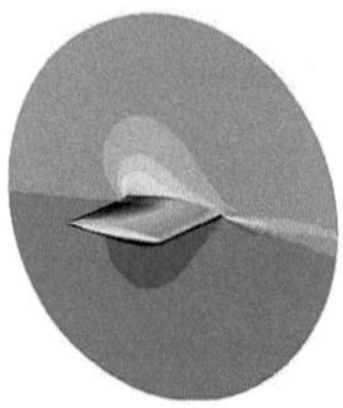

	discrete gradient	adjoint gradient
$\frac{d C_D}{d \Lambda}$		0,000011
$\frac{d C_L}{d \Lambda}$		0,009731
$\frac{d C_D}{d \alpha}$	0,00252	0,00249
$\frac{d C_L}{d \alpha}$	0,04760	0,04710

Discrete Values	
$\Lambda = 2.000$	$\Lambda = 2.275$
$C_L = 0{,}1581_{\alpha=2}$	$C_L = 0{,}1581_{\alpha=1.77}$
$C_D = 0{,}0099_{\alpha=2}$	$C_L = 0{,}0094_{\alpha=1.77}$
$\Delta C_D = 5$ dc $= 0{,}0005$	
Adjoint Values	
$\left.\frac{d C_D}{d \Lambda}\right\|_{CL = const.} = -0{,}000515 = 5$ dc	

$$\left.\frac{d C_D}{d \Lambda}\right|_{CL = const.} = \frac{d C_D}{d \Lambda} - \frac{d C_D}{d \alpha}\,\frac{1}{\frac{d C_L}{d \alpha}}\,\frac{d C_L}{d \Lambda}$$

Resolution	Fine	Medium	Coarse	Blunt/Coarse
Δ CD $_{CL = const}$	-0,0005	-0,000500	-0,00045	+0,0007
dCD / dΛ $_{CL = const}$	-0,000515	-0,000524	-0,00045	+0,0004

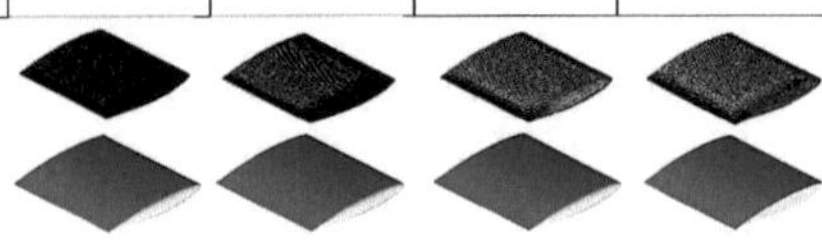

Fig. 13. Comparison of adjoint gradient and discrete aspect ratio sensitivity

All sensitivities which are formed under constant lift conditions and/or may affect lift distributions have to account for the aerodynamic down-wash induction via the equation highlighted in Figure 13 and derived below. Here the basic sensitivities of drag, e.g. with the parameter aspect-ratio has to include the corresponding lift sensitivities. The latter being adjusted by the sensitivities due to the angle of attack required to correct for constant lift as shown in the following equations.

$$dC_L = \delta C_{L\,(parameter)} + \frac{\delta C_L}{\delta \alpha}\Delta\alpha \;;\; dC_D = \delta C_{D\,(parameter)} + \frac{\delta C_D}{\delta \alpha}\Delta\alpha$$

$$C_L = \text{const.} \Rightarrow \quad dC_L = 0.0 \Rightarrow \quad \Delta\alpha = \frac{\delta CL}{\delta \alpha}\;\delta C_{L\,(parameter)}$$

$$_{(parameter)} = \Lambda$$

$$\left.\frac{d C_D}{d \Lambda}\right|_{CL = const.} = \frac{d C_D}{d \Lambda} - \frac{d C_D}{d \alpha}\,\frac{1}{\frac{d C_L}{d \alpha}}\,\frac{d C_L}{d \Lambda}$$

In Figure 13 the comparison of the discrete and adjoint sensitivities at constant lift conditions underlines that the adjoint method performs very well. Only very crude meshes impair the evaluation For example in the case of blunt/coarse geometry-mesh: in this example the flow is not resolved properly and the simulation produces wrong results, which is also shown in the absolutely wrong drag/aspect-ratio sensitivity of both the discrete and adjoint evaluation.

It can be concluded that the adjoint gradient method is very reliable and will reduce optimization efforts with many parameters considerably. Lift dependent effects have account for wake induced interferences by the adjustment described above. The integration of the discontinuous RANS method is almost identical with the continuous one and can be rated as assured.

10 Conclusions

modeFRONTIER, a framework for the scheduling, optimization and analysis of results has been presented. Several features of the software have been introduced and the attention has been focused on the main aspects of the optimization. The most important optimization algorithms have been explained, together with their pros and cons.

Through the analysis of two optimizations of military aircraft configurations also an idea of the practical use of modeFRONTIER has been given. Two very different optimizations have been presented: one using the genetic algorithm and a multiobjective strategy and another one with a gradient based algorithm which use the sensitivities provided by an adjoint simulation code.

References

1. Poles: MOGA-II, An improved Multi-Objective Genetic Algorithm. Technical Report 2003-6, Esteco (December 2003)
2. Silvia, P.: The simplex method. Technical Report 2003-5, Esteco (October 2003)
3. Schittkowski, K.: NLPQLP: A New Fortran Implementation of a Sequential Quadratic Programming Algorithm - Users Guide, Report, Department of Mathematics, University of Bayreuth (2001)
4. Nash, J.: Equilibrium points in n-person games. Proceedings of the National Academy of Sciences 36(1), 48–49 (1950)
5. Clarich, Pediroda, Padovan, Poloni, Periaux: Application of game strategy in multiobjective robust design optimisation implementing selfadaptive search space decomposition by statistical analysis. In: European Congress on Computational Methods in Applied Sciences and Engineering (July 2004)
6. Clarich, Rigoni, Poloni: A new algorithm based on game theory for robust and fast multi-objective optimisation
7. Clarich, Pediroda, Poloni, Periaux: Comparison between different game theory methodologies in robust design optimisation. In: International Conference on Computational Methods for Coupled Problems in Science and Engineering (2005)
8. Dwight, R., Brezillon, J.: Adjoint Algorithms for the Optimization of 3D Turbulent Configurations. In: STAB-Symposium, Darmstadt, November 2

One-Shot Methods for Aerodynamic Shape Optimization

Volker Schulz and Ilia Gherman

Department of Mathematics, University of Trier, 54296 Trier, Germany
volker.schulz@uni-trier.de

1 Introduction

The flow simulation tools in aerodynamics have reached a level of sophistication such that realistic flow simulations can be done in a reasonable computational time. Hence, the use of computational fluid dynamics (CFD) in the aerodynamic research as well as in the industry increases. When already using simulation tools, a natural extension is also to apply numerical shape optimization methods in the design process of an aircraft. With gradient free optimization methods, hundreds or even thousands of flow simulations are required during the optimization process. Even though the simulation tools are well developed and fast, this is still numerically very costly.

Our aim is to develop a gradient based optimization method, which allows the forward and the adjoint equation to be solved inexactly during the optimization. The algorithmic paradigm is derived from the reduced SQP methods. One can also view such a strategy as a simultaneous solution of the equations involved, hence the notation *one-shot approach*, i.e solve the equations (necessary optimality conditions) simultaneously. The aim is to reduce the overall cost of the optimization to just a few flow simulations.

When solving the forward and the adjoint equation inexactly within a reduced SQP-type optimization method, an important issue is the approximation of the reduced Hessian. A further step is – like in partially reduced SQP methods – to handle additional state constraints.

The idea of solving the necessary optimality conditions for problems without additional state constraints was already used by Ta'asan in a pseudo-time embedding in [24], where only the forward and the adjoint equations were solved inexactly. Batterman and Heinkenschloss [1] use an one-shot kind method based on trust-region techniques for similar problems. Müller et al. [21] uses a pseudo-time embedded one-shot method which can be seen as a variant of the inexact Newton method since the adjoint and the forward equations are solved more exactly when the iterates come closer to the solution. Andreas Griewank [13] proposed a one-shot optimization method, which he calls a piggy-back method. Since the piggy-back optimization is based on the automatic differentiation, the

N. Kroll et al. (Eds.): MEGADESIGN and MegaOpt, NNFM 107, pp. 207–220.
springerlink.com

gradient information available is used only in the next iteration, whereas we want to use the available information as soon as possible – resulting in a different kind of iteration. Regarding linear-quadratic problems (linear objective function and quadratic constraints), related preconditioners for Krylov methods are proposed by Biros and Ghattas in [4,5,6] and Battermann and Sachs in [2,3]. For linear-quadratic problems, our method results in the construction of a good preconditioner for the linear system of the necessary optimality conditions, too. The difference is that Biros and Ghattas as well as Battermann and Sachs consider a preconditioner for a Krylov-type iterative solver and the convergence is proved by considering the eigenvalues of the preconditioned system. We want to solve the linear system with a defect correcting type iteration which means that we consider a different type of iteration.

A very important issue is to extend the method to include additional state constraints as it is necessary for realistic optimization problems. So far, no theoretical considerations of inclusion of the additional state constraints in aerodynamic shape optimization have been made in the literature.

This article is structured as follows. In Sect. 2, the one-shot method for linear-quadratic problems is developed and its convergence is proved. Firstly, we consider a linear-quadratic optimization problem with a quadratic objective function and reducable linear constraints. The linear system resulting from the necessary optimality conditions is solved by a defect correcting iteration. For this iteration, a preconditioner based on reduced SQP methods is considered. The difference to standard reduced SQP methods is that the forward and the adjoint matrices are inexact. In this case, also the reduced Hessian has to be the consistent reduced Hessian, according to the approximations of the forward and the adjoint matrices. The convergence of the iteration is proven and also upper bounds for the convergence rate are given. Following that, we consider again a linear-quadratic optimization problem, now containing additional state constraints which *cannot* be reduced. For the resulting linear system, again a preconditioner for a defect correcting iteration is constructed based on partially reduced SQP methods. Again, the forward and the adjoint matrices are only approximations. Then, additionally to the consistent reduced Hessian, also the consistent reduced gradient of the additional state constraints is part of the preconditioner. The convergence of the method is proven and also upper bounds for the linear convergence rate are given. Eventually, generalizations of this additionally constrained method are given to include cross-terms and several (not only one) additional constraints.

Section 3 deals with the generalization of the developed methods to non-linear optimization problems. First, the unconstrained version of the method is generalized. Aerodynamic optimization without additional state constraints can lead to physically not meaningful results. Hence, the next step is to generalize the additionally constrained version of the one-shot method to non-linear optimization problems with additional state constraints. However, one has to modify the developed method at first to be able to generalize it such that the non-linear version of the method can be performed using existing adjoint solvers. We will also show how the original and the modified methods are connected and that they converge to the same limit.

The developed one-shot optimization method is eventually applied to aerodynamic shape optimization problems in Sect. 4. We present 2D results for transonic wing optimization with FLOWer and for optimizing a three-element high-lift configuration with TAU. Further applications of the algorithms, also in 3D cases, can be looked up in [14, 15].

2 One-Shot Method for Linear-Quadratic Problems

2.1 Approximate Reduced SQP Preconditioner for a Defect Correcting Iteration

We consider the following linear-quadratic optimization problem (QP):

$$\min_{u,q} \frac{1}{2}u^{\mathrm{T}}H_u u + \frac{1}{2}q^{\mathrm{T}}H_q q + f_u^{\mathrm{T}} u + f_q^{\mathrm{T}} q \tag{1}$$

$$\text{subject to } C_u u + C_q q + c = 0 \,. \tag{2}$$

$u \in \mathbb{R}^{n_u}$ represents the state variable, and $q \in \mathbb{R}^{n_q}$ represents the design variable (parameters) of the optimization problem. Furthermore, $H_u \in \mathbb{R}^{n_u \times n_u}$ and $H_q \in \mathbb{R}^{n_q \times n_q}$ are assumed to be symmetric, $f_u \in \mathbb{R}^{n_u}$, $f_q \in \mathbb{R}^{n_q}$, $C_u \in \mathbb{R}^{n_u \times n_u}$ and $C_q \in \mathbb{R}^{n_q \times n_u}$. C_u is assumed to be invertible. In aerodynamical context, the cost function of this QP may represent a quadratic approximation of an aerodynamic coefficient, e.g. C_{D}. The constraint (2) represents the linearized flow equations. Since the matrix C_u is invertible, the nullspace of (2) can be expressed explicitly. This results in an unconstrained optimization problem. Furthermore constraints on the state variables u are not posed – hence, the problem (1),(2) is referred to as an unconstrained problem.

The solution of the (QP) is given by the solution of the linear system:

$$\begin{bmatrix} H_u & 0 & C_u^{\mathrm{T}} \\ 0 & H_q & C_q^{\mathrm{T}} \\ C_u & C_q & 0 \end{bmatrix} \begin{pmatrix} u \\ q \\ \lambda \end{pmatrix} = - \begin{pmatrix} f_u \\ f_q \\ c \end{pmatrix} ,$$

where λ is the Lagrange multiplier. For the solution of the above linear system we want to use a defect correcting-type iteration, with a preconditioner based on the ideas of reduced SQP methods (rSQP) (see e.g. [22, 23])

$$\begin{pmatrix} u^{k+1} \\ q^{k+1} \\ \lambda^{k+1} \end{pmatrix} = \begin{pmatrix} u^{k} \\ q^{k} \\ \lambda^{k} \end{pmatrix} - \begin{bmatrix} 0 & 0 & A_a \\ 0 & B & C_q^{\mathrm{T}} \\ A_f & C_q & 0 \end{bmatrix}^{-1} \left(\begin{bmatrix} H_u & 0 & C_u^{\mathrm{T}} \\ 0 & H_q & C_q^{\mathrm{T}} \\ C_u & C_q & 0 \end{bmatrix} \begin{pmatrix} u^{k} \\ q^{k} \\ \lambda^{k} \end{pmatrix} + \begin{pmatrix} f_u \\ f_q \\ c \end{pmatrix} \right) . \tag{3}$$

When looking at the preconditioner, it strikes out that in contrast to traditional rSQP methods as in [23], there appear the matrices A_a instead of C_u^{T} and A_f instead of C_u in the preconditioner. A_a and A_f are supposed to be just approximations of the matrices C_u^{T} and C_u respectively. Hence, the resulting preconditioner will be denoted as *approximate reduced SQP* preconditioner. Note also that we do not require $A_a^{\mathrm{T}} = A_f$. This assumption is important especially with regard to aerodynamic applications to come. A_f can be interpreted as a linearization of the flow equations, A_a as a linearization of the adjoint equations.

Furthermore, in classical rSQP methods, B is an approximation of the reduced Hessian of the Lagrangian, which for the problem (1),(2) is given by

$$S = H_q + C_q^{\mathrm{T}} C_u^{-\mathrm{T}} H_u C_u^{-1} C_q \,,$$

and which we assume to be positive definite in order to obtain a well posed problem. It should be also noted that the reduced Hessian S can be interpreted as the Schur complement of the KKT-matrix in (3) with respect to the variables u and λ. However, since C_u and C_u^{T} are approximated by A_f and A_a respectively, one can consider the reduced Hessian consistent with the choice of A_f and A_a as approximations to C_u and C_u^{T} in the form

$$S_A = H_q + C_q^{\mathrm{T}} A_a^{-1} H_u A_f^{-1} C_q \,. \tag{4}$$

This "reduced Hessian" will be referred to as *consistent reduced Hessian.* Compared to the classical reduced Hessian S, S_A is a projection of the Hessian of the Lagrangian to a different subspace. Refer for details to [11, 19]. Here, only the main results are presented.

Theorem 2.1 *Define the numerical spectral norms*

$$r_A^f := \left\| I - A_f^{-1} C_u \right\|_2, \quad r_A^a := \left\| I - A_a^{-1} C_u^T \right\|_2, \quad r_S := \left\| I - B^{-1} S_A \right\|_2.$$

There is an $\eta < 1$, such that the iteration (3) *converges, provided.*

$$\max\{r_A^f, r_A^a, r_S\} < \eta \,.$$

In the case that $A_a = A_f^{\mathrm{T}} =: A$ and additionally C_u symmetric and one uses the following spectral norm

$$\left\| \begin{pmatrix} u \\ q \\ \lambda \end{pmatrix} \right\|_r = \left\| \begin{pmatrix} A^{1/2} u A^{-1/2} \\ B^{1/2} q B^{-1/2} \\ A^{1/2} \lambda A^{-1/2} \end{pmatrix} \right\|_2$$

the formulation of the theorem 2.1 can be refined:

Theorem 2.2 *Let $A_a = A_f = A$ and $\|.\| := \|.\|_r$, and define*

$$\varrho_A := \varrho\left(I - A^{-1} C_u\right), \quad \varrho_S := \varrho\left(I - B^{-1} S_A\right) \,.$$

There is an $\eta < 1$, such that the iteration (3) *converges, provided* $\max\{\varrho_A, \varrho_S\} < \eta$.

The important conclusion is that B should approximate the consistent reduced Hessian S_A. This fact was already observed in practical examples in [12] and was verified once again at numerical examples in [19].

The developed convergence theory can be easily generalized to the case of non-zero cross-terms, i.e. for the objective function of the form:

$$\min_{u,q} \frac{1}{2}\left(u^{\mathrm{T}} H_u u + u^{\mathrm{T}} H_{uq} q + q^{\mathrm{T}} H_{qu} u + q^{\mathrm{T}} H_q q\right) + f_u^{\mathrm{T}} u + f_q^{\mathrm{T}} q.$$

The consistent reduced Hessian for this case is then

$$S_A = H_q - H_{qu} A_f^{-1} C_q - C_q^{\mathrm{T}} A_a^{-1} H_{uq} + C_q^{\mathrm{T}} A_a^{-1} H_u A_f^{-1} C_q,$$

see [11, 19] for details. Using the Lagrangian

$$\begin{aligned}\mathcal{L}(u,q,\lambda) =& \frac{1}{2}\left(u^{\mathrm{T}} H_u u + u^{\mathrm{T}} H_{uq} q + q^{\mathrm{T}} H_{qu} u + q^{\mathrm{T}} H_q q\right) + f_u^{\mathrm{T}} u + f_q^{\mathrm{T}} q \\ &+ \lambda^{\mathrm{T}}(C_u u + C_q q + c).\end{aligned}$$

Iteration (3) can now be rewritten in a more compact form as

$$\lambda^{k+1} = \lambda^k - A_a^{-1} \nabla_u \mathcal{L}(u^k, q^k, \lambda^k)\,, \tag{5}$$

$$q^{k+1} = q^k - B^{-1} \nabla_q \mathcal{L}(u^k, q^k, \lambda^{k+1})\,, \tag{6}$$

$$u^{k+1} = u^k - A_f^{-1} \nabla_\lambda \mathcal{L}(u^k, q^{k+1}, \lambda^{k+1})\,. \tag{7}$$

Note the early usage of information as soon as it is available (e.g. λ^{k+1} is used to compute q^{k+1}). This resembles the Gauss–Seidel iteration for linear systems. The piggy-back method by Andreas Griewank [13] uses the gained information only in the next step thus resembling the Jacobian iteration for linear systems.

The next important step is to extend the method to be able to handle addtional state constraints.

2.2 Approximate prSQP Preconditioner for a Defect Correcting Iteration

Here, we want to add to the optimization problem (1)-(2) a further constraint which at first should be a linear scalar one:

$$\ell_u^{\mathrm{T}} u + \ell_q^{\mathrm{T}} q + \ell = 0$$

with $\ell_u \in \mathbb{R}^{n_u}$ and $\ell_q \in \mathbb{R}^{n_q}$ representing an additional state constraint. We want to use the same technique as above and solve the resulting linear system by a defect correcting iteration using an appropriate preconditioner. Now, the preconditioner is based on the ideas of the partially reduced SQP methods (see e.g. [23]). The iteration considered is

$$\begin{pmatrix} u^{k+1} \\ q^{k+1} \\ \mu^{k+1} \\ \lambda^{k+1} \end{pmatrix} = \begin{pmatrix} u^{k} \\ q^{k} \\ \mu^{k} \\ \lambda^{k} \end{pmatrix} - \begin{bmatrix} 0 & 0 & 0 & A_a \\ 0 & B & \gamma_A & C_q^{\mathrm{T}} \\ 0 & \gamma_A^a & 0 & 0 \\ A_f & C_q & 0 & 0 \end{bmatrix}^{-1} \left(\begin{bmatrix} H_u & 0 & \ell_u & C_u^{\mathrm{T}} \\ 0 & H_q & \ell_q & C_q^{\mathrm{T}} \\ \ell_u^{\mathrm{T}} & \ell_q^{\mathrm{T}} & 0 & 0 \\ C_u & C_q & 0 & 0 \end{bmatrix} \begin{pmatrix} u^{k} \\ q^{k} \\ \mu^{k} \\ \lambda^{k} \end{pmatrix} + \begin{pmatrix} f_u \\ f_q \\ \ell \\ c \end{pmatrix} \right), \tag{8}$$

where the preconditioner will be referred to as *approximate partially reduced SQP preconditioner*. Same notations as in (3) are used with additionally $\gamma \in \mathbb{R}^{n_q}$ and $\gamma^a \in \mathbb{R}^{1\times n_q}$. In the same manner as the consistent reduced Hessian was introduced, one can also define the consistent reduced gradient of the additional constriant (see [11] for details):

$$g_A := \ell_q - C_q^{\mathrm{T}} A_a^{-1} \ell_u \quad \text{and} \quad g_A^a := \ell_q^{\mathrm{T}} - \ell_u^{\mathrm{T}} A_f^{-1} C_q \,.$$

We assume further that

$$n_g := -g_A^a B^{-1} g_A \neq 0 \,.$$

One obtains similar convergence results as above ([11]):

Theorem 2.3 *Define the numerical spectral norms*

$$r_A^f := \left\| I - A_f^{-1} C_u \right\|_2, \quad r_A^a := \left\| I - A_a^{-1} C_u^T \right\|_2, \quad r_S := \left\| I - B^{-1} S_A \right\|_2 .$$

There is an $\eta < 1$, such that the iteration (8) *converges, provided*

$$\max\{r_A^f, r_A^a, r_S\} < \eta \,. \tag{9}$$

Theorem 2.4 *Suppose that $A_f = A_a^T = A$, C_x and B all symmetric and positive definite. Define*

$$\varrho_A := \varrho(I - A^{-1} C_x) \,, \qquad \varrho_S := \varrho(I - B^{-1} S_A) \,.$$

There is an $\eta < 1$, such that the iteration (8) *converges, if* $\max\{\varrho_A, \varrho_S\} < \eta$.

Just as before, we want to generalize the developed iteration: First, to include the cross-terms in the objective function and second to allow for multiple additional constraints:

$$\begin{aligned} \min_{u,q} \quad & \frac{1}{2}\left(u^{\mathrm{T}} H_u u + u^{\mathrm{T}} H_{uq} q + q^{\mathrm{T}} H_{qu} u + q^{\mathrm{T}} H_q q\right) + f_u^{\mathrm{T}} u + f_q^{\mathrm{T}} q \\ \text{subject to} \quad & C_u u + C_q q + c = 0 \,, \\ & L_u^{\mathrm{T}} u + L_q^{\mathrm{T}} q + l = 0 \end{aligned}$$

with $L_u \in \mathbb{R}^{n_u \times n_\ell}$, $L_q \in \mathbb{R}^{n_q \times n_\ell}$ and $l \in \mathbb{R}^{n_\ell}$, n_ℓ being the number of additional constraints. Instead of the (consistent) reduced gradient of the additional constraints one has now the (consistent) reduced Jacobian of the additional constraints (the matrix which columns are consistent reduced gradients of the repective additional constraints):

$$G^{\mathrm{T}} = L_q^{\mathrm{T}} - L_u^{\mathrm{T}} C_u^{-1} C_q \quad \in \mathbb{R}^{n_q \times n_\ell}$$

and correspondingly

$$G_A^a = L_q^{\mathrm{T}} - L_u^{\mathrm{T}} A_f^{-1} C_q \,.$$

Theorem 2.5 *The result of theorems 2.3 is also valid for the case of n_ℓ additional constraints, and non-zero crossterms, provided that additionally $G_A^a S_A^{-1} G_A$, $G_A^a B^{-1} G_A$ are invertible.*

To be able to generalize the iteration for non-linear problems, we again make use of the Lagrangian

$$\begin{aligned}\mathcal{L}(u,q,\lambda,\mu) =& \frac{1}{2}\left(u^{\mathrm{T}}H_u u + u^{\mathrm{T}}H_{uq}q + q^{\mathrm{T}}H_{qu}u + q^{\mathrm{T}}H_q q\right) + f_u^{\mathrm{T}}u + f_q^{\mathrm{T}}q \\ &+ \lambda^{\mathrm{T}}(C_u u + C_q q + c) + \mu^{\mathrm{T}}(L_u^{\mathrm{T}}u + L_q^{\mathrm{T}}q + l)\,.\end{aligned}$$

The above iteration can be viewed as (refer to [11] for details) *Compute the update in the Lagrange variable λ:*

$$\lambda^{k+1} = \lambda^k - A_a^{-1}\nabla_u \mathcal{L}(u^k, q^k, \lambda^k, \mu^k)\,. \tag{10}$$

Then, the updates for the variables q and μ are computed from the following linear system

$$\begin{bmatrix} B & \Gamma_A \\ \Gamma_A^a & 0 \end{bmatrix} \begin{pmatrix} \Delta q \\ \Delta\mu \end{pmatrix} = \begin{pmatrix} -\nabla_q \mathcal{L}(u^k, q^k, \lambda^{k+1}, \mu^k) \\ -\nabla_\mu \mathcal{L}(u^k, q^k, \lambda^{k+1}, \mu^k) \end{pmatrix} \tag{11}$$

and setting $q^{k+1} = q^k + \Delta q$, $\mu^{k+1} = \mu^k + \Delta\mu$. Notice that here λ^{k+1} is already used in the right-hand side, i.e., like in the unconstrained case, the information is used as soon as it is available.

Eventually, the update in u is computed by

$$u^{k+1} = u^k - A_f^{-1}\nabla_\lambda \mathcal{L}(u^k, q^{k+1}, \lambda^{k+1}, \mu^{k+1}) \tag{12}$$

again using the available information immediately.

3 Generalization to Non-linear Problems

We now turn to general nonlinear problems of the form

$$\begin{aligned} &\min\ f(u,q) \\ &\text{subject to } c(u,q) = 0 \end{aligned} \tag{13}$$

with invertible Jacobian $C_u := \frac{\partial c}{\partial u}$. First, no additional state constraints are considered and we want to generalize the iteration (5)-(7), refer to [11] for details.

Considering (5) in non-linear context it can be interpreted as obtaining the new adjont variable by *inexact* solution of the adjoint equation, starting in the current adjoint variable. If an iterative solver for the adjoint equation is available, that means, that the new adjoint variable is obtained by starting in the current adjoint variable and performing several, say n_a steps of the iterative ajoint solver.

Algorithm 1. One-Shot method for unconstrained shape optimization

$k \leftarrow 0$, Start with initial guess u^0, q^0, λ^0
repeat
  Starting in λ^k, perform n_a steps in the adjoint solver $\rightarrow \lambda^{k+1}$
  Compute the reduced gradient γ^k:

$$\gamma^k := \nabla_q f(u^k, q^k) + C_q^{\mathrm{T}}(u^k, q^k) \cdot \lambda^{k+1}$$

  Approximate the consistent reduced Hessian $\rightarrow B_k$
  $q^{k+1} \leftarrow q^k - \tau_k B_k^{-1} \gamma^k$, $\tau_k \in (0, 1]$ step length
  Compute the according geometry and adjust the computational mesh
  Starting in u^k, perform n_{f} steps in the forward solver $\rightarrow u^{k+1}$
until "convergence"

The update of the design variable as in (6) is interpreted in the following way: Using currently available (inexact) adjoint variable, compute the consistent reduced gradient γ_{d}^k of the objective function. Approximate the consistent reduced Hessian by B_k and compute the update of the design variables by solving the linear system

$$B_k \Delta q = -\gamma_{\mathrm{d}}^k \,,$$

and set $q^{k+1} = q^k + \tau_k \Delta q$, where τ_k is the step length.

The state equation (7) is interpreted in a very similar way as the adjoint equation, i.e., as performing several, say n_{f} steps of the iterative forward solver, starting in the current state variable.

These steps result in the unconstrained version of the one-shot algorithm, presented in algorithm 1. However, the unconstrained formulation of the aerodynamic shape optimization problem can lead to solutions which are physically not meaningful, see [11]. Hence, it is important to have the possibility to include additional state constraints in the shape optimization problem. This leads us to the iteration (10)-(12) which means that we extend the nonlinear optimization problem (13) by an additional scalar (but nonlinear) constraint

$$\ell(u, q) = 0 \,.$$

When generalizing the iteration analogously to the unconstrained case, one unfourtunately obtains an iteration which cannot be performed in practice due to the adjoint equation which cannot be solved by an existing adjoint solver, see [11]. Hence, we have to modify the iteration first to be able to generalize it in the way that it can be solved by existing solvers. The adjoint solvers are such that the right-hand side is the derivative of C_{D}, C_{L} or C_{M} with respect to u. We assume that the adjoint equation with the right-hand side $\nabla_u \ell$ can be solved (think of the additional constraint $\ell(u, q)$ as a restriction of the lift coefficient C_{L}). The idea is to solve (inexactly) the adjoint equation with $\nabla_u f$ as the right-hand side – the corresponding adjont variable is say σ_{d} – and then the adjoint equation with $\nabla_u \ell$ as right-hand side – with the corresponding adjoint σ_ℓ. Combining both adjoint variables σ_{d} and σ_ℓ one obtains an artificial adjoint variable wich corresponds to the adjoint varable λ in the defect-correcting iteration (10), and one can show that they converge to the same limit, see [11].

All this is summarized in the following algorithm.

Algorithm 2. One-shot method for state constrained shape optimization

---

Initialization: $k \leftarrow 0$, Start with initial guess u^0, q^0, σ_{d}^0, σ_ℓ^0
repeat
  Starting in σ_{d}^k, perform n_a steps in the adjoint solver ($\nabla_u f$ as r.h.s.) $\rightarrow \sigma_{\mathrm{d}}^{k+1}$
  Starting in σ_ℓ^k, perform n_a steps in the adjoint solver ($\nabla_u \ell$ as r.h.s.) $\rightarrow \sigma_\ell^{k+1}$
  Compute the reduced gradients:

$$\gamma_{\mathrm{d}}^{k+1} = \nabla_q f(u^k, q^k) + C_q^{\mathrm{T}}(u^k, q^k) \cdot \sigma_{\mathrm{d}}^{k+1}$$
$$\gamma_\ell^{k+1} = \nabla_q \ell(u^k, q^k) + C_q^{\mathrm{T}}(u^k, q^k) \cdot \sigma_\ell^{k+1}$$

  Approximate the consistent reduced Hessian of the Lagrangian $\rightarrow B_k$
  Compute $\mu^{k+1} = \frac{\ell - (\gamma_\ell^{k+1})^{\mathrm{T}} B_k^{-1} \gamma_{\mathrm{d}}^{k+1}}{(\gamma_\ell^{k+1})^{\mathrm{T}} B_k^{-1} \gamma_\ell^{k+1}}$
  $q^{k+1} = q^k - \tau_k B_k^{-1} \gamma_{\mathrm{d}}^{k+1} - B_k^{-1} \gamma_\ell^{k+1} \mu^{k+1}$, $\tau_k \in (0, 1]$ damping factor.
  Compute the corresponding geometry and adjust the computational mesh.
  Starting in u^k, perform n_{f} steps in the forward solver $\rightarrow u^{k+1}$
until "convergence"

---

4 Numerical Applications

4.1 RAE2822 Airfoil in Transonic Euler-Flow with FLOWer

Here, we optimize the RAE2822 airfoil in Euler-flow at transonic flight conditions:

$$\mathrm{Ma}_\infty = 0.73, \qquad \alpha = 2°.$$

The objective function is the drag coefficient C_{D}. Additional (scalar) restriction is that the initial lift coefficient shall be preserved. We treat this constraint explicitly by means of the constrained one-shot method and not by adjusting the angle of attack.

The airfoil is decomposed into thickness and camber distribution. Then, only the camber of the airfoil is parameterized by 21 Hicks–Henne functions, see [9,17]. The thickness is not changed guaranteeing that the thickness of the airfoil will not change during the optimization process, and hence the optimization will not result in a flat airfoil.

The flow region is discretized by a 133×33 C-type grid.The structured flow solver FLOWer provided by DLR in forward and adjoint mode will be used to solve the state and the adjoint equations. Both, the forward and the adjoint solvers are used with the multi-grid option with three multi-grid levels.

After each update of the design variables, also the grid has to be moved. For the structured grid, the so called "grid-moving technique" based on Reuther's

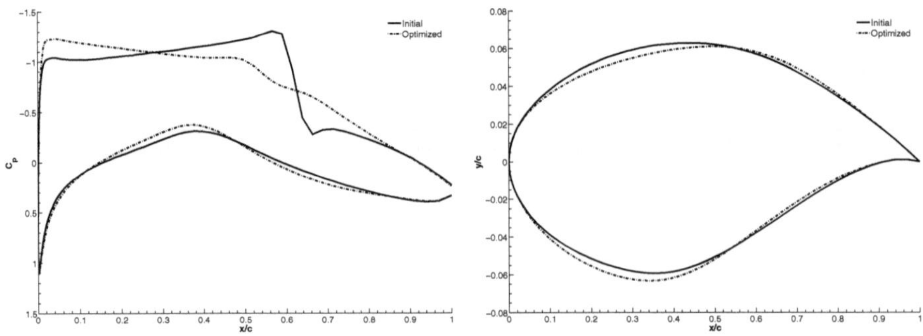

Fig. 1. Distribution of the pressure coefficient over the initial and optimized airfoils (left) and the shape of the initial and optimized airfoils (right)

approach [20] is used, see e.g. [9,16]. The computation of the reduced gradients for Euler-flow based on such grid-moving strategy can be found in [9].

The reduced Hessian B of the Lagrangian is approximated by limited-memory BFGS updates. We use three stages for the L-BFGS updates, i.e. the update is computed based on the gradient information from the three preceding steps.

The optimization is stopped, when the ∞-norm of the reduced gradient of the Lagrangian

$$\gamma_{\mathcal{L}}^k := \gamma_{\mathrm{d}}^k + \mu^k \gamma_{\ell}^k ,$$

based on the consistent reduced gradients of the objective function and of the additional constraint is small, i.e.

$$\|\gamma_{\mathcal{L}}^k\|_\infty < \varepsilon .$$

We choose $\varepsilon = 5\times10^{-3}$. For the step length selection, we use a simplistic strategy, $\tau_k \equiv 0.8$, i.e. a constant step length.

The optimization stops after 320 optimization iterations with the values

$$C_{\mathrm{D}} = 37.2 \times 10^{-4} \qquad C_{\mathrm{L}} = 0.8261 ,$$

whereas the initial values are

$$C_{\mathrm{D}} = 86.88 \times 10^{-4} \qquad C_{\mathrm{L}} = 0.8263$$

see also Fig. 1. The total numerical cost of the optimization equals to less than 8 flow simulations, see [11] for details.

4.2 High-Lift Configuration with TAU

Within the EUROLIFT II project [25], the problem of optimizing the shape of the flap device of a three-element airfoil at take-off conditions is formulated, see also [7,8]. For this high-lift configuration, the flow considered is the Navier–Stokes flow, the flow conditions are:

$$\mathrm{Ma}_\infty = 0.17146\,, \quad \mathrm{Re} = 14.7 \times 10^6, \quad \alpha = 9.6°,$$

Again, the cost function is the drag coefficient and we formulate the additional constraint of constant lift explicitly.

Only the shape and the position of the flap device are free for optimization. Neither other parts of the airfoil nor the angle of attack are allowed to be changed by the optimization. The flap is parameterized by 10 design variables, see [7, 11] for details. Several issues arise here. Firstly, one needs to scale the design variables and secondly also box-constriants on the design variables have to be formulated for geometrical reasons on one hand and in order to be able preserve the topology of the mesh when building new mesh for the changed airfoil on the other hand.

Here, the unstructured flow solver TAU provided by DLR is used. The adjoint solver is only used as a direct solver, i.e. the adjoint equations are always solved "exactly". The computational space is discretized by an unstructred mesh consisting of about 90 000 points. As an implementational issue, the problem arises that one has to recompute the mesh each time the geometry (design variables) is changed. Also, the mesh sensitivity has to be computed by finite differences which is numerically enormously, cf. [11].

In [7] the same optimization problem is considered. However, there, the constraint of constant lift is enforced by adjusting the angle of attack. We treat the constraint explicitly and hence, first implement the linearized gradient projection method (LPG) as presented in [10] or in [11] as a black-box method to be able to compare the one-shot optimization with a black-box optimization.

Within the LGP method, all equations are solved "exactly". The state equation is solved by an iterative time-integration method until a steady state is reached. Here, we perform 7000 iterations in the forward equation at each optimization step, so that the residual of the forward equation is always less than 10^{-6}. We stop the optimization when the update Δq of the design variables yields

$$\|\Delta q\|_2^2 \leq 10^{-3}.$$

and we use a constant step length $\tau_k \equiv 0.5$. It takes 21 optimization iterations of LGP method to converge. The total wallclock run-time of this optimization on a Linux PC, Intel 4 with 3 GHz and 2 GB ROM, is about 27 hours. The main numerical cost stems from the "exact" solution of the state equation, which takes approximately one hour in each optimization iteration.

The first variant of the one-shot optimization can be designated as one-shot linearized gradient projection (OS-LGP). The idea is to perform the LGP optimization just as above, however, now *not* solving the state equation exactly, but only performing several, n_f, steps in the iterative time-integrator of the forward equation. This OS-LGP method can be viewed also as a "usual" one-shot method. The update of the design variables computed in that way corresponds to the one-shot method and L-BFGS updates for the reduced Hessian with trivial history choice $m = 0$. Compared to the LGP method, we reduce the damping factor to $\tau_k \equiv 0.1$ due to the inexactness of the forward solution. Furthermore, we choose $n_{\mathrm{f}} = 100$, i.e. we perform 100 (time) steps in the forward solver at each optimization iteration (LGP method: 7000 steps).

In this way, computation of the (inexact) solution of the state equation takes less than one minute and so it is not the driving cost of the optimization process any more. Now, the computation of the mesh sensitivities by finite differences and the direct solution of the adjoint equations carry the most weight. Hence, we parallelize the computation of the mesh derivatives. Also, both adjoint equations are solved simultaneously. That means, that more resources (10 computer/processors) are now needed for the computation. The process stops after 103 optimization iterations. The wallclock run-time of the OS-LGP optimization is (only) 9 hours (However, 10 computers are needed. Without parallelization, the OS-LGP optimization takes 20 hours, which is still an improvement compared to 27 hours needed by LGP method).

The next logical step is to perform "proper" one-shot optimization, using L-BFGS updates for the reduced Hessian approximation, now with $m > 0$. We choose $m = 3$. Furthermore, we use more information to build up the reduced Hessian and solve the state equation less exactly. Now we perform 20 iterations within the state solver at each optimization iteration, i.e. $n_{\mathrm{f}} = 20$. All other parameters remain unchanged.

In this case, the optimization needs only 33 iterations to converge. The runtime is 4.5 hours. The solution obtained, however, differs from the one obtained by both LGP and OS-LGP methods. The values of the drag and of the lift coefficients are:

$$\begin{aligned} \text{Initial:} \quad & C_{\mathrm{D}} = 811.3 \times 10^{-4} \quad & C_{\mathrm{L}} = 2.7300 \\ \text{Optimized LGP and OS-LGP:} \quad & C_{\mathrm{D}} = 741.5 \times 10^{-4} \quad & C_{\mathrm{L}} = 2.7291 \\ \text{Optimized one-shot L-BFGS:} \quad & C_{\mathrm{D}} = 743.1 \times 10^{-4} \quad & C_{\mathrm{L}} = 2.7256 \end{aligned}$$

There is again a negligible drop in the lift coefficient and a slightly worse reduction of the drag coefficient. Although the values of the aerodynamic coefficients are similar, the corresponding geometries differ distinctly as one can see from Fig. 2, showing the airfoils and C_{p} distribution of all optimization results.

The apparent explanation is that using L-BFGS updates for reduced Hessian approximation leads to a different local minimum as when using LGP-kind of design update. It is well known that gradient-based optimization "only" finds a local solution, and changing the way of the reduced Hessian approximation can apparently lead to another local solution. Aerodynamically, both obtained solutions are comparable since the drag and the lift values are similar. For a "unique" solution, further constraints have to be formulated.

One can also overcome the problem of the computation of the mesh sensitivities by finite differences, see [11]. Even when using finite differences one can use the matrix of the mesh derivatives constant over several optimization iterations. Even when computing the matrix only once and then using it through the whole optimization thus making the computation of the mesh sensitivities during the optimization obsolete delivers acceptable results, see [11]. In this case, the whole

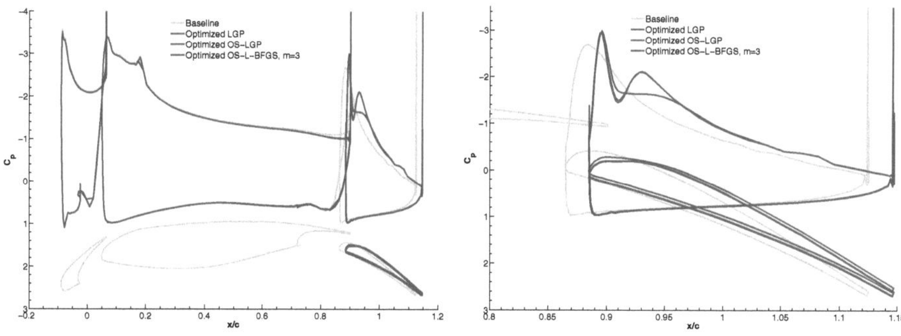

Fig. 2. Comparison of different solutions: LGP, one-shot LGP and one-shot method with L-BFGS-updates for reduced Hessian. The airfoil and the distribution of the pressure coefficient C_{p} over the airfoil (left) and zoom to the flap (right).

optimization runs on *one* PC (no parallelization is needed) with a wall clock time of about 4 hours.

References

1. Battermann, A., Heinkenschloss, M.: Preconditioners for Karush–Kuhn–Tucker matrices arising in optimal control of distributed systems. In: Desch, W., Kappel, F., Kunisch, K. (eds.) Optimal Control of Partial Differential Equations, Vorau 1997, pp. 15–32. Birkhäuser Verlag (1998)
2. Battermann, A., Sachs, E.W.: Block preconditioner for KKT systems in PDE-governed optimal control problems. In: Hoppe, R.H.W., Hoffmann, K.-H., Schulz, V. (eds.) Workshop on Fast Solutions of Discretized Optimization Problems, pp. 1–18. Birkhäuser, Basel (2001)
3. Battermann, A., Sachs, E.W.: An indefinite preconditioner for KKT systems arising in optimial control problems (2003)
4. Biros, G., Ghattas, O.: Parallel preconditioners for KKT systems arising in optimal control of viscous incompressible flows. In: Keyes, D.E., Ecer, A., Periaux, J., Satofuka, N. (eds.) Parallel Computational Fluid Dynamics 1999. North-Holland, Amsterdam (1999)
5. Biros, G., Ghattas, O.: Parallel Newton–Krylov algorithms for PDE-constrained optimization. In: Proceedings of SC 1999, The SCxy Conference Series, Portland. ACM/IEEE (1999)
6. Biros, G., Ghattas, O.: Parallel Lagrange–Newton–Krylov–Schur methods for PDE-constrained optimization. Part I: The Krylov–Schur solver. SIAM Journal on Scientific Computing 27(2), 687–713 (2005)
7. Brezillon, J., Wild, J.: Evaluation of different optimization strategies for the design of a high-lift flap device. In: EUROGEN 2005 (September 2005)
8. Dwight, R.P., Brezillon, J.: Effect of various approximations of the discrete adjoint on gradient-based optimization. In: 44th AIAA Aerospace Sciences Meeting and Exhibit, Reno, Nevada, January 9–12, pp. 1–22 (2006)
9. Gauger, N.: Das Adjungiertenverfahren in der aerodynamischen Formoptimierung. PhD thesis, TU Braunschweig (2003)

10. Gauger, N., Brezillon, J.: DLR's technical annex to the 2nd year progress report of the european project AEROSHAPE. DLR, Institut für Aerodynamik und Strömungstechnik, Institutionsbericht (2002)
11. Gherman, I.: Approximate Partially Reduced SQP Approaches for Aerodynamic Shape Optimization Problems. PhD thesis, Universität Heidelberg (2007)
12. Gherman, I., Schulz, V.: Preconditioning of one-shot pseudo-timestepping methods for shape optimization. PAMM 5(1), 741–742 (2005)
13. Griewank, A.: Projected Hessians for preconditioning in one-step one-shot design optimization. Large Scale Nonlinear Optimization (2005) (to appear)
14. Hazra, S.B., Schulz, V.: Simultaneous pseudo-timestepping for aerodynamic shape optimization problems with state constraints. SIAM Journal on Scientific Computing 28(3), 1078–1099 (2006)
15. Hazra, S.B., Schulz, V., Brezillon, J.: Simultaneous pseudo-time stepping for 3D aerodynamic shape optimization. Forschungsbericht Nr. 05-2, Universität Trier (2005)
16. Hazra, S.B., Schulz, V.H., Brezillon, J., Gauger, N.R.: Aerodynamic shape optimization using simultaneous pseudo-timestepping. Journal of Computational Physics 204, 46–64 (2005)
17. Hicks, R.M., Henne, P.A.: Wing design by numerical optimization. Journal of Aircraft 15(7), 407–412 (1978)
18. Hintermüller, M., Schulz, V.: On the convergence of approximate reduced space iterations for optimization problems (in preparation)
19. Ito, K., Kunisch, K., Schulz, V., Gherman, I.: Approximate nullspace iterations for KKT systems in model based optimization. Forschungsbericht, Universität Trier, 06-5 (2006)
20. Jameson, A., Reuther, J.: Control theory based airfoil design using the Euler equations. In: AIAA Proceedings 94-4272-CP (1994)
21. Jaworski, A., Cusdin, P., Müller, J.-D.: Uniformly converging simultaneous time-stepping methods for optimal design. In: Schilling, R., Haase, W., Periaux, J., Baier, H., Bugeda, G. (eds.) Evolutionary and Deterministic Methods for Design, Optimization and Control with Applications to Industrial and Societal Problems, EUROGEN 2005, FLM, Munich (2005)
22. Nocedal, J., Wright, S.J.: Numerical Optimization. Springer Series in Operations Research. Springer, Heidelberg (1999)
23. Schulz, V.H.: Solving discretized optimization problems by partially reduced SQP methods. Computing and Visualization in Science 1, 83–96 (1998)
24. Ta'asan, S.: Pseudo-time methods for constrained optimization problems governed by PDE. ICASE Report No. 95-32 (1995)
25. Wild, J., Mertins, R., Quagliarella, D., Brezillon, J., Quest, J., Amoignon, O., Moens, F.: Applying numerical optimization to realistic high-lift design of transport aircraft – An overview of the aerodynamic design optimization investigation with the EUROLIFT II project. In: EUROGEN 2005 (September 2005)

Automatic Differentiation of FLOWer and MUGRIDO

Ralf Giering[1], Thomas Kaminski[1], Bernhard Eisfeld[2], Nicolas Gauger[2], Jochen Raddatz[2], and Lars Reimer[3]

[1] FastOpt, Hamburg
http://FastOpt.com
[2] DLR, Inst. for Aerodynamics and Flow Technology, Braunschweig
http://DLR.de
[3] Mechanics Department (LFM), RWTH Aachen
http://www.lufmech.rwth-aachen.de

1 Introduction

This chapter addresses the efficient computation of accurate sensitivity information in the aerodynamic design process. Mathematically, this sensitivity information is expressed by a derivative of a function that is defined via the numerical model of the aerodynamic system. This function links a number of independent variables to relevant target quantities such as lift, drag, or pitching moment. Figure 1 illustrates such a function definition via the process chain for a wing simulation. Independent parameters are the coordinates of the wing contours (see, e.g., [56]). In an automated design procedure, the space of wing coordinates can then be searched for a point that yields the optimum of the target quantity. The sensitivity of the target quantity with respect to the wing coordinates enables the use of powerful gradient algorithms for optimisation.

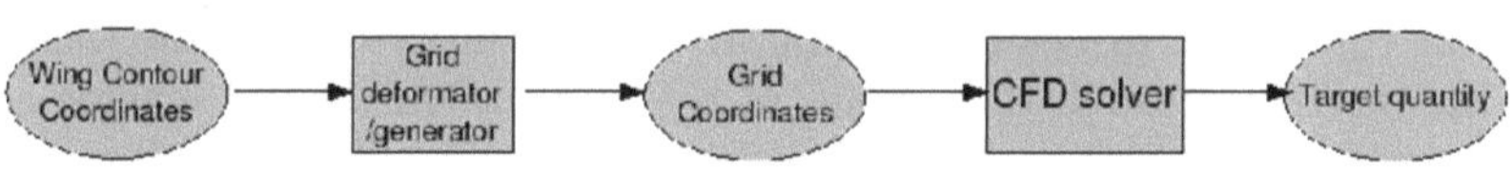

Fig. 1. Aerodynamic design chain. Oval boxes denote data and rectangular boxes numerical operations.

For extracting such sensitivity information there are two major approaches. The first approach, often referred to as continuous approach, applies perturbation theory [38] to the equations underlying the numerical model. This results in linearised equations, which are discretised and then numerically integrated by the so-called tangent code. The computational cost of running the tangent code is roughly proportional to the number of independent variables. For most problems this number is much larger than that of the target quantities. For computationally demanding numerical models, sensitivity computation is then only feasible

N. Kroll et al. (Eds.): MEGADESIGN and MegaOpt, NNFM 107, pp. 221–235.
springerlink.com

by means of adjoint code. In the continuous approach, adjoint code numerically integrates the discretisation of the adjoint version of the linearised model equations [39]. Lacking adjoint code, the inverse problem can only be tackled by reducing the complexity of the numerical model or the number of independent variables. The reduction of the design space is typically achieved via a parametrisation of the contour with only a few parameters. This would extend the chain of figure 1 by prepending the parametrisation, and the parameters would take the role of the independent variables and span the reduced design space. This regularisation of the inverse problem will typically yield a suboptimal solution, as the solution is to some degree prescribed by the parametrisation [37, 55].

The alternative route to sensitivity information applies Automatic Differentiation (AD, [27]). directly to the code of the original model: To generate the derivative code (tangent or adjoint code), the model code is decomposed into elementary functions (such as $+, -, \sin(\cdot)$), which (more or less) correspond to the individual statements in the code. These elementary functions are differentiated; this derivative is called local Jacobian. According to the chain rule, the product of the local Jacobians yields the derivative of the composite function. As opposed to derivative approximation by finite differences (also known as numerical differentiation), AD provides sensitivity information that is accurate within round-off error.

Like the continuous approach, AD can construct both tangent and adjoint codes. The tangent code uses the order in which the original model code evaluates the statements to evaluate the product of their local Jacobians. The adjoint code performs this evaluation in reverse order. In AD terminology, the tangent code operates in forward mode, and the adjoint code operates in reverse mode. Similar to the finite difference approximation, the computational resources needed in forward mode increase with the number of independent variables. In reverse mode, they are roughly proportional to the number of target quantities, but virtually independent of the number of independent variables.

The continuous approach involves the choice of a discretisation scheme for the adjoint equations. Typically it is not trivial to identify the scheme that yields, on the discretised level, the adjoint of the tangent. Any other scheme risks to produce sensitivity information that is inconsistent with the original code. This is of particular concern [50], when the adjoint sensitivity is used by an optimisation algorithm together with the target quantities provided by the original code.

AD can be carried out by an AD tool (for an overview see http://www.autodiff.org), but it is also common to apply the basic principles of derivative code construction [19, 52] by hand (see, e.g., [26, 58, 61]). The present paper describes AD of two Fortran codes that cover the design chain in figure 1: the DLR's RANS solver FLOWer [44], including a number of turbulence models, and the RWTH's flow grid deformator MUGRIDO [7, 33, 45]. For a number of CFD codes in Fortran, tangent (e.g. [5, 8, 9, 16, 35]) adjoint (e.g. [43]) and Hessian (e.g. [53]) codes have been generated by the AD tool ADIFOR [6], and adjoint codes (e.g. [31, 41]) by the AD tool Odyssée [46] and its successor

TAPENADE [30]. Cusdin and Müller [13] compare the performance of tangent and adjoint versions simple CFD codes that can be handled by three AD tools.

For the present application we use the commercial AD tool Transformation of Algorithms in Fortran (TAF [19]). TAF and its predecessor TAMC generated tangent, adjoint, and Hessian codes, for a long list of applications (over 150 papers), primarily for large codes from Earth Sciences (up to 300,000 lines of Fortran excluding comments). For the feasibility of most of these applications, flexibility and computational efficiency of the derivative code are crucial and were, thus, in the focus of TAF development. Applications to CFD codes started this decade and include design of aircraft [54], turbo machinery [29], or cabin ventilation [42] as well as aeroacoustics [12]. Industrial CFD applications encompass design of race cars and aircraft. Generation of efficient second derivative code [20] for an airfoil configuration has been demonstrated by [24].

Use of an AD tool such as TAF is also favourable regarding the maintenance of the derivative code for models that are under development. Once the model code is TAF-compliant, the maintenance of the derivative code can be automated: At least after small changes of the model, the corresponding adjoint, tangent, and Hessian code can generally be updated automatically. TAF-compliance means that the derivative code is both correct and efficient as generated by TAF without any user intervention after the generation process. The effort of achieving this compliance typically pays off rapidly via the automated derivative code maintenance. For examples of this new concept that uses TAF as integrated component of the modelling system see [1,28,36,42].

The remainder of this chapter is arranged as follows. Section 2 introduces the AD tool TAF and is followed by the descriptions of the AD process for FLOWer in section 3 and for MUGRIDO in section 4. Finally, section 5 draws conclusions.

2 TAF

Transformation of Algorithms in Fortran (TAF, [19]) is an AD tool for programmes written in Fortran 77-95. TAF operates as a source-to-source transformation tool. That is, from a given Fortran programme that evaluates a function, TAF generates a second Fortran programme that evaluates the function's derivative (gradient or Jacobian). TAF generates both forward and reverse mode derivative codes, i.e. tangent and adjoint models. In each mode, TAF can generate code to evaluate Jacobian times vector products or the full Jacobian. Second order derivative (Hessian) code is generated by invoking TAF twice. Typically, the most efficient strategy of obtaining second derivative information for a scalar-valued target quantity is the so-called forward over reverse mode of AD: TAF is invoked to generate the adjoint code, which afterwards is resubmitted to TAF to be differentiated in forward mode [20]. TAF is accessed via a secure shell connection to the FastOpt servers.

Another TAF feature [23] is Automatic Sparsity Detection (ASD), i.e. efficient determination of the sparsity structure of the Jacobian. This sparsity information can be important, because the Jacobian's sparsity pattern can be exploited to render the evaluation of the Jacobian more efficient (see, e.g., [36] for an application). In a CFD context ASD is of particular interest for evaluation of the sparse Jacobian representing the linearisation of a single solver iteration [14].

Recent TAF enhancements include basic support of parallel programming, namely the Message Passing Interface (MPI) and OpenMP (see [25,32] for large-scale applications) as well as a mode for generation of a divided adjoint, which allows interruption and restart of the adjoint model run (see [32] for details).

TAF performs an analysis of the data flow in the code to be differentiated, which determines the active/passive variables. Active variables [4, 19] are all variables that depend on the independent variables and have influence on the target quantities. All non-active variables are called passive variables. Derivative information needs only be propagated and stored for active variables.

Required variables are all variables whose values are needed to evaluate the local Jacobian. For example, in integrations of non-linear systems the trajectory is part of the required variables. Their values can be provided by recomputation or by storing them on disk or in memory in an initial integration and reading them in the course of the adjoint integration. Most efficient adjoint code uses a combination of both [19,21]. By default TAF inserts recomputations; automatic generation of a store/read scheme is triggered by TAF store directives. TAF can also generate a so-called checkpointing scheme [27] for particularly efficient use of disk/memory at the cost of an additional model integration.

For converging iterations, Christianson [10,11] suggests an efficient alternative adjoint (based on the implicit function theorem), which only uses the required values from the last iteration and, thus, compared to the general adjoint considerably reduces the resources required for storing/recomputing. TAF implements automatic generation of the Christianson scheme, triggered by a TAF loop directive [24]. This is another feature of high interest in aerodynamic simulations, as these often address steady problems.

In case there are pieces of source code missing (black box routines), e.g. library routines, the user can provide the relevant data flow information via TAF flow directives [22, 24]. TAF flow directives are also applied to include available derivative code into TAF-generated code, which is useful, e.g., in case of self-adjoint routines, as demonstrated by [25, 34].

3 Automatic Differentiation of FLOWer

FLOWer is a Reynolds-averaged Navier-Stokes (RANS) solver [44] developed and maintained by DLR and used by its scientific and industrial partners. Excluding comments, the FLOWer code comprises over 100,000 lines of Fortran 77 (see Tab. 1) and can be run in a large variety of configurations [2], with a suite of algebraic, one-, and two-equation turbulence models [15]. For an Euler configuration, an adjoint version derived via the continuous approach [17] was available to the project. The initial strategy was to couple TAF-generated adjoint turbulence code with the continuous adjoint of the FLOWer core. In the course of the project, however, it turned out to be favourable to apply TAF to the entire FLOWer code.

Table 1. Performance of FLOWer's derivative code

Component	# of code lines	memory	CPU	rel. accuracy
Primal	166,000	1	1	
TLM	268,000	≈ 2	≈ 3	$\approx 10^{-8}$
ADM steady	310,000	2–3	6–10	$\approx 10^{-5}$
ADM general	310,000	variable	<10	$\approx 10^{-8}$

When rendering the FLOWer code TAF-compliant, we met a number of challenges. One is the use of large super-arrays for an implementation of (pseudo) dynamic memory management, a typical feature of legacy codes designed in the pre-Fortran-90 period. Another challenge is the implementation of an error-exit procedure via goto statements in every routine, which considerably complicates the programme's control flow structure. For the adjoint, an efficient store/read scheme has been devised. From the TAF-compliant solver code, a tangent and two adjoint versions were generated. The tangent code is mainly an intermediate result and is used for verification of the two adjoints. The first adjoint version (general adjoint) uses a flexible checkpointing scheme (see section 2) that stores required values on disk and in memory. It provides the exact gradient for steady and unsteady computations. The second version of the adjoint (steady adjoint) assumes convergence of the solver to a steady flow (see section 2) and stores this flow in memory. As an example of TAF-generated code, the Appendix shows the adjoint of FLOWer's LEA k-ω routine [47].

The generated code has been verified for a 2d test configuration simulating the turbulent flow around a NACA 12 airfoil with 2000 iterations on a single fine grid. We evaluate the derivative of lift with respect to angle of attack. Figure 2 shows the relative difference of the general adjoint to the finite difference approximations for a range of finite difference intervals. Since we are running the evaluation in double precision with about 16 significant digits, a relative accuracy of the best finite difference approximation in the order of 10^{-8} is all we can expect. Lower accuracies usually indicate errors in the derivative code. The inaccuracy of the steady adjoint (see Tab. 1) is probably due to insufficient convergence of the primal integration. The relative difference between tangent and standard adjoint is in the order of 10^{-12}.

Tab. 1 lists the performances of the tangent and both adjoint versions for the test configuration with k-ω turbulence scheme [59, 60]. Owing to the flexibility of the checkpointing scheme (see section 2) the memory requirement for the general adjoint is variable. The CPU time is listed in multiples of primal solver integrations, and refers to the evaluation of the target function plus its derivative. For the adjoints this derivative refers to the full gradient, and for the tangent this refers to a directional derivative. CPU times vary with platform, compiler, and compiler options.

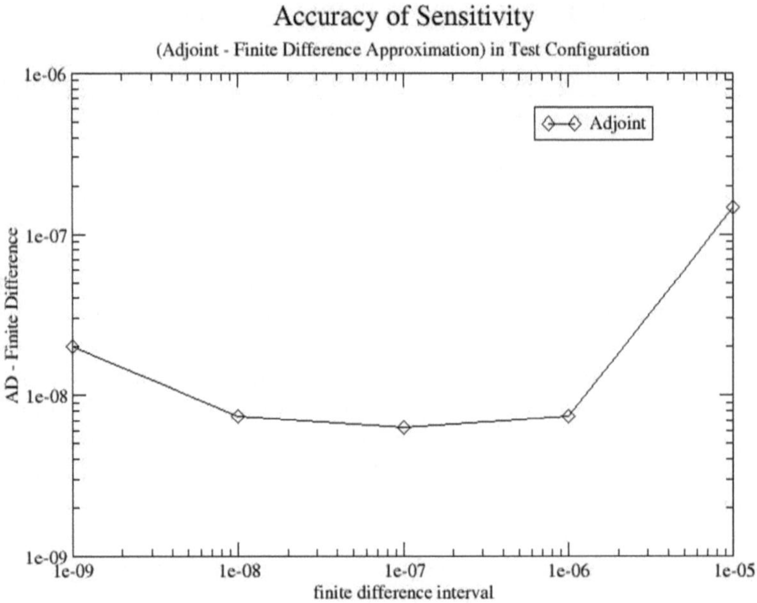

Fig. 2. Comparison of the derivative evaluated by the adjoint to finite difference approximations for a range of finite difference intervals

In addition to the Euler configuration, We have verified the derivative for the following five turbulence models:

- Baldwin and Lomax [3]
- Wilcox k-ω [59, 60]
- LLR k-ω [48]
- SST k-ω [40]
- LEA k-ω [47]

Curiously, for the one equation model of Spallart and Allmaras [51] the generated adjoint code produced a wrong gradient. We did not look into details but expect the problem is not too hard to identify and correct.

As an example of an adjoint sensitivity evaluation, figure 3 displays the sensitivity of lift with respect to angle of attack over the number of iterations in the adjoint solver. The computation uses the turbulence formulation according to Wilcox [59, 60].

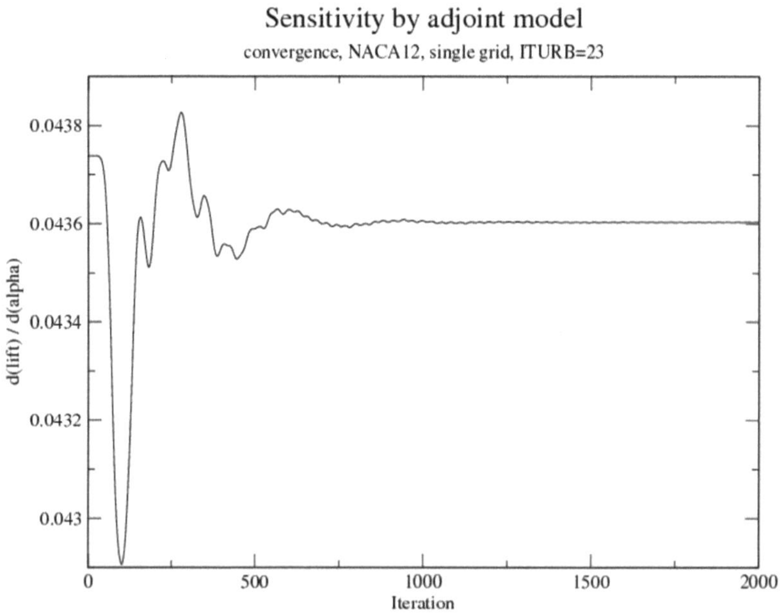

Fig. 3. Sensitivities of lift with respect to angle of attack computed by the adjoint of FLOWer with Wilcox [59,60] turbulence formulation

4 Automatic Differentiation of MUGRIDO

The Multiblock Grid Deformation Tool (MUGRIDO, [7,33,45]), was developed at RWTH Aachen. It can handle block-structured grid topologies, especially those used by FLOWer, and thus is well-suited for the design chain depicted in figure 1. MUGRIDO generates a fictitious beam framework by modelling block boundaries of the flow grid and a given percentage of additional grid lines as massless Timoshenko beams. After applying the Finite Element method the resulting linear system of equations is solved using the SPARSKIT utility for sparse matrices [49]. The right hand side to this system is supplied by deflections of the wetted surface relative to the undeformed grid. A well-shaped flow grid is finally reconstructed from the deformed beam framework by transfinite interpolation. MUGRIDO is written entirely in Fortran 77, and its coding concept is similar to FLOWer's, with the same super-array stucture for pseudo dynamic memory managment.

To demonstrate the applicability of TAF to a grid deformation tool, we rendered MUGRIDO TAF-compliant and generated its tangent in a fully automated procedure. For more efficient use of the tangent in the design chain, in addition to the standard (scalar) tangent, we also provided a vector mode version of the tangent, which simultaneously evaluates multiple directional derivatives. Tab. 2 lists the performance for the scalar tangent and the vector tangent for

Table 2. Performance of FLOWer's derivative code

Component	# of code lines	memory	CPU	rel. accuracy
Primal	25,000	1.0	1.0	
TLM scalar	42,800	2.1	1.0	$\approx 10^{-9}$
TLM 5 columns	43,400	3.4	1.2	$\approx 10^{-9}$
TLM 10 columns	43,400	5.8	1.5	$\approx 10^{-9}$
TLM 15 columns	43,400	8.3	2.2	$\approx 10^{-9}$

different numbers of directional derivatives. The CPU time refers to evaluation of the function plus the derivative. Increasing the number of directional derivatives does only marginally increase the CPU time. The agreement with the best finite difference approximation (last column) is excellent.

5 Conclusions

We demonstrated the feasibility of adjoint code generation for the CFD code FLOWer including a number of advanced turbulence models. The adjoint has been generated in two forms, one for steady simulations and a general one, which is suitable, e.g., for time-dependent simulations. The generated code is ready for applications, without any posterior modifications. The TAF-compliant FLOWer version is an excellent basis for further development of FLOWer, minimising the effort for updating the adjoint. The generated adjoint is efficient both in terms of memory usage and CPU time.

We also generated tangent code of the grid deformation tool MUGRIDO. The tangent is available in two forms, a scalar version for evaluation of a single directional derivate and a vector version for evaluation of multiple directional derivatives. The derivative code is highly efficient.

Many of the TAF algorithms can be ported without or with little modification to other programming languages. TAC++ [57] is the equivalent to TAF for differentiation of codes written in C(++). For a routine in the simplified Euler version of the DLR's RANS solver TAU [18], the tool generates highly efficient adjoint code in a fully automated procedure [57].

Appendix: Adjoint Code Example

Below we show the adjoint of LEA k-ω model [47], as an example of adjoint code generated by TAF. The declaration block, comment and blank lines are removed, to save space. Adjoint variables are denoted by the suffix `_ad`. The adjoint subroutine takes the sensitivity of `fmuet` (held in `fmuet_ad`) with respect to the target variable (e.g. lift) as input and propagates it back to the sensitivities of `r` (held in `r_ad`) and `shearvar`) (held in `shearvar_ad` with respect to the target variable. Note the recomputations before the nested loop and at the beginning of its kernel. For details in the generated code consult [19].

```
      subroutine turb26_ad( r, r_ad, swshear, shearvar, shearvar_ad,
     $fmuet_ad )
... (declarations, comments removed)
      help_h = epsma*1.e+8
      if (help_h .lt. 9.9999999999999e-31) then
        tolepsma = 9.9999999999999e-31
      else
        tolepsma = help_h
      endif
      twothird = 2./3.
      fothird = 4./3.
      cmu = 0.09
      c3 = 1.25
      c4 = 0.45
      do k = k2, 2, -1
        do j = j2, 2, -1
          do i = i2, 2, -1
            rho = r(i,j,k,1)
            help_j = r(i,j,k,it1)/rho
            if (help_j .ge. 0) then
              help_i = help_j
            else
              help_i = -help_j
            endif
            ka = help_i+tolepsma
            help_l = r(i,j,k,it2)/rho
            if (help_l .ge. 0) then
              help_k = help_l
            else
              help_k = -help_l
            endif
            om = help_k+tolepsma
            s = shearvar(i,j,k,1)/cmu/om
            st = shearvar(i,j,k,2)/cmu/om
            help_m = 1.5*st**1.7/(17.1+1.875*st**1.7)
            if (0.4 .lt. help_m) then
              c2 = help_m
            else
              c2 = 0.4
            endif
            arg1 = sqrt(0.8*s*s+0.2*st*st)
            arg2 = st*st/4.6225
            if (arg2 .lt. 1000.) then
              fact1 = 1.+0.95*(1.-tanh(arg2))
            else
              fact1 = 1.+0.95
            endif
            gr = 1./(1.6*fact1+st*st/(4.+1.83*arg1))
            beta1 = (fothird-c2)*gr*0.5
            beta2 = (2.-c4)*gr*0.5
            beta3 = (2.-c3)*gr
            xi2 = 0.5*beta2*beta2*s*s
            eta2 = 0.125*beta3*beta3*st*st
            cmust = beta1/(1.-twothird*eta2+2.*xi2)
            if (0.12 .gt. cmust) then
              help_n = cmust
            else
              help_n = 0.12
            endif
            if (0.04 .lt. help_n) then
              cmust = help_n
            else
              cmust = 0.04
            endif
            cmust_ad = cmust_ad+fmuet_ad(i,j,k)*(rho*ka/om/cmu)
            ka_ad = ka_ad+fmuet_ad(i,j,k)*(rho*cmust/om/cmu)
            om_ad = om_ad-fmuet_ad(i,j,k)*(rho*ka*cmust/(om*om)/cmu)
            rho_ad = rho_ad+fmuet_ad(i,j,k)*(ka*cmust/om/cmu)
```

```
      fmuet_ad(i,j,k) = 0.
      if (0.04 .lt. help_n) then
        help_n_ad = help_n_ad+cmust_ad
        cmust_ad = 0.
      else
        cmust_ad = 0.
      endif
      cmust = beta1/(1.-twothird*eta2+2.*xi2)
      if (0.12 .gt. cmust) then
        cmust_ad = cmust_ad+help_n_ad
        help_n_ad = 0.
      else
        help_n_ad = 0.
      endif
      beta1_ad = beta1_ad+cmust_ad/(1.-twothird*eta2+2.*xi2)
      eta2_ad = eta2_ad+cmust_ad*(beta1*twothird/((1.-twothird*
$eta2+2.*xi2)*(1.-twothird*eta2+2.*xi2)))
      xi2_ad = xi2_ad-cmust_ad*(2*beta1/((1.-twothird*eta2+2.*xi2)
$*(1.-twothird*eta2+2.*xi2)))
      cmust_ad = 0.
      beta3_ad = beta3_ad+0.25*eta2_ad*beta3*st*st
      st_ad = st_ad+0.25*eta2_ad*beta3*beta3*st
      eta2_ad = 0.
      beta2_ad = beta2_ad+xi2_ad*beta2*s*s
      s_ad = s_ad+xi2_ad*beta2*beta2*s
      xi2_ad = 0.
      gr_ad = gr_ad+beta3_ad*(2.-c3)
      beta3_ad = 0.
      gr_ad = gr_ad+0.5*beta2_ad*(2.-c4)
      beta2_ad = 0.
      c2_ad = c2_ad-0.5*beta1_ad*gr
      gr_ad = gr_ad+0.5*beta1_ad*(fothird-c2)
      beta1_ad = 0.
      arg1_ad = arg1_ad+gr_ad*(1.*(1.83*st*st/((4.+1.83*arg1)*(4.+
$1.83*arg1)))/((1.6*fact1+st*st/(4.+1.83*arg1))*(1.6*fact1+st*st/(
$4.+1.83*arg1))))
      fact1_ad = fact1_ad-gr_ad*(1.6/((1.6*fact1+st*st/(4.+1.83*
$arg1))*(1.6*fact1+st*st/(4.+1.83*arg1))))
      st_ad = st_ad-gr_ad*(1.*(2*st/(4.+1.83*arg1))/((1.6*fact1+
$st*st/(4.+1.83*arg1))*(1.6*fact1+st*st/(4.+1.83*arg1))))
      gr_ad = 0.
      if (arg2 .lt. 1000.) then
        arg2_ad = arg2_ad-0.95*fact1_ad*(1./cosh(arg2)**2)
        fact1_ad = 0.
      else
        fact1_ad = 0.
      endif
      st_ad = st_ad+arg2_ad*(2*st/4.6225)
      arg2_ad = 0.
      s_ad = s_ad+1.6*arg1_ad*1./(2.*sqrt(0.8*s*s+0.2*st*st))*s
      st_ad = st_ad+0.4*arg1_ad*1./(2.*sqrt(0.8*s*s+0.2*st*st))*st
      arg1_ad = 0.
      if (0.4 .lt. help_m) then
        help_m_ad = help_m_ad+c2_ad
        c2_ad = 0.
      else
        c2_ad = 0.
      endif
      st_ad = st_ad+help_m_ad*(2.55*st**0.7/(17.1+1.875*st**1.7)-
$3.1875*1.5*st**1.7*st**0.7/((17.1+1.875*st**1.7)*(17.1+1.875*st**
$1.7)))
      help_m_ad = 0.
      om_ad = om_ad-st_ad*(shearvar(i,j,k,2)/cmu/(om*om))
      shearvar_ad(i,j,k,2) = shearvar_ad(i,j,k,2)+st_ad*(1/cmu/om)
      st_ad = 0.
      om_ad = om_ad-s_ad*(shearvar(i,j,k,1)/cmu/(om*om))
      shearvar_ad(i,j,k,1) = shearvar_ad(i,j,k,1)+s_ad*(1/cmu/om)
      s_ad = 0.
```

```
        help_k_ad = help_k_ad+om_ad
        om_ad = 0.
        if (help_l .ge. 0) then
          help_l_ad = help_l_ad+help_k_ad
          help_k_ad = 0.
        else
          help_l_ad = help_l_ad-help_k_ad
          help_k_ad = 0.
        endif
        r_ad(i,j,k,it2) = r_ad(i,j,k,it2)+help_l_ad/rho
        rho_ad = rho_ad-help_l_ad*(r(i,j,k,it2)/(rho*rho))
        help_l_ad = 0.
        help_i_ad = help_i_ad+ka_ad
        ka_ad = 0.
        if (help_j .ge. 0) then
          help_j_ad = help_j_ad+help_i_ad
          help_i_ad = 0.
        else
          help_j_ad = help_j_ad-help_i_ad
          help_i_ad = 0.
        endif
        r_ad(i,j,k,it1) = r_ad(i,j,k,it1)+help_j_ad/rho
        rho_ad = rho_ad-help_j_ad*(r(i,j,k,it1)/(rho*rho))
        help_j_ad = 0.
        r_ad(i,j,k,1) = r_ad(i,j,k,1)+rho_ad
        rho_ad = 0.
      end do
    end do
  end do
  end subroutine turb26_ad
```

References

1. Adcroft, A., Campin, J.M., Heimbach, P., Hill, C., Marshall, J.: The MITgcm. Online documentation. Massachusetts Institute of Technology, USA (2002)
2. Aumann, P., Bartelheimer, W., Bleecke, H., Eisfeld, J., Lieser, J., Heinrich, R., Kroll, N., Kuntz, M., Monsen, E., Raddatz, J., Reisch, U., Roll, B.: FLOWer Installation and USER Handbook Release 116. Technical Report MEGAFLOW-1001, DLR (2000)
3. Baldwin, B., Lomax, H.: Thin-layer approximation and algebraic model for separated turbulent flows. IAAA Paper 1978-0257, AIAA, Reston Va, USA (1978)
4. Bischof, C., Carle, A., Khademi, P., Mauer, A.: ADIFOR 2.0: Automatic differentiation of Fortran 77 programs. IEEE Computational Science & Engineering 3(3), 18–32 (1996)
5. Bischof, C.H., Bücker, H.M., Lang, B., Rasch, A., Slusanschi, E.: Efficient and accurate derivatives for a software process chain in airfoil shape optimization. Technical Report RWTH-CS-SC-02-06, Institute for Scientific Computing, Aachen University of Technology, Aachen (2002)
6. Bischof, C.H., Carle, A., Corliss, G.F., Griewank, A., Hovland, P.D.: ADIFOR: Generating derivative codes from Fortran programs. Scientific Programming 1, 11–29 (1992)
7. Boucke, A.: Kopplungswerkzeuge für aeroelastische Simulationen. PhD thesis, RWTH Aachen (2003)
8. Bücker, H.M., Lang, B., Rasch, A., Bischof, C.H.: Computation of sensitivity information for aircraft design by automatic differentiation. In: Sloot, P.M.A., Tan, C.J.K., Dongarra, J.J., Hoekstra, A.G. (eds.) ICCS-ComputSci 2002. LNCS, vol. 2330, pp. 1069–1076. Springer, Heidelberg (2002)

9. Carle, A., Green, L., Bischof, C.H., Newman, P.: Applications of automatic differentiation in CFD. In: Proceedings of the 25th AIAA Fluid Dynamics Conference, AIAA Paper 94-2197, American Institute of Aeronautics and Astronautics (1994)
10. Christianson, B.: Reverse accumulation and attractive fixed points. Optimization Methods and Software 3, 311–326 (1994)
11. Christianson, B.: Reverse accumulation and implicit functions. Optimization Methods and Software 9(4), 307–322 (1998)
12. Collis, S.S., Ghayour, K., Heinkenschloss, M., Ulbrich, M., Ulbrich, S.: Towards Adjoint-Based Methods for Aeroacoustic Control. IAAA Paper 2001-0821, AIAA, Reston Va, USA (2001)
13. Cusdin, P., Müller, J.-D.: Improving the performance of code generated by automatic differentiation. Technical Report QUB-SAE-03-04, QUB School of Aeronautical Engineering (2003)
14. Dwight, R., et al.: Development of Adjoint Methods for Hybrid RANS Solver TAU. In: Kroll, N., et al. (eds.) MEGADESIGN and MegaOpt. NNFM, vol. 107. Springer, Heidelberg (2008)
15. Eisfeld, B.: Turbulence Models in FLOWer. In: Kroll, N., Fassbender, J.K. (eds.) MEGAFLOW- Numerical Flow Simulation for Aircraft Design. Notes on Numerical Fluid Mechanics and Multidisciplinary Design, vol. 89, pp. 63–77. Springer, Heidelberg (2005)
16. Forth, S.A., Evans, T.P.: Aerofoil Optimisation via AD of a Multigrid Cell-Vertex Euler Flow Solver. In: Corliss, G., Faure, C., Griewank, A., Hascoët, L., Naumann, U. (eds.) Automatic Differentiation: From Simulation to Optimization. Computer and Information Science, pp. 153–160. Springer, New York (2001)
17. Gauger, N.R.: Das Adjungiertenverfahren in der aerodynamischen Formoptimierung. PhD thesis, TU Braunschweig (2004)
18. Gerhold, T.: Overview of the hybrid rans code tau. In: Kroll, N., Fassbender, J.K. (eds.) MEGAFLOW- Numerical Flow Simulation for Aircraft Design. Notes on Numerical Fluid Mechanics and Multidisciplinary Design, vol. 89, pp. 81–92. Springer, Heidelberg (2005)
19. Giering, R., Kaminski, T.: Recipes for Adjoint Code Construction. ACM Trans. Math. Software 24, 437–474 (1998)
20. Giering, R., Kaminski, T.: Using TAMC to generate efficient adjoint code: Comparison of automatically generated code for evaluation of first and second order derivatives to hand written code from the minpack-2 collection. In: Faure, C. (ed.) Automatic Differentiation for Adjoint Code Generation, INRIA, Sophia Antipolis, France, pp. 31–37 (1998)
21. Giering, R., Kaminski, T.: Recomputations in reverse mode AD. In: Corliss, G., Griewank, A., Fauré, C., Hascoet, L., Naumann, U. (eds.) Automatic Differentiation of Algorithms: From Simulation to Optimization, ch. 33, pp. 283–291. Springer, Heidelberg (2002)
22. Giering, R., Kaminski, T.: Applying TAF to generate efficient derivative code of Fortran 77-95 programs. PAMM 2(1), 54–57 (2003)
23. Giering, R., Kaminski, T.: Automatic sparsity detection implemented as a source-to-source transformation. In: Alexandrov, V.N., van Albada, G.D., Sloot, P.M.A., Dongarra, J. (eds.) ICCS 2006. LNCS, vol. 3994, pp. 591–598. Springer, Heidelberg (2006)
24. Giering, R., Kaminski, T., Slawig, T.: Generating Efficient Derivative Code with TAF: Adjoint and Tangent Linear Euler Flow Around an Airfoil. Future Generation Computer Systems 21(8), 1345–1355 (2005)

25. Giering, R., Kaminski, T., Todling, R., Errico, R., Gelaro, R., Winslow, N.: Generating tangent linear and adjoint versions of NASA/GMAO's Fortran-90 global weather forecast model. In: Bücker, H.M., Corliss, G., Hovland, P., Naumann, U., Norris, B. (eds.) Automatic Differentiation: Applications, Theory, and Tools. Lecture Notes in Computational Science and Engineering. Springer, Heidelberg (2005)
26. Giles, M.B., Duta, M.C., Mueller, J.D., Pierce, N.: Algorithm developments for discrete adjoint methods. AIAA Journal 41(2), 198–205 (2003)
27. Griewank, A.: Achieving logarithmic growth of temporal and spatial complexity in reverse automatic differentiation. Optimization Methods and Software 1, 35–54 (1992)
28. Griffies, S.M., Harrison, M.J., Pacanowski, R.C., Rosati, A.: The FMS MOM4-beta User Guide. Technical report, NOAA/Geophysical Fluid Dynamics Laboratory (2002)
29. Hall, K.C., Thomas, J.P.: Sensitivity analysis of coupled aerodynamic/structural dynamic behavior of blade rows. In: Extended Abstract for the 7th National Turbine Engine High Cycle Fatigue (HCF) Conference, Palm Beach Gardens, Florida, May 14-17 (2002)
30. Hascoët, L., Pascual, V.: TAPENADE 2.1 user's guide. Rapport technique 300, INRIA, Sophia Antipolis (2004)
31. Hascoët, L., Vázquez, M., Dervieux, A.: Automatic differentiation for optimum design, applied to sonic boom reduction. In: Kumar, V., Gavrilova, M.L., Tan, C.J.K., L'Ecuyer, P. (eds.) ICCSA 2003, Part II. LNCS, vol. 2668, pp. 85–94. Springer, Heidelberg (2003)
32. Heimbach, P., Hill, C., Giering, R.: An efficient exact adjoint of the parallel MIT general circulation model, generated via automatic differentiation. Future Generation Computer Systems 21(8), 1356–1371 (2005)
33. Hesse, M.: Entwicklung eines automatischen Gitterdeformationsalgorithmus zur Stroemungsberechnung um komplexe Konfiguration auf Hexaeder-Netzen. PhD thesis, RWTH Aachen (2006)
34. Hinze, M., Slawig, T.: Adjoint gradients compared to gradients from algorithmic differentiation in instataneous control of the Navier-Stokes equations. Optimization Methods & Software 18(3), 299–315 (2003)
35. Hovland, P.D., Mohammadi, B., Bischof, C.H.: Automatic differentiation of Navier-Stokes computations. Technical Report MCS-P687-0997, Argonne National Laboratory (1997)
36. Kaminski, T., Giering, R., Scholze, M., Rayner, P., Knorr, W.: An example of an automatic differentiation-based modelling system. In: Kumar, V., Gavrilova, M.L., Tan, C.J.K., L'Ecuyer, P. (eds.) ICCSA 2003, Part II. LNCS, vol. 2668, pp. 95–104. Springer, Heidelberg (2003)
37. Kaminski, T., Heimann, M.: Inverse modeling of atmospheric carbon dioxide fluxes. Science 294(5541), 259 (2001)
38. Kato, T.: Perturbation theory for linear operators. Springer, Berlin (1966)
39. Marchuk, G.I.: Adjoint Equations and Analysis of Complex Systems. Kluwer, Dordrecht (1995)
40. Menter, F.R.: Two-equation eddy-viscosity turbulence models for engineering applications. AIAA Journal 32(18), 1598–1605 (1994)

41. Mohammadi, B., Malé, J.M., Rostaing-Schmidt, N.: Automatic differentiation in direct and reverse modes: Application to optimum shapes design in fluid mechanics. In: Berz, M., Bischof, C.H., Corliss, G.F., Griewank, A. (eds.) Computational Differentiation: Techniques, Applications, and Tools, pp. 309–318. SIAM, Philadelphia (1996)
42. Othmer, C., Kaminski, T., Giering, R.: Computation of topological sensitivities in fluid dynamics: Cost function versatility. In: Wesseling, P., nate, E.O., Périaux, J. (eds.) ECCOMAS CFD 2006, TU Delft (2006)
43. Park, M.A., Green, L.L., Montgomery, R.C., Raney, D.L.: Determination of Stability and Control Derivatives Using Computational Fluid Dynamics and Automatic Differentiation. IAAA Paper 1999-3136, AIAA, Reston Va, USA (1999)
44. Raddatz, J., Fassbender, J.: Block Structured Navier-Stokes Solver FLOWer. In: Kroll, N., Fassbender, J.K. (eds.) MEGAFLOW- Numerical Flow Simulation for Aircraft Design. Notes on Numerical Fluid Mechanics and Multidisciplinary Design, vol. 89, pp. 27–44. Springer, Heidelberg (2005)
45. Reimer, L., Hesse, M.: Kurzdokumentation des Mehrblock-Gitterdeformationsverfahrens MUGRIDO. Technical report, RWTH Aachen (2006)
46. Rostaing, N., Dalmas, S., Galligo, A.: Automatic differentiation in Odyssée. Tellus 45A, 558–568 (1993)
47. Rung, T., Luebcke, H., Franke, M., Xue, L., Thiele, F., Fu, S.: Assessment of explicit algebraic stress models in transonic flows. In: Proceedings of the 4th International Symposium on Engineering Turbulence Modelling and Measurements, Ajaccio, France, May 24-26, pp. 659–668 (1999)
48. Rung, T., Thiele, F.: Computational modelling of complex boundary-layer flows. In: Proceedings of the 9th Int. Symp. on Transport Phenomena in Thermal-Fluid Engineering, Singapore (1996)
49. Saad, Y.: Sparskit: A Basic Tool Kit for Sparse Matrix Computation (1994)
50. Shah, P.: Application of adjoint equations to estimation of parameters in distributed dynamic systems. In: Griewank, A., Corliss, G.F. (eds.) Automatic Differentiation of Algorithms: Theory, Implementation, and Application, pp. 181–190. SIAM, Philadelphia (1991)
51. Spallart, P.R., Allmaras, S.R.: A One-Equation Model for Aerodynamic Flows. AIAA Journal 92(2) (1992)
52. Talagrand, O.: The use of adjoint equations in numerical modelling of the atmospheric circulation. In: Griewank, A., Corliss, G.F. (eds.) Automatic Differentiation of Algorithms: Theory, Implementation, and Application, pp. 169–180. SIAM, Philadelphia (1991)
53. Taylor III, A.C., Green, L.L., Newman, P.A., Putko, M.M.: Some Advanced Concepts in Discrete Aerodynamic Sensitivity Analysis. IAAA Paper 2001-2529, AIAA, Reston Va, USA (2001)
54. Thomas, J.P., Hall, K.C., Dowell, E.H.: A discrete adjoint approach for modeling unsteady aerodynamic design sensitivities. AIAA Journal 43(9), 1931–1936 (2005)
55. Trampert, J., Snieder, R.: Model estimations biased by truncated expansions: Possible artifacts in seismic tomography. Science 271, 1257–1260 (1996)
56. Ulbrich, S.: Optimal Control of Nonlinear Hyperbolic Conservation Laws with Source Terms, Habilitationsschrift. Fakultät für Mathematik, Technische Universität München, Germany (2002)

57. Voßbeck, M., Giering, R., Kaminski, T.: Development and First Applications of TAC++. In: Bischof, C., Bücker, H.M., Hovland, P.D., Naumann, U., Utke, J. (eds.) Advances in Automatic Differentiation, Berlin. Lecture Notes in Computational Science and Engineering. Springer, Heidelberg (2008) (to appear)
58. Weaver, A.T., Vialard, J., Anderson, D.L.T.: Three-and Four-Dimensional Variational Assimilation with a General Circulation Model of the Tropical Pacific Ocean. Part I: Formulation, Internal Diagnostics, and Consistency Checks. Monthly Weather Review 131(7), 1360–1378 (2003)
59. Wilcox, D.C.: Reassessment of the scale-determining equation for advanced turbulence models. AIAA, Aerospace Sciences Meeting 26, 1299–1310 (1988)
60. Wilcox, D.C.: Turbulence Modeling for CFD, DCW Industries. Inc., La Canada, California (1993)
61. Zhu, J., Kamachi, M.: The Role of Time Step Size in Numerical Stability of Tangent Linear Models. Monthly Weather Review 128(5), 1562–1572 (2000)

Adjoint Methods for Coupled CFD-CSM Optimization

Nicolas R. Gauger[1,2] and Antonio Fazzolari[1]

[1] German Aerospace Center (DLR),
Institute of Aerodynamics and Flow Technology,
Lilienthalplatz 7, 38108 Braunschweig, Germany
Nicolas.Gauger@dlr.de
[2] Humboldt University Berlin, Department of Mathematics,
Unter den Linden 6, 10099 Berlin, Germany

Summary. Multi-disciplinary analysis is necessary to reach physically meaningful optimum designs. For aero-structural shape optimization this means coupling two disciplines – aerodynamics and structural mechanics. In this paper the sensitivity evaluation for aerodynamic shape optimization is considered, while taking into account the static aeroelastic effects introduced by the variations in the aerodynamic forces, which are associated with changes in the aerodynamic shape. Due to the high computational cost of a finite difference evaluation step for such a coupled problem, an extension of the adjoint approach to aeroelasticity is necessary for an efficient calculation of the sensitivities. The theory, implementation, validation and application of such a method in the multi-disciplinary design optimization (MDO) context described above, is presented.

1 Introduction

The use of successively performed single-disciplinary optimizations in case of a multi-disciplinary optimization problem is not only inefficient but in some cases has been shown to lead to wrong, non-optimal designs [9]. On the computational structural mechanics (CSM) side we use the commercial solver MSC Nastran and on the computational fluid dynamics (CFD) side the flow solver FLOWer [8] from DLR. Although multi-disciplinary optimization is possible with classical approaches for sensitivity evaluation by means of finite differences, this method is extremely expensive in terms of calculation time, requiring the reiterated solution of the coupled problem for every design variable.

A new approach that allows the evaluation of the gradient with low computational cost takes advantage of the so-called adjoint formulation of the multi-disciplinary optimization problem [9,10]. Therefore, the FLOWer adjoint option has been coupled with the structure solver MSC Nastran for an efficient coupled aero-structure adjoint solver. In the following section we will derive the needed cross terms for the aero-structure coupling. Here, we exploit the symmetry of the structural stiffness matrix.

Finally, we will show the implementation, validation and application of the presented adjoint aero-structure approach.

N. Kroll et al. (Eds.): MEGADESIGN and MegaOpt, NNFM 107, pp. 237–246.
springerlink.com

2 Adjoint Formulation for Aero-Structure Coupling

The derivation of the adjoint equations in case of a multi-disciplinary problem is similar to what has been carried out for the pure aerodynamic case [1, 5, 6, 7], with the difference that we will end up with a dual adjoint variable for each set of state variables of the problem. An adjoint formulation is possible for any problem involving the calculation of the gradient of a function of one or more sets of variables obeying one or more constraint equations. We will restrict ourselves to the case of two sets: one represents the flow field variables w, the other the structure nodal displacement field Z. Furthermore, $I(X, w, Z)$ denotes the aerodynamic cost function of the aero-structure coupled optimization problem. Then, the gradient with respect to the vector of design variables X, which describes the aerodynamic shape, takes the form

$$\frac{\mathrm{d}I}{\mathrm{d}X} = \frac{\partial I}{\partial X} + \frac{\partial I}{\partial w}\frac{\partial w}{\partial X} + \frac{\partial I}{\partial Z}\frac{\partial Z}{\partial X} , \tag{1}$$

or, in terms of differentials

$$\delta I = \frac{\partial I}{\partial X}\delta X + \frac{\partial I}{\partial w}\delta w + \frac{\partial I}{\partial Z}\delta Z . \tag{2}$$

The fields (w, Z) are the solution of the system of partial differential equations

$$R(X, w, Z) = 0 , \tag{3}$$

$$S(X, w, Z) = 0 , \tag{4}$$

being (3) the flow and (4) the structural equations. We take the first variation of the PDEs. This yields

$$\delta R = \frac{\partial R}{\partial X}\delta X + \frac{\partial R}{\partial w}\delta w + \frac{\partial R}{\partial Z}\delta Z = 0 , \tag{5}$$

$$\delta S = \frac{\partial S}{\partial X}\delta X + \frac{\partial S}{\partial w}\delta w + \frac{\partial S}{\partial Z}\delta Z = 0 . \tag{6}$$

We multiply (5) and (6) with the Lagrange multipliers ψ and ϕ respectively and add the result to the expression for the differential increment of I in terms of the differentials of the independent set (X, w, Z), obtaining

$$\begin{aligned}\delta I = &\left(\frac{\partial I}{\partial X} + \psi^{\mathrm{T}}\frac{\partial R}{\partial X} + \phi^{\mathrm{T}}\frac{\partial S}{\partial X}\right)\delta X + \\ &\left(\frac{\partial I}{\partial w} + \psi^{\mathrm{T}}\frac{\partial R}{\partial w} + \phi^{\mathrm{T}}\frac{\partial S}{\partial w}\right)\delta w + \left(\frac{\partial I}{\partial Z} + \psi^{\mathrm{T}}\frac{\partial R}{\partial Z} + \phi^{\mathrm{T}}\frac{\partial S}{\partial Z}\right)\delta Z .\end{aligned} \tag{7}$$

Since we want to avoid recalculation of the (w, Z) fields, we cancel the terms in δw and δZ from δI by imposing the fields ϕ and ψ to be the solution of the equations

$$\left(\frac{\partial I}{\partial w} + \psi^{\mathrm{T}}\frac{\partial R}{\partial w} + \phi^{\mathrm{T}}\frac{\partial S}{\partial w}\right) = 0 , \tag{8}$$

$$\left(\frac{\partial I}{\partial Z} + \psi^{\mathrm{T}}\frac{\partial R}{\partial Z} + \phi^{\mathrm{T}}\frac{\partial S}{\partial Z}\right) = 0 . \tag{9}$$

These are the adjoint equations for the problem of coupled aeroelasticity. After their solution, the gradient can be recovered from the expression

$$\delta I = \left(\frac{\partial I}{\partial X} + \psi^{\mathrm{T}}\frac{\partial R}{\partial X} + \phi^{\mathrm{T}}\frac{\partial S}{\partial X}\right)\delta X . \tag{10}$$

We can assume the cost function to be a functional in the form

$$I(X, w, Z) = \int_V i(X, w, Z)\mathrm{d}V . \tag{11}$$

For convenience reasons we restrict ourselves to 2D drag (inviscid) reduction cases. The extension to 3D and other optimization cases is straight forward. In the case of drag reduction we now have

$$i(X, w, Z) = \frac{C_{\mathrm{p}}}{C_{\mathrm{ref}}}(n_x \cos\alpha + n_y \sin\alpha)\delta(\eta) , \tag{12}$$

with $\delta(\eta)$ being the Dirac delta function and $\eta = 0$ the equation defining the airfoil shape in the body fitted coordinates (ξ, η). For the Dirac delta function under integration the following equation holds

$$\int \delta(\eta) f(\eta)\mathrm{d}\eta = f(0) . \tag{13}$$

In the context of (11), it reduces the volume integral to a surface integral. We suppose that the fluid obeys the Euler equations, which in body fitted coordinates take the form

$$\frac{\partial F}{\partial \xi} + \frac{\partial G}{\partial \eta} = 0 , \tag{14}$$

where the transformed F, G are appropriate combinations of f and g, e.g.

$$F = J\frac{\partial \xi}{\partial x} f + J\frac{\partial \xi}{\partial y} g = J\begin{bmatrix} \varrho U \\ \varrho u U + \frac{\partial \xi}{\partial x} p \\ \varrho v U + \frac{\partial \xi}{\partial y} p \\ \varrho H U \end{bmatrix} . \tag{15}$$

Since our cost function I is of the form shown in (11), we have to, as a first step, formulate (8) and (9) in an appropriate way, using the following property

$$\delta I(X,w,Z) = \int\limits_V \delta i(X,w,Z)\mathrm{d}V$$
$$\int\limits_V \left(\frac{\partial i(X,w,Z)}{\partial X}\delta X + \frac{\partial i(X,w,Z)}{\partial w}\delta w + \frac{\partial i(X,w,Z)}{\partial Z}\delta Z \right) \mathrm{d}V \,. \tag{16}$$

The derivation is identical to what has already been seen, and gives the adjoint equations

$$\int\limits_V \left(\frac{\partial i}{\partial w} + \psi^{\mathrm{T}} \frac{\partial R}{\partial w} + \phi^{\mathrm{T}} \frac{\partial S}{\partial w} \right) \mathrm{d}V = 0 \,, \tag{17}$$

$$\int\limits_V \left(\frac{\partial i}{\partial Z} + \psi^{\mathrm{T}} \frac{\partial R}{\partial Z} + \phi^{\mathrm{T}} \frac{\partial S}{\partial Z} \right) \mathrm{d}V = 0 \,. \tag{18}$$

And for the gradient we get

$$\delta I(X,w,Z) =$$
$$\int\limits_V \left(\frac{\partial i(X,w,Z)}{\partial X}\delta X + \psi^{\mathrm{T}} \frac{\partial R(X,w,Z)}{\partial X}\delta X + \phi^{\mathrm{T}} \frac{\partial S(X,w,Z)}{\partial X}\delta X \right) \mathrm{d}V \,. \tag{19}$$

It can be shown that (17) is equivalent to the equation

$$\int\limits_V \left(\left(\frac{\partial \psi}{\partial \xi} \right)^{\mathrm{T}} \frac{\partial F}{\partial w} + \left(\frac{\partial \psi}{\partial \eta} \right)^{\mathrm{T}} \frac{\partial G}{\partial w} \right) \mathrm{d}V = 0 \tag{20}$$

and the boundary condition (in the case of the drag)

$$\psi_2 n_x + \psi_3 n_y + n_x \cos(\alpha) + n_y \sin(\alpha) - \mathbf{n}^{\mathrm{T}} \phi = 0 \,. \tag{21}$$

Note that the structural adjoint variables appear only in the boundary condition (21), while the adjoint flow equation (20) is unchanged. This implies that in order to implement the coupling, only the boundary condition treatment in the FLOWer code has to be modified. Equation (9) represents the structural adjoint equation and its boundary conditions. The structural equation reads, in the case of linear elasticity

$$S(X,w,Z) = K \cdot Z - a = 0 \,, \tag{22}$$

where K is the symmetric stiffness matrix and a is the aerodynamic force. The derivative ${\partial S}/{\partial Z}$ in (9) can thus be replaced by K and the product $\phi^{\mathrm{T}} K$ by $K\phi$. In this way, the same solver can be used for the structural direct and adjoint equation, with different boundary conditions, given by the first and second term in (9). The first term is reduced to a surface integral by the presence of the Dirac delta function, giving a vector defined by

$$V_i = \frac{\partial \int_S I(X,w,Z)\mathrm{d}S}{\partial Z_i} \,, \tag{23}$$

that is the derivative of the cost function with respect to a structural degree of freedom. The second term, namely

$$\int_V \left(\psi^{\mathrm{T}} \frac{\partial R}{\partial Z}\right) \mathrm{d}V \,, \tag{24}$$

represents the integral of the scalar product of the adjoint field ψ and the partial derivative of the flow operator $R(X, w, Z)$ with respect to a structural degree of freedom, thus keeping the flow field and the design variables constant. It is evaluated by making use of the finite-volume formulation implemented in FLOWer. A similar term appears in the expression for the gradient (19), which explicitly becomes

$$\frac{\mathrm{d}I}{\mathrm{d}X} = \frac{\partial I}{\partial X} + \int_V \left(\psi^{\mathrm{T}} \frac{\partial R}{\partial X}\right) \mathrm{d}V + \int_V \left(\phi^{\mathrm{T}} \frac{\partial S}{\partial X}\right) \mathrm{d}V \,. \tag{25}$$

The third term reduces to the surface integral of the adjoint field ϕ multiplied by the term

$$\frac{\partial S}{\partial X} = \frac{\partial K}{\partial X} Z - \frac{\partial a}{\partial X} \,. \tag{26}$$

Of the two terms on the right hand side, the first has been neglected, which is equivalent to assuming that shape deformations do not act on the structural mesh and thus on the stiffness matrix.

More details on the adjoint aero-structure approach and its derivation can be found in [2, 3, 4].

3 Implementation

In order to solve the coupled equations of the aero-structural system, a sequential staggered method has been implemented, where forces are transferred from the flow mesh to the structure mesh and give the nodal loads, and deflections are transferred back from the structure mesh to the flow mesh which is consequently deformed. The flow around the body described by the Euler equation is solved by the DLR solver FLOWer, while the structural problem is solved by MSC Nastran. The transfer of information between the two meshes is managed by a module developed in-house based on B-spline volume interpolation. Typically, 20 exchanges of information between the two codes are more than enough to reach a converged aeroelastic solution, as shown in Fig. 1.

The same staggered scheme has been used to solve the systems of the coupled adjoint equations, with the difference that now only adjoint deflections are interpolated from the structural mesh to the flow mesh, in order to evaluate the boundary condition (21) for the new adjoint flow computation. Each 100 steps of the adjoint flow solver, boundary conditions coming from the coupling are exchanged and updated, as shown in Fig. 2.

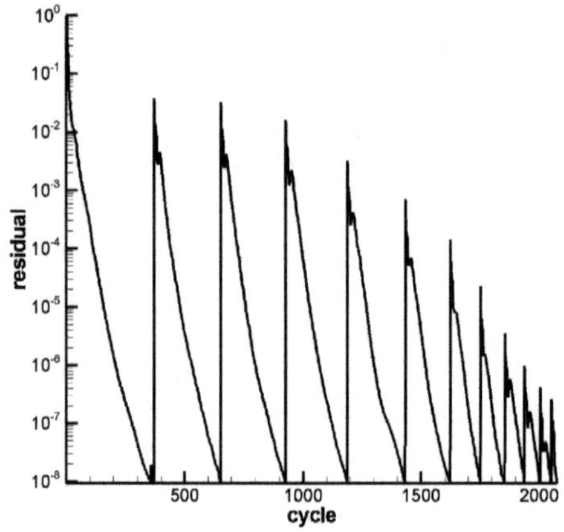

Fig. 1. Plot of residual (log scale) of flow equation during coupled computation (multigrid is used): AMP wing, $Ma_\infty = 0.78$, $\alpha = 2.83°$, 2-block structured grid of about 140,000 nodes each

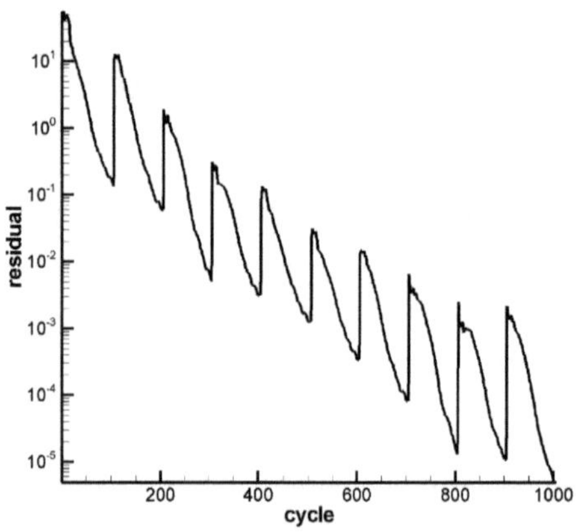

Fig. 2. Plot of residual (log scale) of adjoint flow equation during coupled computation: AMP wing, $Ma_\infty = 0.78$, $\alpha = 2.83°$, 2-block structured grid of about 140,000 nodes each. Each 100 iterations, the boundary conditions of the adjoint flow solver are updated.

4 Validation and Application

The validation of both the theory and the implementation of the adjoint formulation for the aeroelastic system has been achieved by comparison with the finite difference method.

As test case for the validation the AMP wing has been chosen. The structure has been modelled with a simplified model of 126 nodes, all lying on the wing

Fig. 3. Wing structure model

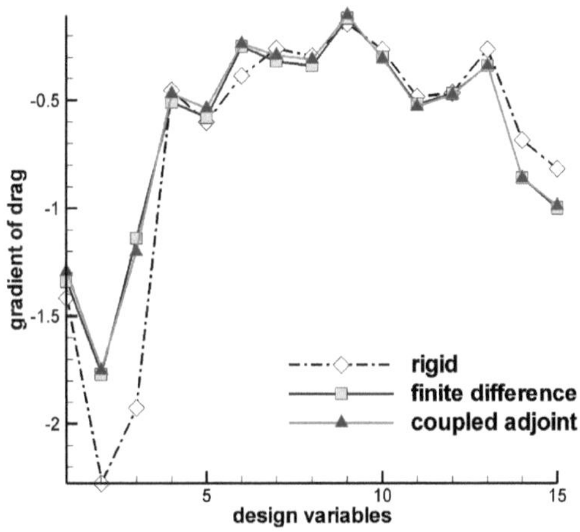

Fig. 4. Validation of the aero-structural coupled adjoint with finite differences (AMP wing, $\mathrm{Ma}_\infty = 0.78$ and $\alpha = 2.83°$)

surface, connected by 422 tria/quad shell and 198 beam elements (see Fig. 3). Such a model, unlike its fluid counterpart, is not state of the art, but is sufficient to demonstrate the features of the method. In order to underline the effect of aeroelasticity, the thickness of the beam elements of the wing has been tuned to reach a deflection of about 10% of the wing span at the wing tip.

Making use of the finite difference method, the gradient of the drag with respect to the shape parameters has been calculated, this time including the effect of aeroelastic interaction. This means that after a deformation of the jig shape (undeflected shape), an aeroelastic coupling was carried out and a stationary state was reached, as shown in Fig. 1. This operation was repeated for every design parameter.

On the other hand, after the solution of the coupled adjoint equations, both the flow and structural adjoint fields have been used to reconstruct the gradient

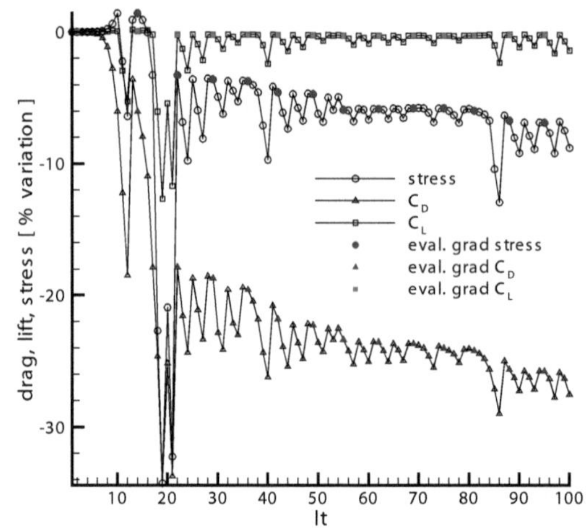

Fig. 5. Optimization history for the drag reduction by constant lift while taking into account the static deformation of the AMP wing (Ma = 0.78, $\alpha = 2.83°$). Free-form deformation with 240 design variables was used for parameterization and feasible directions was used as optimization strategy.

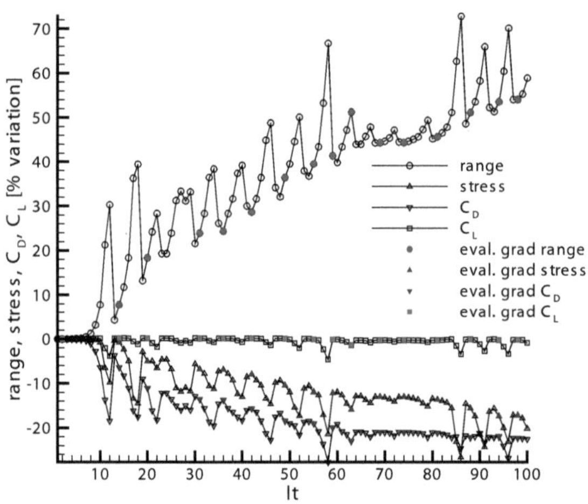

Fig. 6. Optimization history for the range maximization by constant lift while taking into account the static deformation of the AMP wing (Ma = 0.78, $\alpha = 2.83°$). Free-form deformation with 240 design variables was used for parameterization and feasible directions was used as optimization strategy.

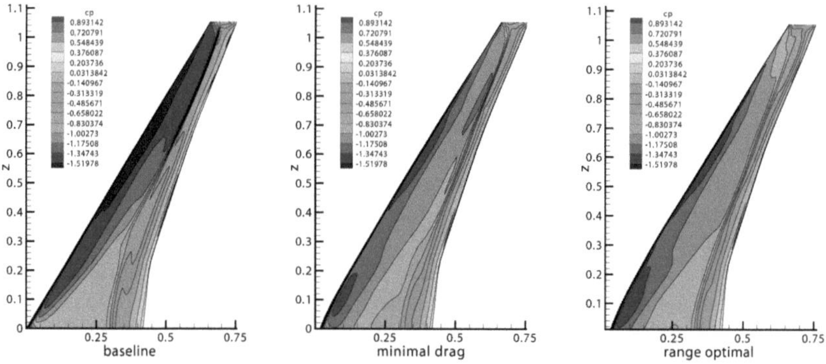

Fig. 7. Pressure distribution for the baseline AMP wing shape and for the optimal wing shapes for drag minimization and range maximization (Ma = 0.78, $\alpha = 2.83°$)

according to (25). The comparison of both methods is shown in Fig. 4, together with the gradient obtained when neglecting the aeroelastic coupling (rigid).

Finally, Figs. 5, 6 and 7 illustrate the application of the coupled aero-structural adjoint approach to the drag reduction of the AMP wing by constant lift, while taking into account the static deformation of this wing caused by the aerodynamic forces as well as for the Breguet formula of aircraft range, where in addition to the lift to drag ratio the weight of the wing is taken into account.

5 Conclusion

A coupled aero-structure adjoint approach for the efficient evaluation of sensitivities, needed for gradient-based optimization methods in MDO context, has been derived. Its implementation and validation has been presented. Finally, as a first application, the drag reduction as well as range maximization of the AMP wing has been presented.

References

1. Brezillon, J., Gauger, N.R.: 2D and 3D aerodynamic shape optimization using the adjoint approach. Aerospace Science and Technology 8(8), 715–727 (2004)
2. Fazzolari, A.: An aero-structure adjoint formulation for efficient multi-disciplinary wing optimization, PhD Thesis, TU Braunschweig (2006)
3. Fazzolari, A., Gauger, N.R., Brezillon, J.: An aero-structure adjoint formulation for efficient multi-disciplinary wing optimization. In: Proceedings of EUROGEN 2005, Munich (2005)
4. Fazzolari, A., Gauger, N.R., Brezillon, J.: Efficient aerodynamic shape optimization in MDO context. Journal of Computational and Applied Mathematics 203, 548–560 (2007)

5. Gauger, N.R.: Aerodynamic shape optimization using the adjoint Euler equations. In: Proceedings of the GAMM Workshop on Discrete Modelling and Discrete Algorithms in Continuum Mechanics, pp. 87–96. Logos Verlag, Berlin (2001)
6. Gauger, N.R.: Das Adjungiertenverfahren in der aerodynamischen Formoptimierung, DLR-Report No. DLR-FB–2003-05 (2003) ISSN 1434-8454
7. Gauger, N.R., Brezillon, J.: Aerodynamic shape optimization using adjoint method. Journal of the Aeronautical Society of India 54(3), 247–254 (2002)
8. Kroll, N., Rossow, C.-C., Schwamborn, D., Becker, K., Heller, G.: MEGAFLOW – A Numerical Flow Simulation Tool for Transport Aircraft Design, ICAS-2002-1105.20 (2002)
9. Martins, J.R., Alonso, J.J., Reuther, J.J.: High-Fidelity aero-structural design optimization of a supersonic business jet, AIAA 2002-1483 (2002)
10. Martins, J.R., Alonso, J.J., Reuther, J.J.: Complete configuration aero-structural optimization using a coupled sensitivity analysis method, AIAA 2002-5402 (2002)

Part V

Aerodynamic and Multidisciplinary Optimization of 3D-Configurations

Aerodynamic Optimization for Cruise and High-Lift Configurations

Joël Brezillon, Richard P. Dwight, and Markus Widhalm

German Aerospace Center (DLR), Institute of Aerodynamics and Flow Technology, Lilienthalplatz 7, 38108 Braunschweig, Germany
Joel.Brezillon@dlr.de, Richard.Dwight@dlr.de, Markus.Widhalm@dlr.de

Summary. Within the next few years, numerical shape optimization based on high fidelity methods is likely to play a strategic role in future aircraft design. In this context, suitable tools have to be developed for solving aerodynamic shape optimization problems, and the adjoint approach – which allows fast and accurate evaluations of the gradients with respect to the design parameters – is seen as a promising strategy. Based on the unstructured RANS solver TAU, a continuous as well as a discrete adjoint have been developed and applied to cruise and high-lift configuration optimization problems. This paper describes investigations of planform optimizations for a flying wing transport aircraft with an Euler continuous adjoint method, the wing design of the DLR-F6 wing-body aircraft configuration and the flap and slat settings of the DLR-F11 high-lift wing-body aircraft with a viscous discrete adjoint method.

1 Introduction

Numerical shape optimization is playing an increasing strategic role in aerodynamic aircraft design. It offers the possibility of designing or improving aircraft components with respect to a given objective, subject to geometrical and physical constraints. However, Computational Fluid Dynamics (CFD) still suffers from high computational effort for flow simulations around realistic 3d configurations which limits its use in design process. Consequently, worldwide a large effort is being devoted to developing efficient optimization strategies for industrial aerodynamic aircraft design.

At the DLR, activities focus on developing several key technologies relating to the establishment of an efficient and flexible numerical optimization capability based on high fidelity methods. These include suitable techniques for geometry parameterization, meshing and mesh movement methods, efficiency and accuracy improvements of the flow solvers, as well as robust and efficient optimizers. One of the most promising strategies is the use of the adjoint approach of a flow solver for efficient and accurate computation of gradients in high-dimensional design spaces, which can then be applied within gradient-based optimizers.

The paper will give an overview of the work performed at the German Aerospace Center's Institute of Aerodynamics and Flow Technology, on the application of the continuous as well as the discrete adjoint approach for solving

N. Kroll et al. (Eds.): MEGADESIGN and MegaOpt, NNFM 107, pp. 249–262.
springerlink.com

various aerodynamic shape optimization problems. The paper introduces first the strategy developed in the unstructured TAU code [1,2,3] to solve the adjoint problem and to compute the gradients. In the second part, the paper focuses on the application to 3d design in cruise and take-off conditions.

2 Gradients via Adjoint Approach

2.1 Primal Approach

Let the optimization problem be stated as

$$\min_{D} I(W, X, D), \tag{1}$$

subject to the constraint

$$R(W, X, D) = 0, \tag{2}$$

where I is a cost function such as lift or drag, D is a vector of design variables that control the shape of aircraft subject to aerodynamic design, $W(X, D)$ the vector of flow variables, $X(D)$ the computational mesh and $R(W, X, D)$ the residuals of the flow.

For a gradient based optimization strategy, the search for the minimum requires the total derivative of the cost function I with respect to the design variables D. This total derivative - also called here the sensitivity - can be written as:

$$\frac{dI}{dD} = \frac{\partial I}{\partial X}\frac{dX}{dD} + \frac{\partial I}{\partial W}\frac{dW}{dD}. \tag{3}$$

The first term of (3) expresses the direct effect of the geometry perturbation and the second term contains the effect of the flow alteration caused by the geometry perturbation on the cost function I. Solving the above equation can be done by applying finite differences which requires evaluations of the flow solver on n perturbed geometries, with n the number of design parameter. Alternatively, the adjoint approach allows a rapid evaluation of dI/dD for a large number of design variables D, without computing the flow solution on the perturbed geometry.

2.2 Dual Approach

Instead of applying the chain rule to I, apply it to the Lagrangian:

$$\mathcal{L}(W, X, D, \Lambda) = I(W, X, D) + \Lambda^T R(W, X, D) \tag{4}$$

where Λ are known as the adjoint variables. Since (2) holds for all D, $\mathcal{L} = I$ for all Λ and all D. Hence,

$$\frac{d\mathcal{L}}{dD} = \frac{dI}{dD} \quad \forall \Lambda, D, \tag{5}$$

and so, applying the chain rule to $\mathcal{L}$, the total derivative of I becomes:

$$\frac{dI}{dD} = \left\{ \frac{\partial I}{\partial X}\frac{dX}{dD} + \frac{\partial I}{\partial W}\frac{dW}{dD} \right\} + \Lambda^T \left\{ \frac{\partial R}{\partial X}\frac{dX}{dD} + \frac{\partial R}{\partial W}\frac{dW}{dD} \right\} \tag{6}$$
$$= \left\{ \frac{\partial I}{\partial X} + \Lambda^T \frac{\partial R}{\partial X} \right\} \frac{dX}{dD} + \left\{ \frac{\partial I}{\partial W} + \Lambda^T \frac{\partial R}{\partial W} \right\} \frac{dW}{dD}.$$

The unknown quantity dW/dD may be eliminated by choosing Λ such that

$$\left(\frac{\partial R}{\partial W}\right)^T \Lambda = -\left(\frac{\partial I}{\partial W}\right)^T. \tag{7}$$

This is the adjoint equation, and must be solved only once to evaluate the gradient of a single I with respect to any number of design variables. The resulting Λ allows rapidly computing the total derivative using:

$$\frac{dI}{dD} = \left\{ \frac{\partial I}{\partial X} + \Lambda^T \frac{\partial R}{\partial X} \right\} \frac{dX}{dD}. \tag{8}$$

2.3 The Continuous Formulation

In this approach the adjoint of the continuous governing equations with respect to a given cost function is derived, before being discretized. Its implementation in the TAU-Code is related to previous work [4,5,6]. We derive the adjoint using a variational formulation which emphasises the generality of the approach [7].

The steady compressible Euler equations on the domain Ω may be written for 2d flow:

$$\nabla \cdot F(w) = 0 \quad \text{in } \Omega, \tag{9}$$

where $w = (\rho, \rho u, \rho v, \rho E)$ is the vector of conserved quantities, and the flux tensor F may be written

$$F = \begin{bmatrix} \rho u & \rho v \\ \rho u^2 + p & \rho uv \\ \rho uv & \rho v^2 + p \\ \rho H u & \rho H v \end{bmatrix} \tag{10}$$

and ρ, u, v, E, p and H are the density, Cartesian components of velocity, total energy, pressure and enthalpy respectively, and the ideal gas relations are assumed. This equation is subject to slip boundary conditions on solid walls $\Gamma_w \subset \Gamma$

$$U \cdot n = 0 \quad \text{on } \Gamma_w, \tag{11}$$

where $U = (u, v)$, and n is the surface normal vector. Furthermore we are interested in a cost function I given by

$$I(w) = \int_\Gamma g(w)\, \mathrm{d}\Gamma. \tag{12}$$

To derive the adjoint we multiply (9) by $\psi \in V$, where V is a Sobolev space containing the solution of (9), linearize about w_0, and integrate by parts

$$\int_{\Omega} \nabla \cdot F'(\phi)\psi \, \mathrm{d}\Omega = \int_{\Gamma} n \cdot F'(\phi)\psi \, \mathrm{d}\Gamma - \int_{\Omega} F'(\phi)\nabla\psi \, \mathrm{d}\Omega, \tag{13}$$

where $F' = F'[w_0]$ is the derivative of F with respect to w evaluated at w_0. Therefore the variational formulation of the adjoint problem is given by: find adjoint solution ψ such that $\forall \phi \in V$

$$\int_{\Gamma} \phi(n \cdot F')^T \psi \, \mathrm{d}\Gamma - \int_{\Omega} \phi F'^T \nabla\psi \, \mathrm{d}\Omega = I'[w_0](\phi). \tag{14}$$

The continuous adjoint problem is therefore

$$\begin{aligned} -F'^T \nabla \psi &= 0 \quad \text{in } \Omega, \\ (n \cdot F')^T \psi &= g' \quad \text{on } \Gamma. \end{aligned} \tag{15}$$

The singularity of F' on slip walls leads to the well-known result that not all choices of g result in a well-posed adjoint problem.

Given the adjoint field ψ the derivative of I with respect to any design variable D may be written

$$\frac{\mathrm{d}I}{\mathrm{d}D} = \int_{\Gamma} \frac{\partial g}{\partial D} \mathrm{d}\Gamma + \int_{\Omega} \psi \frac{\partial}{\partial D} \nabla \cdot F \, \mathrm{d}\Omega, \tag{16}$$

which is notable for not containing any total deriatives of the flow solution $\mathrm{d}w/\mathrm{d}D$.

In the TAU code, the inviscid continuous adjoint equations are discretized in space in an analogous way to the flow equations. As the adjoint equations are not conservative, care was taken that the discretization is non-conservative when applying the finite volume method. The resulting equations are solved using similar iterative methods to those used for the Euler flow equation.

2.4 The Discrete Adjoint Approach

Here the discrete variant of the adjoint equation (7) is considered. Its implementation requires the ability to evaluate the quantities $(\partial R/\partial W)^T \Lambda$ - the adjoint residual - and $\partial I/\partial W$. The Jacobian $(\partial R/\partial W)$ is evaluated by hand, which is a straightforward exercise as R may be written explicitly in terms of W, while being time-consuming as R is often extremely complex. As R is a sum of convective fluxes, viscous fluxes, boundary conditions etc., each of these may be handled independently, and by application of the chain rule may be further subdivided into manageable chunks. The derivatives are further simplified by choosing primitive variables as working variables. Because the equations remain in conservative form this choice has no effect on the final solution. A more detailed description of the implementation in the TAU code can be found in [10]. A wide range of the spatial discretizations available in TAU have been differentiated, including the Spalart-Allmaras-Edwards one-equation turbulence model. The effect of various approximations of the Jacobian was investigated and their impacts on the efficiency of the optimization process has been demonstrated on several 2d optimization problems [10, 11]. In the present study, the viscous 3d adjoint computations have been performed by considering frozen turbulent effects.

Despite the guarantees regarding convergence provided by the theory of adjointed fixed-point iterations (FPIs) [12] there are regularly situations in which it is possible to obtain a reasonably converged solution of the non-linear equations, but not of the corresponding adjoint equations. This can occur for three reasons: either a) the non-linear solution is not sufficiently converged, or b) there is a discrepancy between the linear and non-linear problem due to some approximation of the Jacobian, or c) the FPI applied to the non-linear problem does not converge *asymptotically* itself. All three cases appear regularly in practice. An engineer may reasonably consider a computation converged when the integrated forces that interest her no longer vary significantly, though this may occur prior to the asymptotic regime.

In an effort to understand and mitigate these phenomena, we consider the Recursive Projection Method (RPM), originally developed by Schroff and Keller in 1993 as a means of stabilizing unstable procedures [13]. The main idea can be described briefly as follows: let the (linear) adjoint system be written $Ax = b$, and regard the transient solution of the linear problem as a sum of eigenvectors of the relaxation operator $\Phi = (I - M^{-1}A)$ where M is some iteration operator, e.g. LU-SGS with multigrid. The application of Φ to an approximate solution then corresponds to a product of each eigenvector with its corresponding eigenvalue. Divergence of the iteration implies that there is at least one eigenvalue of Φ with modulus greater than unity. Assuming that the number of such eigenvalues is small, and that the space spanned by their eigenvectors is known, call it $\mathcal{P}$, then it must be possible to solve the projection of the problem onto this low dimensional subspace using some expensive but stable method, while solving the projection onto the complimentary subspace $\mathcal{Q}$ using the original FPI iteration, which is known to be stable there.

Newton-Raphson is used on $\mathcal{P}$. The space of dominant eigenvectors is determined as the calculation progresses, by applying the principle that the difference between successive applications of the FPI on $\mathcal{Q}$ form a power iteration on the dominant eigenvalues of Φ restricted to $\mathcal{Q}$.

This procedure has been successfully applied for the design of DLR-F6 configuration that features flow separation in the junction between the upper surface of the wing and the fuselage, see Part 4.

Further investigations revealed that the robustness of RPM for the viscous adjoint problem was limited in the case that the base iteration diverges too rapidly for $\mathcal{P}$ to be well approximated. In this case applying the well-known Generalized Minimum Residual (GMRes) method in it's restarted form [14], with 10-50 iterations of LU-SGS with multigrid as a preconditioner has been seen to be an exceptionally robust alternative. This was the stabilization using to converge the adjoint problem for the high-lift configurationin Section 5.

2.5 The Metric Terms

In order to compute the total derivatives of the cost function I as given in (8), the metric term variation is computed by using finite differences, which is in case of R computed like:

$$\frac{\partial R}{\partial X}\frac{\partial X}{\partial D_k} \approx \frac{(R(W, X(D + \Delta D_k)) - R(W, X))}{\Delta D_k} = \frac{\Delta R}{\Delta D_k}. \tag{17}$$

After obtaining Λ, the sensitivities can be evaluated with a single point deformation and yields for each design variable D_k to a variation of the cost function due to the perturbed geometry and we get a scalar difference for the direct variation and a matrix-vector product for the dependency of the residual,

$$\frac{dI}{dD_k} \approx \frac{\Delta I}{\Delta D_k} + \Lambda^T \frac{\Delta R}{\Delta D_k}. \tag{18}$$

3 Planform Optimization of a Very Efficient Large Aircraft (VELA)

The adjoint approach in an optimization context is first applied to a blended wing body aircraft. The initial geometry was originally developed within the European Project VELA (Very Efficient Large Aircraft) and was provided by Strüber and Hepperle [15, 16]. The aircraft is designed as a future passenger transport aircraft with long-haul capabilities, as depicted in Figure 1.

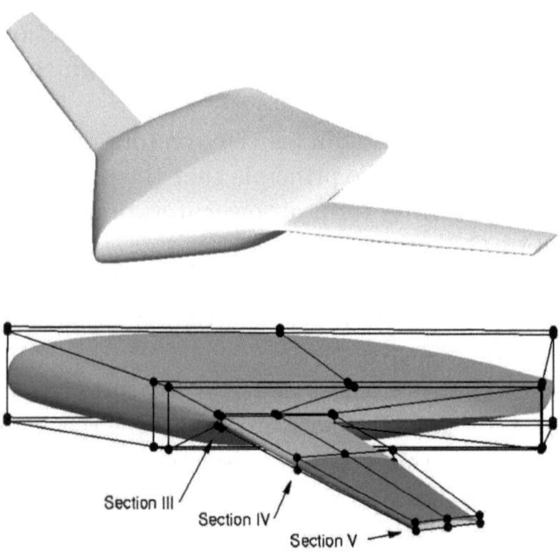

Fig. 1. VELA configuration (above) and initial lattice box used for deformation

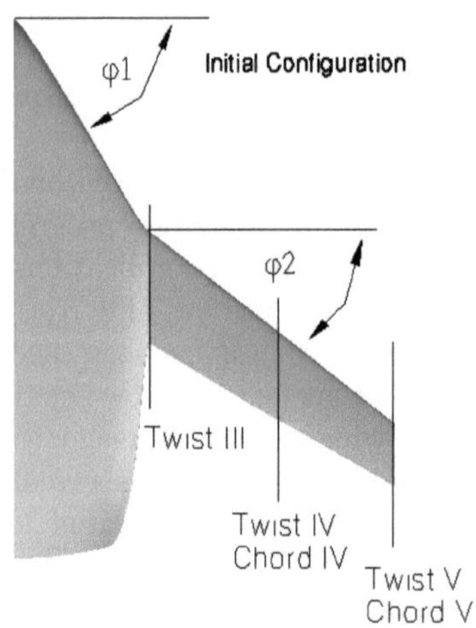

Fig. 2. VELA configuration depicting the 7 design parameters

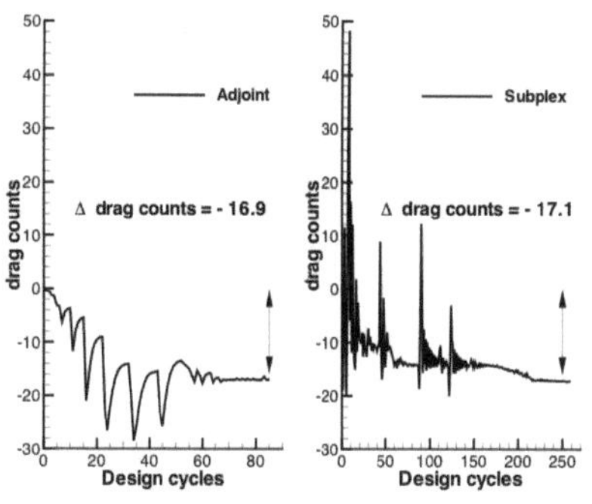

Fig. 3. Convergence of the VELA drag-minimization, Adjoint (left) and Subplex (right)

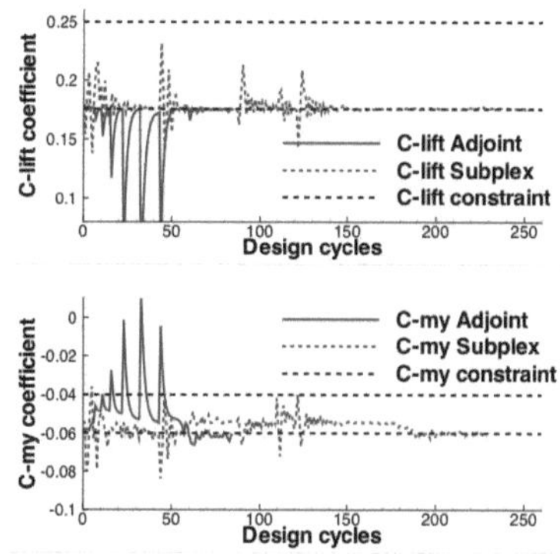

Fig. 4. Lift (above) and Pitching moment (bottom) history during the optimization

The geometry is parameterized by the coordinates of the lattice box points controlling the free-form deformation (FFD) [17,18,19,20]. The FFD technique allows a broad range of deformations with a low number of parameters and ensures a smooth deformation. The lattice boxes were generated with DLR's parametric grid generator MegaCads [21], and are directly applied on the unstructured surface grid points, bottom left Figure 1. The design parameters are the sweep angles φ_1 and φ_2, the variations of the chord length in the middle and at the tip of the outboard wing and three twist angles again for the outboard wing as depicted in Figure 2. Once the surface grid is deformed, the TAU deformation module is used to propagate the deformation to the surrounding volume grid. The initial grid was generated with the software package Centaur [22], was designed for inviscid flow, and contains approximately 1 million points.

The optimization test case concerns the drag minimization for a prescribed lift and pitching moment at a free-stream Mach number of 0.85 and fixed angle of attack of 1.8 degrees. The allowed deviation in lift and pitching moment is set to 1% of the initial configuration during the optimization process. The gradient based optimization strategy used is the modified method for feasible directions (MMFD) [23] and the gradients for the goal function and the constraints are computed from to the continuous adjoint. It is here intended to demonstrate that the adjoint approach can be used to perform efficient planform optimization. For this purpose, the same optimization is performed a second time using the SubPlex approach [24], a gradient free optimizer based on the Nelder-Mead method [25].

The convergences of the optimizations are shown in Figure 3 for the drag reduction and Figure 4 for the lift and pitching moment. Both methods provide almost the same drag decrease, 17 drag counts, and satisfy the constraints on the lift and pitching moment. As expected, the gradient approach is faster and needs only 82 aerodynamic evaluations while the gradient free requires 260 evaluations

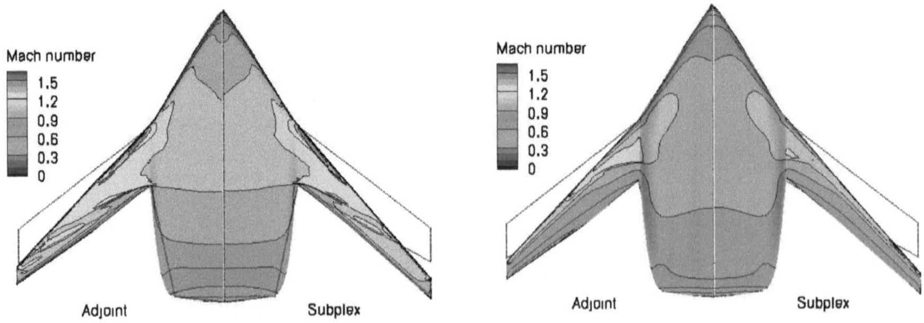

Fig. 5. Comparison of the Mach number distribution on the upper (left figure) and lower (right figure) surfaces, obtained with the Adjoint (left cut) and Subplex (right cut). The initial configuration is depicted with dots.

to converge the optimization process. In terms of CPU time, the use of adjoint allows to complete the optimization within less than 3 days which represents a speed up of 2.7 compared to the gradient free approach.

Figure 5 presents the surface Mach number distribution on the upper and lower side of both optimized geometry. In order to stress the changes of the sweep angles the initial planform is superposed as dotted contour on both pictures. The shock on the upper side decreased at the outward side of the outboard wing but one can still see a strong shock at the inner side on both optimized geometries. On the lower side, near section III, the gradient free optimizer seems to better reduce the shock strength than the gradient based approach, but this has a limited influence on the drag coefficient. Finally, the resulting planforms are very close to each other with almost the same inboard and outboard sweep angles.

4 Wing Shape Optimization of the DLR-F6 Configuration

The adjoint method is now applied to the drag minimization of the DLR-F6 wing-body configuration at Mach 0.75, a Reynolds number of 3×10^6, and $C_L = 0.8$, at which conditions the case has a large region of separated flow in the junction between the upper surface of the wing and the fuselage, as well as along most of the length of the wing. Here, the discrete variant of adjoint formulation is used and the standard method of adjointed LU-SGS with multigrid alone was unconditionally unstable, and applying RPM was necessary to obtain a converged solution.

The optimization algorithm used is conjugate-gradients (CG), as in [26], where the angle-of-attack is automatically set by TAU to maintain the lift constant. The surface of the computational grid is shown in Figure 6, and is coarse, but sufficient to resolve the separation mentioned. The 84 paired nodes of a free-form deformation bounding box are also shown, whose vertical positions, as well as 12

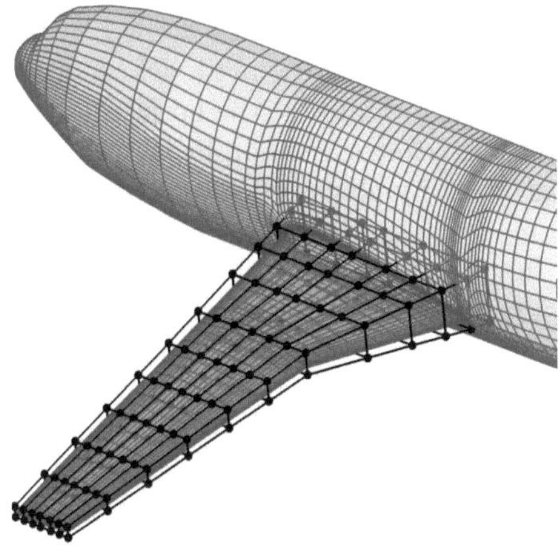

Fig. 6. Parameterization of the wing with a free-form deformation box with 84 paired nodes. Twist is parameterized separately with 12 variables.

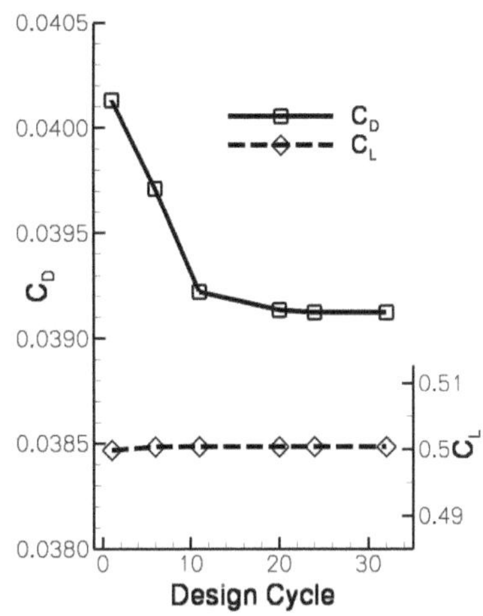

Fig. 7. Convergence of the F6 drag-minimization optimization

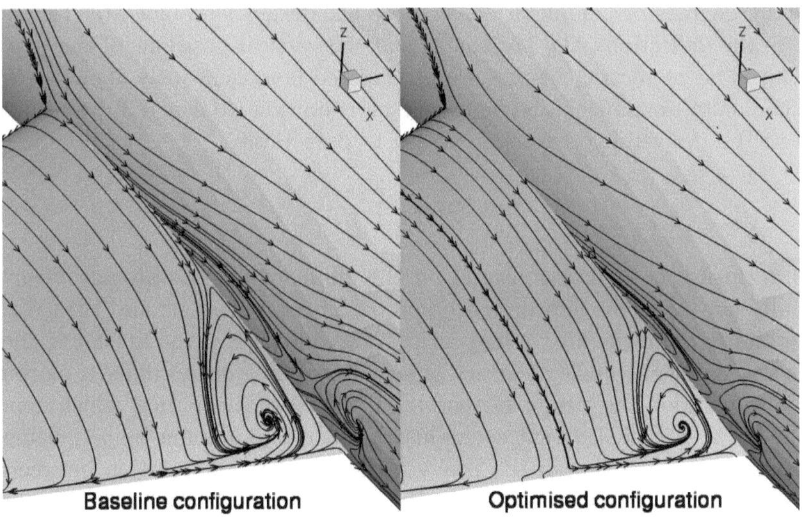

Fig. 8. Comparison of the region of corner separation before and after optimization

additional wing twist variables were used to parameterize the wing. The pairing of nodes constrained the wing thickness. With such a large number of design variables only gradient-based optimization is viable, and only the adjoint method can deliver the gradient efficiently. Note that since the bounding box passes inside the fuselage, the wing-body junction also varies, and this is accounted for by the geometry and grid generation process. The metric sensitivities needed in the gradient calculation are evaluated by finite-differences on the mesh deformation.

The convergence of the optimization is shown in Figure 7, the horizontal axis shows the number of calls to the flow solver (both linear and non-linear), thereby approximately representing computational effort. Symbols indicate gradient evaluations. After 32 solver calls CG was unable to reduce the drag further, giving a final reduction of about 10 drag counts. In contrast a similar optimization with 42 parameters produced a reduction of only 8 counts on this mesh (in a similar CPU time) [26], emphasizing the need for a comprehensive parameterization. The optimization reduced the region of corner separation considerably, Figure 8, while not completely eliminating it, which is unlikely to be possible within the design space considered, as it does not allow deformation of the body.

5 Flap and Slat Settings Optimization of the DLR-F11 Aircraft

The last configuration optimized is the so called DLR-F11 model with full span flap and slat in take-off configuration, see Figure 9. This model is a wide-body Airbus-type research configuration with a half span of 1.4 meter that can feature different degrees of complexity [27]. Here six design variables are selected to modify the deflections, the horizontal and the vertical positions of the flap and the slat. The geometric changes are propagated homogeneously along the span. The goal is to maximize at a single take-off condition (Mach=0.3; Re=20×10^6; AoA=8°) a derived expression of the lift to drag ratio:

$$Obj = \frac{Cl^3}{Cd^2}. \tag{19}$$

This performance indicator, based on the climb index, has already been successfully employed for flap design based on 2d computations and turned to be better suited than the lift to drag ratio [28]. Additionally, the lift is not allowed to decrease and the angle of attack is kept fixed. In order to make a more realistic optimization the weight of the high-lift system kinematics, which depends on the horizontal deployment capability, is taken into account by penalizing the objective function to avoid too heavy a mechanism. The relation between the horizontal displacement and the penalty is set according to industrial specifications [28].

An ICEM-CFD macro has been developed to handle both the parameterization and the mesh procedure. This macro first sets the position of the elements according to the design variables and computes automatically the flap and slat intersection lines with the body. Once the CAD geometry is ready, the meshing part starts and automatically projects the mesh on the moving part and on the updated intersections lines, sets the position and size of the O blocks surrounding the elements. The resulting mesh has in total 2.5 millions points, see Figure 9. The complete process, from reading the design variables to writing the

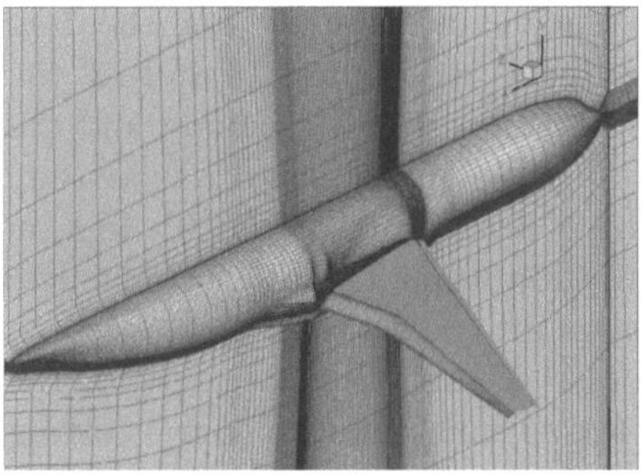

Fig. 9. Mesh around the DLR-F11 model in full span flap and slat configuration

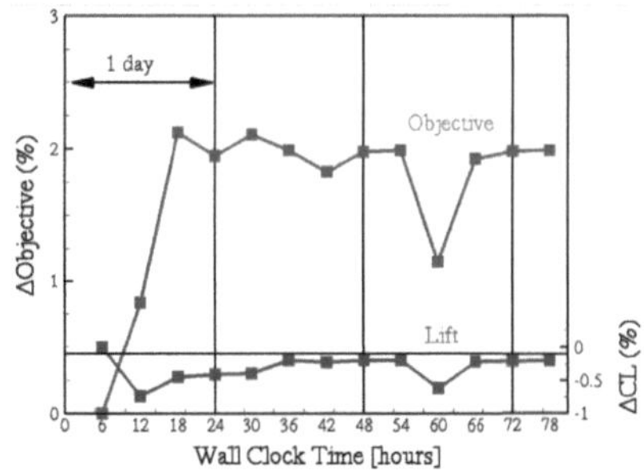

Fig. 10. Evolution of the objective and lift coefficient according to the wall-clock time

mesh in unstructured formats takes about 1 minute on a single AMD Opteron 2.5 GHz processor. The numerical simulations are based on the RANS equations and the Spalart-Allmaras-Edwards turbulence model. For fast convergence, the low Mach number preconditioning approach is adopted and the steady state is reached by a Runge-Kutta scheme using multigrid W-cycles on 3 levels. A fully converged solution with almost 5 orders of density residual decrease is obtained after 5,000 TAU cycles. In order to exploit the parallel capability of the TAU code, the aerodynamic flow is computed on a cluster of 32 AMD Opteron 2.4 GHz processors and the drag and lift adjoints are computed simultaneously on 2 clusters of 16 processors each. Each solution is fully converged after 3 hours wall clock time.

Figure 10 presents the evolution of the optimization process obtained with the NLPQLP optimization strategy [29] coupled to the adjoint approach for the computations of the gradients. After 13 evaluations and 78 hours of simulations the optimization converged to a maximum with a limited deviation on the lift

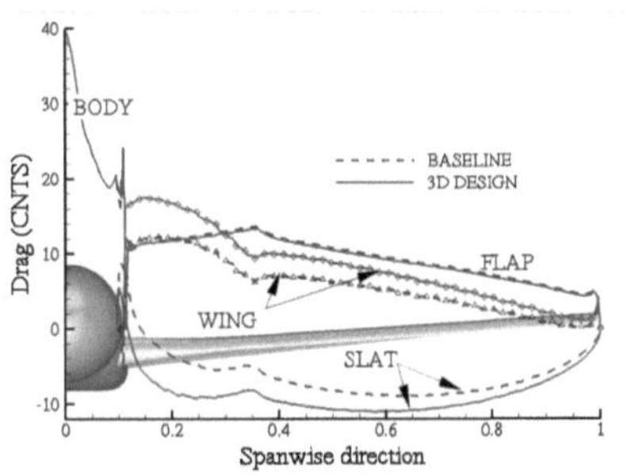

Fig. 11. Drag distribution along the spanwise direction on the baseline and optimised configurations

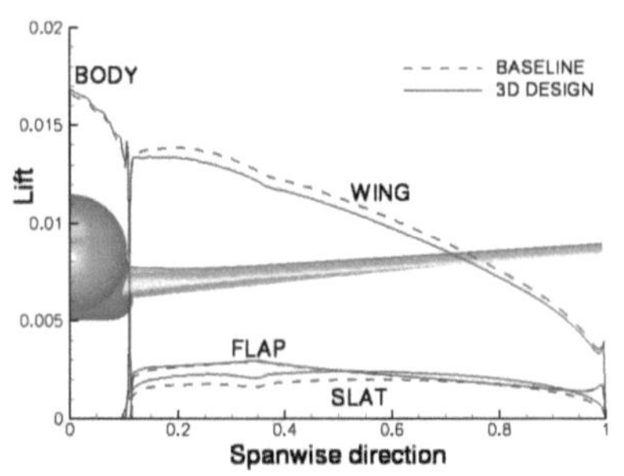

Fig. 12. Lift distribution along the spanwise direction on the baseline and optimised configurations

coefficient. The performance improvement is made evident by plotting the drag distribution in span-wise direction for each element, see Figures 11 and 12: the optimization has almost no influence on the lift and drag of the body and the flap but permits to made further negative the drag on the slat by increasing the lift. This improvement has to be paid by a drag increase and lift loss on the main wing. Finally, the optimized configuration has in total 17.8 counts less drag than on the baseline configuration by same lift coefficient.

Thanks to the adjoint approach, the process is almost independent of the number of design variables and a more complex optimization problem involving more design parameters should require almost the same turn around time.

6 Conclusion

This article presented activities carried out at the DLR for the development of the continuous as well as the discrete adjoint approaches in the unstructured RANS solver TAU code and their application to solving wide range of aerodynamic shape designs.

The capability to perform planform optimization using a gradient-based method coupled with the adjoint has been first demonstrated on a flying wing configuration. The adjoint approach has permitted drastic savings in computation time compared to a gradient-free method while giving comparable results. The capability of the adjoint approach to handle problems with large number of design parameters has been then demonstrated for the optimization of a wing-body configuration in viscous flow. It has been observed that the region of separation was considerably reduced thanks to the fine parameterization of the wing. The MEGADESIGN milestone dedicated to the application of the adjoint approach on unstructured meshes for solving shape optimization problems has been then fully acchieved.

Thanks to the adjoint approach, only a few flow computations were required to converge the 3d high-lift configuration optimization problem and the milestone on the optimization of such a configuration within 2 weeks has been demonstrated as well.

Near future activities will focus on even more complex configurations, such as engine integration problems, and on multi-points design problems.

References

1. Galle, M.: Ein Verfahren zur numerischen Simulation kompressibler, reibungsbehafteter Strïmungen auf hybriden Netzen. DLR-FB 99-04 (1999)
2. Gerhold, T.: Overview of the Hybrid RANS TAU-Code. In: Kroll, N., Fassbender, J. (eds.) MEGAFLOW Numerical Flow Simulation Tool for Transport Aircraft Design, Notes on Multidisciplinary Design, vol. 89. Springer, Heidelberg (2005)
3. Schwamborn, D., Gerhold, T., Heinrich, R.: The DLR TAU-Code: Recent Applications in Research and Industry. In: Proceedings of ECCOMAS CFD 2006, Delft The Netherland (2006)

4. Jameson, A.: Aerodynamic design via control theory. Journal of Scientific Computing 3(3), 233–260 (1988)
5. Giles, M., Pierce, N.: An introduction to the adjoint approach to design. In: Proceedings of ERCOFTAC Workshop on Adjoint Methods (1999)
6. Anderson, W.K., Venkatakrishnan, V.: Aerodynamic Design Optimization on Unstructured Grids with a Continuous Formulation. AIAA Paper 97-0643 (1997)
7. Hartmann, R.: Adjoint consistency analysis of Discontinuous Galerkin discretizations. SIAM J. Numer. Anal. 45(6), 2671–2696 (2007)
8. Jameson, A.: Optimum aerodynamic design using CFD and control theory AIAA 95-1729. In: AIAA 12th Computational Fluid Dynamics Conference, San Diego (1995)
9. Gauger, N.R., Brezillon, J.: Aerodynamic shape optimization using adjoint methods. Journal of Aero. Soc. of India 54(3), 247–254 (2002)
10. Dwight, R.: Efficiency Improvements of RANS-Based Analysis and Optimization using Implicit and Adjoint Methods on Unstructured Grids. School of Mathematics, University of Manchester (2006)
11. Dwight, R., Brezillon, J.: Effect of Various Approximations of the Discrete Adjoint on Gradient-Based Optimization. In: Proceedings of the 44th AIAA Aerospace Sciences Meeting and Exhibit, AIAA Paper 2006-0690, Reno NV (2006)
12. Dwight, R., Brezillon, J., Vollmer, D.B.: Efficient Algorithms for Solution of the Adjoint Compressible Navier-Stokes Equations with Applications. In: Proceedings of the ONERA-DLR Aerospace Symposium (ODAS), Toulouse (2006)
13. Schroff, G., Keller, H.: Stabilization of unstable procedures: The Recursive Projection Method. SIAM Journal of Numerical Analysis 30, 1099–1120 (1993)
14. Saad, Y., Schultz, M.H.: A generalized minimum residual algorithm for solving non-symmetric linear systems. SIAM Journal of Scientific and Statistical Computing 7(3), 856–859 (1988)
15. Strüber, H., Hepperle, M.: Optimization of Flying Wing Transport Aircraft Configurations in the VELA Project. DLR-IB 124-2005/905 (2005)
16. Mialon, B., Hepperle, M.: Flying Wing Aerodynamics Studies at ONERA and DLR. In: CEAS/KATnet Conference on Key Aerodynamic Technologies, Bremen (2005)
17. Samareh, J.A.: Novel Shape Parameterization Approach. NASA TM-1999-209116 (1999)
18. Samareh, J.A.: Aerodynamic Shape Optimization Based on Free-Form Deformation. AIAA 2004-4630 (2004)
19. Ronzheimer, A.: Post-Parametrisation of CAD-Geometries Using Freeform Deformation and Grid generation Techniques. In: Notes on Numerical Fluid Mechanics and multidisciplinary Design, vol. 87, pp. 382–389
20. Ronzheimer, A.: Shape Parameterization in Multidisciplinary Design Optimization Based on Freeform Deformation. In: Evolutionary and Deterministic Methods for Design, Eurogen, Munich (2005)
21. Brodersen, O., Hepperle, H., Ronzheimer, A., Rossow, C.-C., Schöning, B.: The Parametric Grid Generation System Mega Cads. In: 5th International Conference on Numerical Grid Generation in Computational Field Simulation, national Science Foundation (1996)
22. Centaur Soft (2008), `http://www.Centaursoft.com`
23. Vanderplaats, G.N.: Numerical Optimization Techniques for Engineering Design. McGraw-Hill Series in Mechanical Engineering (1984) ISBN 0-07-066964-3
24. Rowan, T.: Functional Stability Analysis of Numerical Algorithms Thesis, Department of Computer Sciences, University of Texas at Austin, USA (1990)

25. Nelder, J.A., Mead, R.: A Simplex Method for Function Minimization. Comput. J. 7, 308–313 (1965)
26. Brezillon, J., Brodersen, O., Dwight, R., Ronzheimer, A., Wild, J.: Development and application of a flexible and efficient environment for aerodynamic shape optimisation. In: Proceedings of the ONERA-DLR Aerospace Symposium (ODAS), Toulouse (2006)
27. Rudnik, R., Thiede, P.: European research on high lift commercial aircraft configurations in the EUROLIFT projects. In: CEASKATnet Conference on Key Aerodynamic Technologies, Bremen, DE (2005)
28. Wild, J., et al.: Advanced high-lift design by numerical methods and wind tunnel verification within the European project EUROLIFT II. In: 25th AIAA Applied Aerodynamics Conference, AIAA-2007-4300, Miami, FL, USA (2007)
29. Schittkowski, K.: NLPQLP: A new Fortran implementation of a sequential quadratic programming algorithm for parallel computing. Report, Department of Mathematics, University of Bayreuth (2001)

Aerodynamic Optimization of an UCAV Configuration

St. M. Hitzel, L. Nardin, K. Sørensen, and H. Rieger

Aerodynamics and Methods, EADS Military Air Systems, Germany

Summary. Advanced aircraft design is characterized by multipoint, multidisciplinary requirements. Optimization techniques probe the aerodynamic, flight mechanical and structural design sensitivities for a balanced vehicle-system. Aircraft optimization exercises were performed in an universal optimization environment, which controls the CAD, robust mesh generation, RANS-flow simulation and the selection of multidiszi-plinary variables. Genetic algorithms, evolutionary strategies and simplex were used. The algorithms applied will be compared.

1 Introduction

Aircraft design more and more is characterized by the intense collaboration of multiple technical disciplines and the economic needs to increase the efficiency of the air-vehicle as a system of systems. The success of integrated configurations depends on the efficient concurrent engineering of all airframe disciplines. It is the only way for a robust design which fits into a set of mission and design requirements. Well posed early risk definitions redirect funding into the most crucial design sensitivities and allows a safe development process plan with growth potential for added value. This is the only way for a well balanced and robust product development in the future. This holds for civil as well as military programs and products. The success and the realisation of highly integrated configurations such as a blended wing body transport aircraft, a flying wing or a low observable unmanned aerial vehicle (LO-UCAV) (Fig. 1) depend on the efficient concurrent engineering of the most important airframe disciplines.

The aircraft design process is an overall multidisciplinary optimization in itself. Usually it begins with a set of design requirements which include parameters such as range, payload, take-off and landing conditions, manoeuvrability, speed requirements and especially in military application special systems to be integrated. Novel mission demands may require non-traditional regimes of flight for which no previous experience may be drawn off.

In the conceptual/preliminary design the basic questions of configuration arrangement, size and weight as well as performance are answered. Usually it will begin with a set of design requirements which include parameters such as range,

N. Kroll et al. (Eds.): MEGADESIGN and MegaOpt, NNFM 107, pp. 263–285.
springerlink.com

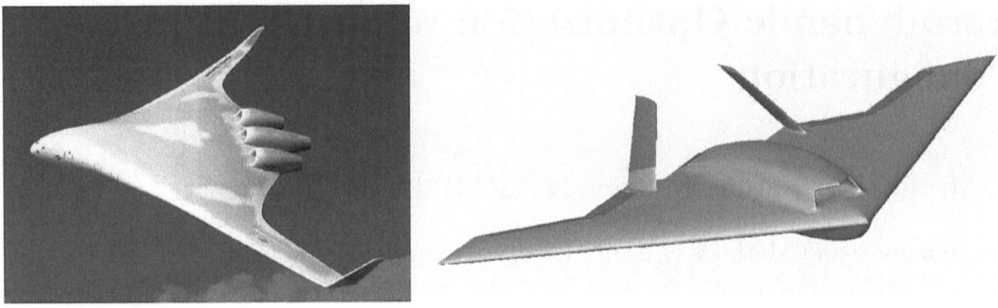

Fig. 1. Highly integrated civil and military aircraft configurations

payload, take-off and landing conditions, manoeuvrability, speed requirements and especially in military application special systems to be integrated.

Nowadays CAD-capabilities allow to define the configuration as a parameterized model which may include the typical features of fuselage shape, wing and control geometries, internal layouts, landing gear, fuel volume etc. These models are ready for first overall checks and sizing processes with load, balance and mission compatibility in mind. The integration of modern simulation techniques allows to regard true aerodynamic, propulsion and structural properties at a very early stage, which is a major advantage when it comes to performance and development risk estimations. The optimization techniques are used to find the lightest or lowest-cost aircraft that will perform its missions and meet all other requirements.

The preliminary design organically can emerge out of the conceptual design by the ongoing refinement of the configurations considered. Here shaping, lofting, sizing and the rearrangement of the details is driven to the confidence that the aircraft can be built at an estimate cost and time. Even later detail design cycles as well as in service design enhancements may utilize optimization when it comes to the adaptation of fairings, stores, modified engines etc. Considering flight performance the key disciplines are aerodynamic shaping together with an early assurance of proper stability margins and efficient control power. The type, shape, size and location of the corresponding – possibly novel - control devices have to be balanced into tightly coupled aircraft configurations. the integration of which is the enabler for widened envelopes, carefree handling and robustness with regard to design changes and in service enhancements.

The nonlinear nature of the aerodynamic properties of these systems such as compressibility, the complexity of controlled, uncontrolled separation as well as unsteady flow phenomena demands emphasis into high quality flow simulations as early as possible [1]. Since almost every other design issue related to performance, flight control systems, loads, operational capabilities and other systems more and more are mutually dependent on the reliable prediction of both flow structures and effective forces and moments high fidelity tools are required to explore the design space at best [2].

High fidelity flow simulation (RANS, URANS) [3], structural analysis together with a very flexible process control systems and optimization procedures open the path towards a future highly efficient design process with integrated optimization cycles for single and multidisciplinary design tasks. Modern CAD-capabilities allow for parameterized aircraft models which can include all typical features of a design. Their object oriented layout helps to integrate modern simulation techniques at an early stage.

The focus of EADS-MAS onto military combat aircraft is challenged by often contradicting requests e.g. high speed manoeuvrability, possibly supersonic cruise, nonlinear lift control, carefree handling quality impacted by flow control, stall and buffet limits, flutter, IR/RCS compatibility, weapons integration, complex intake and nozzle demands as well as special features such as thrust vectoring. Reduced or even discarded vertical fins may sharpen the problem when low observable vehicles, possibly operating in non-traditional regimes of flight become important. The integration of modern sensors and their competition with other vital aircraft-components such as flight controls is only another aspect of tightly coupled system-of-system designs.

In the frame of the Megadesign project EADS-MAS put emphasis in the fields of geometry modelling via CATIA-V5 and its parameterization and the integration of flow simulations into the optimization framework modeFrontier [4]. This also included the automation of the tasks to be integrated into a process-chain suited for industrial needs. A robust meshing technique for unstructured, hybrid grids was selected in the EADS-Mesher tool and combined the DLR Tau RANS / URANS-Solver or the EADS-MAS SimServer environment to provide the means necessary for flight-mechanical and aero-elastic control in future developments.

This paper describes an integrated design-optimization environment from parameterized geometry modelling, the integration of flow simulations into the optimization framework modeFrontier. Python scripts implement new optimization strategies, methods to control aerodynamic and various simplified weight estimations together with autopilot trim-functionalities into the design system. Different applications were tested. They included the aerodynamic optimization of transonic cases for 3-D wing-designs and of complete aircraft configurations.

2 Design Optimization Process

The typical design optimization cycle starts with a first set of “reasonable” geometries. These initial layouts and lofts based on the experience and creativity of a design team address the key requirements of the future product. Those “ideas” then are subject to the analysis by the appropriate disciplinary methods, while the optimization procedure through its sensitivity search strategy determines the next designs to be investigated within the design space. This cycle is repeated until a converged best set of design parameters is determined (Fig. 2). Since an aircraft mission (Fig. 3) may show very different states of operation, multipoint analysis is necessary in order to regard and balance the influence of the different phases accordingly. The design may regard the overall mission as well as specific

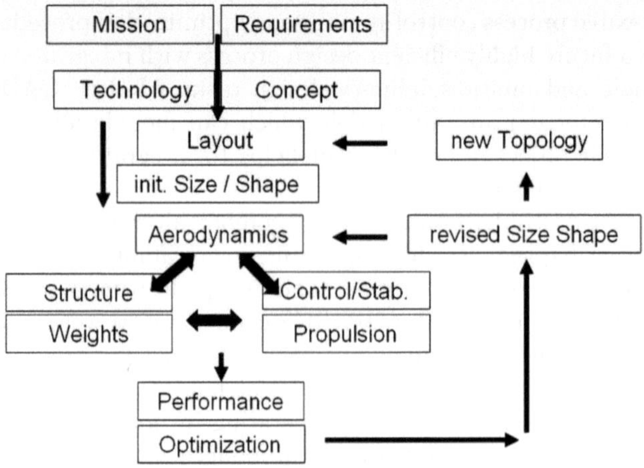

Fig. 2. Design cycle process

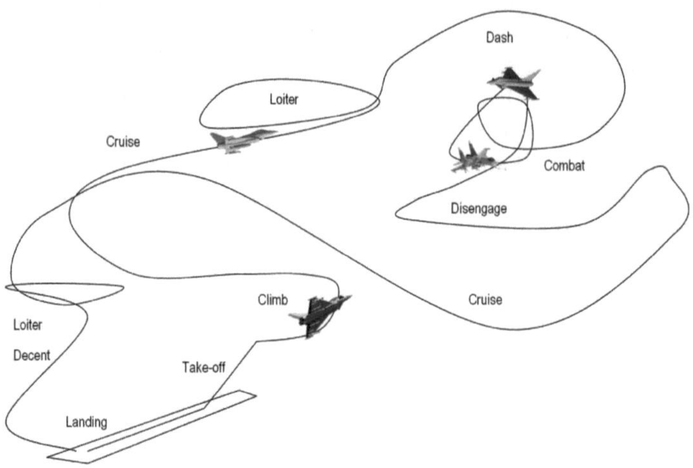

Fig. 3. Typical mission of a combat aircraft

parts of it (e.g. cruise and/or loiter). Other optimizations may aim for typical criteria such as aerodynamic lift and drag at one or multiple conditions.

An optimization environment such as modeFrontier links and controls all steps of the design, evaluation and optimization process itself (Fig. 4). Here, for simplicity emphasis is put onto the aerodynamic steps. modeFrontier [4] provides the overall strategies, selects the CAD-models, controls the demands on the meshing resolution, starts the flow simulation(s), analyses the results and evaluates the objectives. Multidisciplinary connection e.g. for weight estimations and/or by

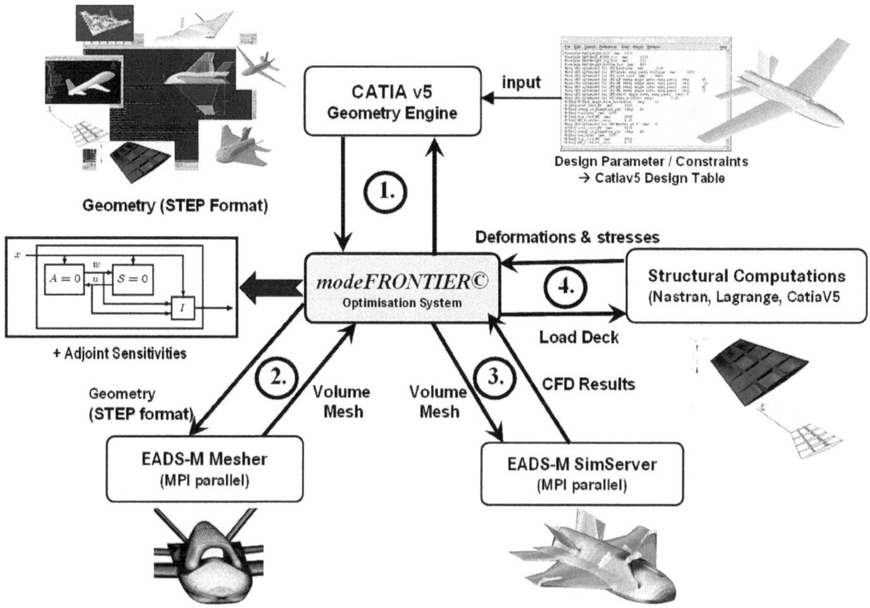

Fig. 4. *modeFrontier* optimization process including aerodynamics and structures

structural sizing through FEM are defined. Depending on the optimization procedure selected, regarding constraints and limits, modeFrontier then determines the next design by redefining the parameters. More details of the modeFrontier framework are given in a companion paper in this book.

3 Tools of the Optimization Environment

Critical to the overall process is the robustness of the tools which provide the geometries, the computational meshes, the solvers and the optimization procedures. Here a review of the properties of the geometry definition through CATIA-V5 models, the meshing of hybrid grids suited for RANS and URANS simulations is given. Also the SimServer-environment [5,6], combining the flight-mechanical and structural analysis into the simulation is described, followed by the optimization procedures applied.

The parameterization of CATIA-V5-models (Fig. 5) provides a variety of configurations of a common topology. Interdependent formulations may be used to change all components and parts in any detail and recombine them into a new shape. The optimization procedure reconfigures the CAD parameter-set selected according to the sensitivities of the objectives evaluated. By the means of knowledge software cross-topology variations are possible, however, experience advises to provide separate models for each topology and select those by the means of the

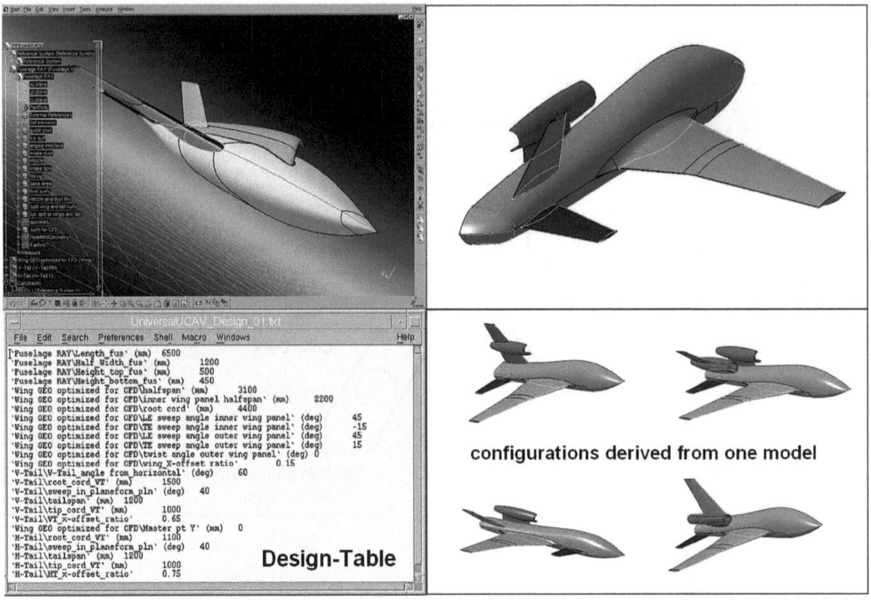

Fig. 5. Parameterized CATIA-V5 aircraft-families of similar topology

process and optimizer control since such a change also effects the optimization variables and the corresponding design sensitivities.

Typical design parameters for an aircraft configuration (Fig. 6) may include airfoil description, planform variables such as span, taper-ratio, wing and tail position, control location and size and any fuselage shape parameters desired.

A very important issue is the provision of feasible geometries. CATIA-V5 only can repeat – script-wise – the basic features of a topology in different shapes. Therefore any infeasible overlap of geometry, e.g. the intersection of a horizontal tail with the wing has to be controlled beforehand by a check on the overall relations provided by the parameters eligible. Before the CATIA-model is run in the automatic mode within an optimization process all extremes of the parameters have to be tested and restricted by control software within the overall process by defining geometric constraints or by procedures which avoid invalid designs beforehand. The construction of complex surfaces within such models such as e.g. wing-fuselage fairings may impose severe restrictions in this sense. Depending on the surface-building method even slight changes in sweep, fuselage cross section, an- or dihedral may negate a reasonable closed surface model. This only can be alleviated by smart, robust CAD-design utilizing universal surface definitions and strategies.

The more complex a geometry, the more difficult the feasiblitiy checks to avoid entanglements and non-unique geometric constellations. Unstructured meshing requirements demand closed surfaces and the avoidance of surface slivers,

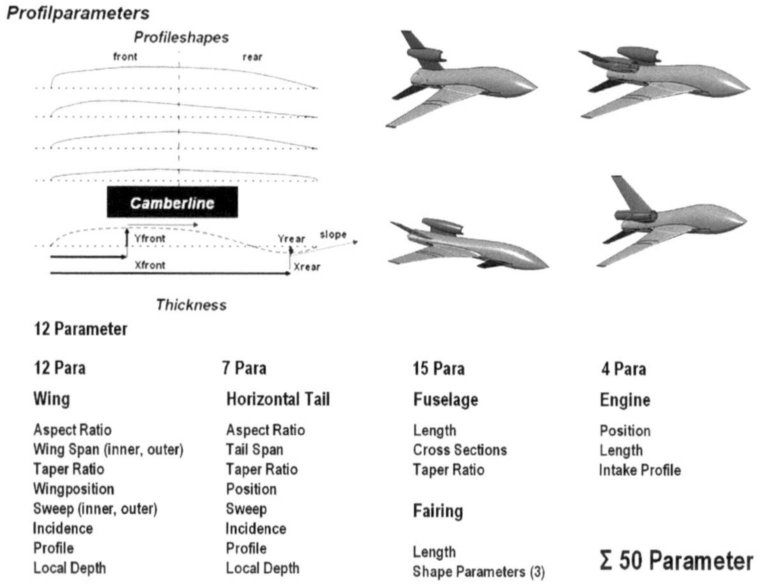

Fig. 6. Possible parameters of a full aircraft configuration

short-comings which in the future may be avoided by surface-patch independent procedures.

The EADS-Mesher [7] procedure is a robust mesh-generator for unstructured, hybrid meshes and very well adapted to CAD input surface definitions. It is based on BREP-surfaces of CATIA-step output and has the provisions to automatically check the surface-geometry for meshing criteria. Fig. 7 gives an overview of the Mesher-tool and an example of the grouping of configuration components.

For Navier-Stokes hybrid meshes prismatic layers are grown into the volume by an advancing front method along averaged normalized directions which are highly flexible and include the effects of surface curvature etc. Fig. 8 shows surface- and hybrid layer-grids for some typical configurations.

The SimServer-environment [5, 6] (Fig. 9) allows the chimera approach for movable controls, stores etc. Here the core aircraft is meshed without the independent control-surfaces which are performed in a similar but independent step. A merging procedure overlays the Chimera meshes according to a describing logic model file. The motion of the Chimera controls or stores then is actuated through a file which may hold all 6 degrees of freedom, while controls may be rotated around their hinge lines.

The SimServer performs the connection of different functions of flow simulations and their interdisciplinary integration. It automatically performs all pre-processing steps such as the partitioning of the dual meshes as well as the hole cutting treatment of Chimera type meshes as well as any relative motion of a Chimera mesh. The data extraction of forces and moments to be provided

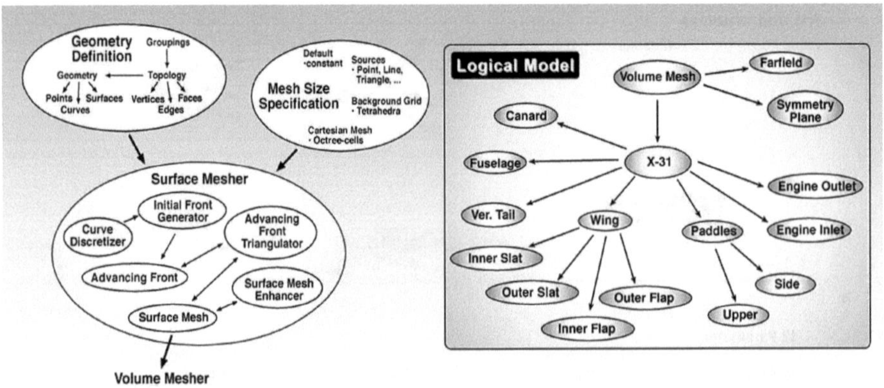

Fig. 7. Scheme of the Mesher tools and example of a configuration component grouping logic

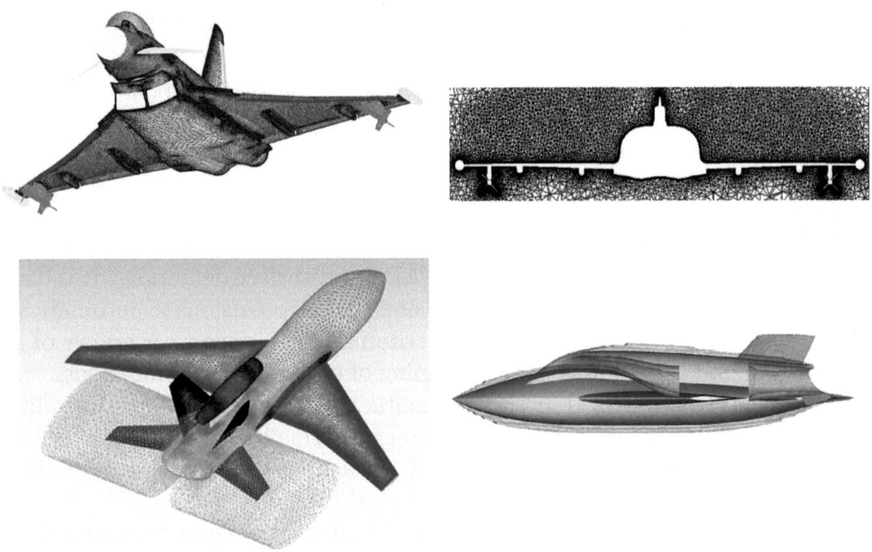

Fig. 8. Surface, hybrid, volume and chimera meshes on typcial aircraft configurations

for optimization objectives and flow investigations through visualisation are applied through the logical hierarchy model. The results also are prepared for post-processing and visualisation.

The data extraction of forces and moments for the evaluation of objectives and/or flow field or flow feature investigations through visualisation also in handled by the appropriate set of the SimServer by using the logical model hierarchy

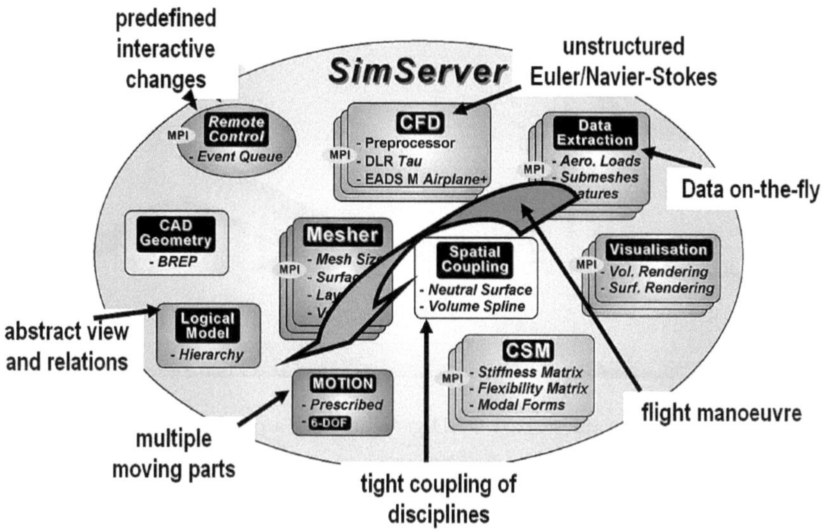

Fig. 9. SimServer incl. motion, aeroelastic effects. Details show CAD, meshing, post-processing and visualisation

and post-processing scripts which provide the output of interest. The results also are prepared for post-processing and visualisation.

The SimServer could be run in steady or unsteady mode according to the flow simulation problem posed. However, it also can be configured for unsteady simulations to perform manoeuvres [7] and store releases [6]. Here only the steady mode was used to adjust the angle of attack and the angle of incidence of the horizontal tail of an aircraft for autopilot purposes to be discussed later. Aeroelastic simulations are possible by the SimServers features to link aerodynamic pressure distributions to the nodes of FEM-models. With this approach steady state evaluations of configurations under inertia and aerodynamic loads may develop the corresponding displacement of the configurations and its parts such as wing bending and twisting.

The backbone of the flow simulation is the second order finite volume Euler and Navier-Stokes DLR-TAU method [3] which is integrated as flow solver into the SimServer environment. This elaborated system of numerical solver tools for RANS and URANS simulations works on structured and unstructured grids.

4 Aerodynamic Optimization of a Combat Aircraft by an Evolutionary Approach

The scenario of an aerodynamic multipoint optimization for a simplified configuration (Fig. 10) without engine-nacelle is depicted in Fig. 11. The simplified mission is characterized by a cruise and loiter phase, the latter of which is to

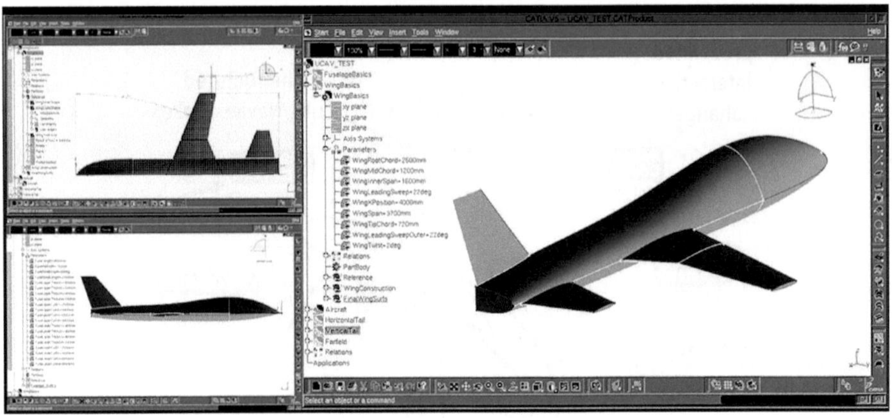

Fig. 10. Parameterized CAD-geometry for a combat aircraft

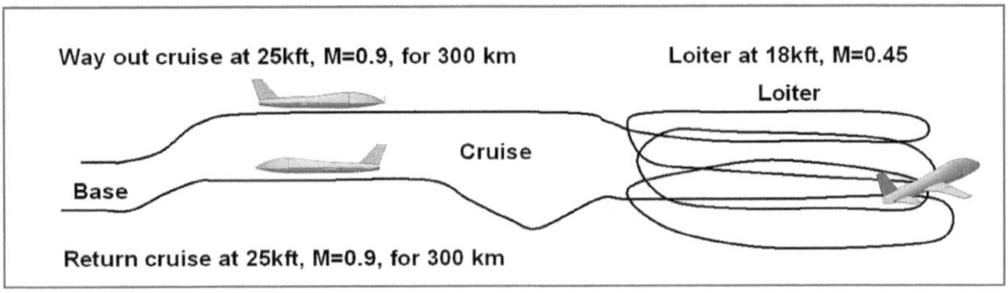

Fig. 11. Scenario for a multipoint optimization of a combat aircraft

be maximized. For simple stealth reasons the leading and trailing edge of the horizontal tail had to become parallel to the wing leading and trailing edges. Furthermore the vertical tail and the fuselage were left unchanged, while a single 'rubber' engine was assumed and not modelled in detail in the CAD. The wing and tail profiles were kept unchanged. The mission weight was assumed to be constant for the cruise and loiter phase at an average of ZFW + 0.5 full fuel weight. The design variables are wing parameters such as sweep, aspect ratio, taper ratio (inner, outer), inner wing kink span and possibly the position of the wing.

The fuel consumption was evaluated via the thrust equals drag condition, the drag being evaluated by the simulation:

$$T_{cruise} = D_{cruise}$$

The max rated sea level thrust as a function of elevation and Mach number is computed using data for a generic high-bypass turbofan engine

$$T_{0,\max} = f(M_{cruise}, h_{cruise}) T_{cruise}$$

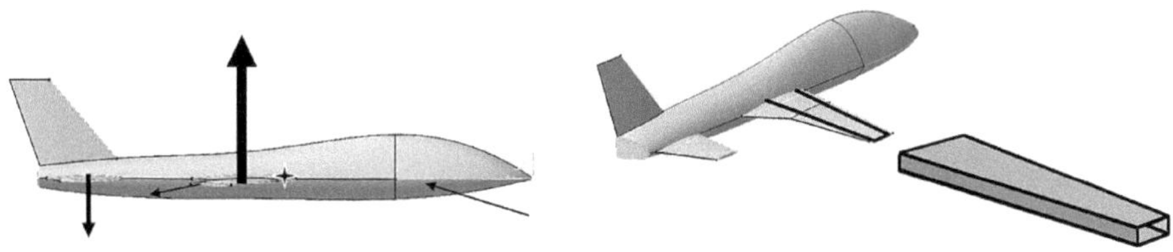

Fig. 12. Longitudinal trim of aircraft (left) and wing box model (right)

The thrust-specific fuel consumption, as a function of elevation, Mach number and throttle setting is found using generic engine data:

$$(TSFC)_{cruise} = g(M_{cruise}, h_{cruise}, \eta = 1), \quad \eta = \frac{T_{cruise}}{T_{0,\max} f^{-1}(M_{cruise}, h_{cruise})}$$

The loiter fuel consumption is found, using the same functions:

$$(TSFC)_{loiter} = g(M_{loiter}, h_{loiter}, \eta_{loiter}),$$
$$\eta_{loiter} = \frac{T_{loiter}}{T_{0,\max} f^{-1}(M_{loiter}, h_{loiter})}$$

The cruise fuel weight is found through:

$$W_{fuel,cruise} = (TSFC)_{cruise} T_{cruise} \frac{s_{cruise}}{v_{cruise}}$$

The loiter thrust required is assumed to depend on the glide ratio:

$$T_{loiter} = W_{0.5} \left(\frac{L}{D}\right)^{-1}_{loiter}$$

Finally, the loiter time is determined:

$$t_{loiter} = (W_{fuel} - W_{fuel,cruise}) \left[(TSFC)_{loiter} T_{loiter}\right]^{-1}$$

Both cruise and loiter had to be evaluated at trimmed conditions (Fig. 12 left). For a simplified version the assumption of linear aerodynamics at low angle-of-attack the angle-of-attack and angle-of-incidence for the horizontal tail trim-position where estimated according to Raymer [8].

$$C_{L_\alpha} = \frac{2\pi A}{2 + \sqrt{4 + \frac{A^2\beta^2}{\eta^2}\left(1 + \frac{\tan^2 \Lambda_{\max,t}}{\beta^2}\right)}} \left(\frac{S_{exposed}}{S_{ref}}\right)(F)$$
$$\beta^2 = 1 - M^2, \; F = 1.07(1 + d/b)^2$$

A very simple structural wing box model (Fig. 13 right) was used to estimate the wing weight. This box-model - being symmetric about its neutral axis - was

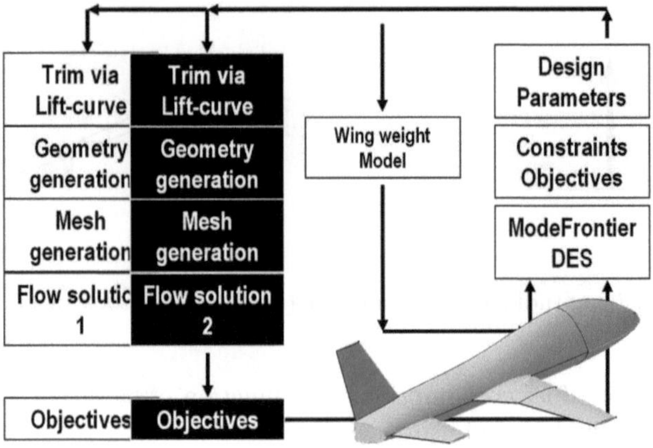

Fig. 13. Evolutionary optimization for cruise and loiter of an aircraft

sized according to the stresses in the upper and lower skin due to the bending moment,

$$M_b = 2\sigma \frac{A}{2}\frac{t}{2} = \sigma A \frac{t}{2}$$

which is produced by the elliptic lift distribution

$$M_b(y) = \int_y^{\frac{b}{2}} l(y) y dy = \frac{L_{total}}{b} \int_y^{\frac{b}{2}} \frac{1}{\left(\frac{L}{b}\right)_{avg}} y dy$$

on a linear chord distributed wing

$$c(y) = \frac{S}{b}\left(\frac{2}{1+\lambda}\right)(1-\eta(1-\lambda))$$

with the weight of its material to be evaluated by

$$W_b = 2\int_0^{\frac{b}{2}} \rho A dy = \frac{4\rho}{\sigma[t/c]} \int_0^{\frac{b}{2}} \frac{M_b}{c} dy$$

the wing weight becomes

$$W_b \propto \frac{\rho b^3 n_{\text{ult}} \sqrt{(ZFW)(TOW)}(1+2\lambda)}{\sigma S[t/c](1+\lambda)\cos^2 \Lambda}$$

with ZFW for the zero fuel weight, TOW for take off weight, b representing the half span, t/c the maximum thickness, λ the taper ratio, Λ the sweep, S the wing surface, n the load factor and ρ the material density.

The flow chart of the evolutionary optimization for maximum loiter time of an aircraft under two point design conditions is shown in Fig. 13. Design variables

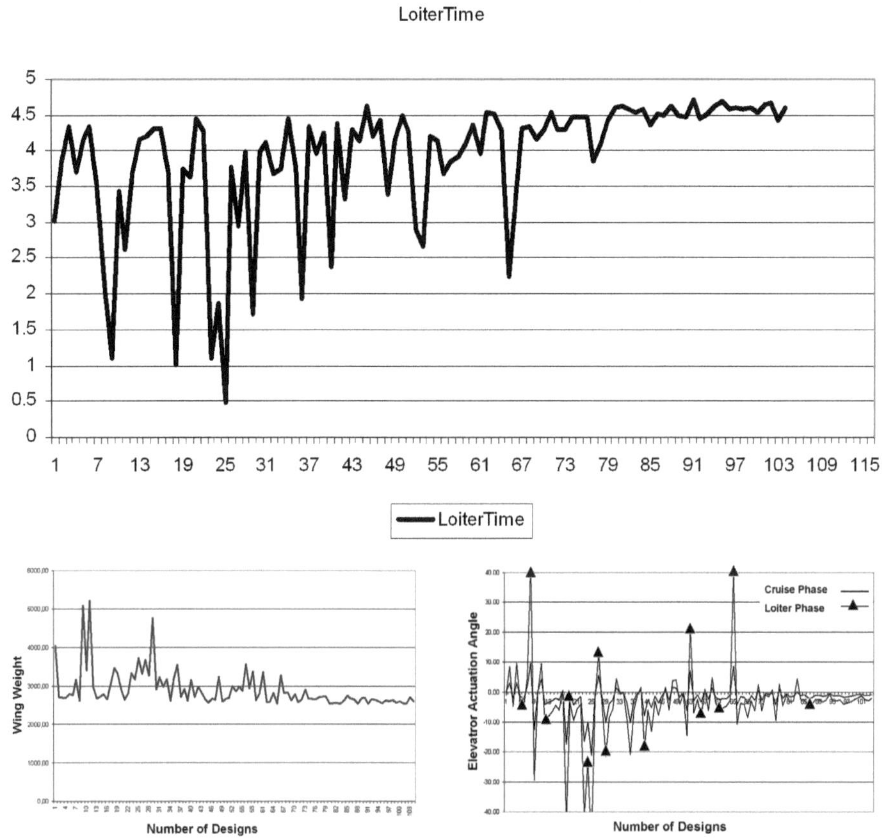

Fig. 14. Loiter time, wing weight and elevator of the optimization with fixed wing position

are wing parameters such as sweep, aspect ratio, taper ratio (inner, outer), inner wing span and possibly the position of the wing.

Two cases, one with fixed wing position and the other with free position were tested. The most important constraints are the trimmed condition and geometric relations of the horizontal tail shape with the wing shape due to RCS-assumptions. The design evolution is shown in Fig. 14 with the maximum loiter time around 4.5 hours, with the development of the wing weight close to 2500 Newton and the elevator trim-angle, the latter being minimized for low trim drag. The shape of the best and worst designs is shown in Fig. 15. The best shape for the fixed wing position conditions is a lambda planform, while highly swept wings suffer from high trim-drag of the horizontal tail visualized in a high pressure (red) which counters the pitch down moment. As soon as the wing position becomes a free parameter this situation is redeemed and the unfavorable lift-trim distribution is resolved. The best design with free wing position is depicted

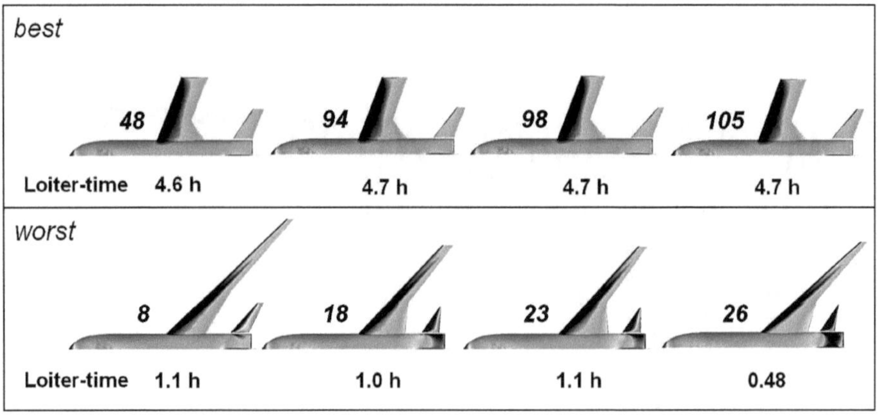

Fig. 15. Some worst and best design of the evolutionary optimization of loiter time

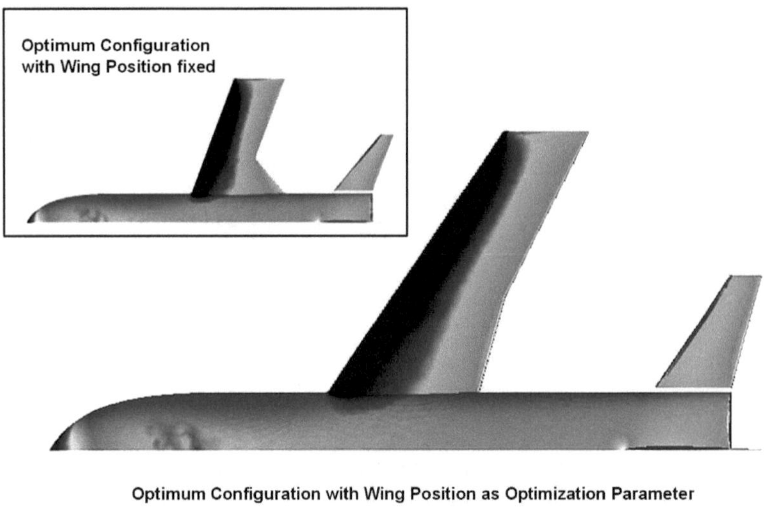

Fig. 16. Best design with free wing position (inset best wing with fixed wing position)

in Fig. 16. Now, located forward, the aspect ratio 5, 35°sweep, almost straight wing pushes the loiter time to almost 5 hours.

5 Optimization of an Aircraft Wing by a SIMPLEX Approach

The SIMPLEX-methodology was used for the optimization of an isolated wing, the very best of which to be integrated into a full aircraft design optimization later on. The effect of different design conditions was checked together with

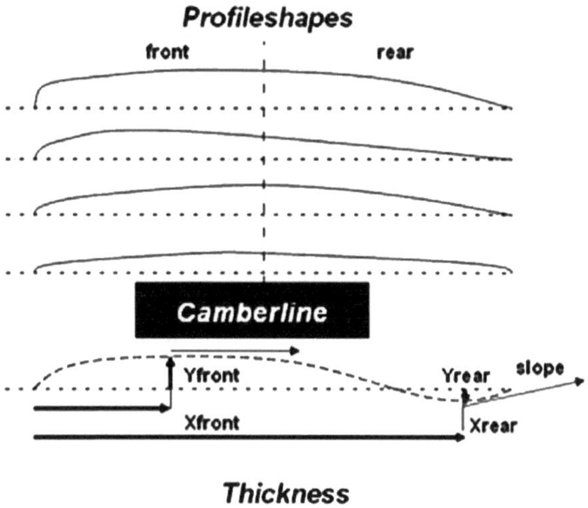

Fig. 17. Airfoil shapes and parametric camber

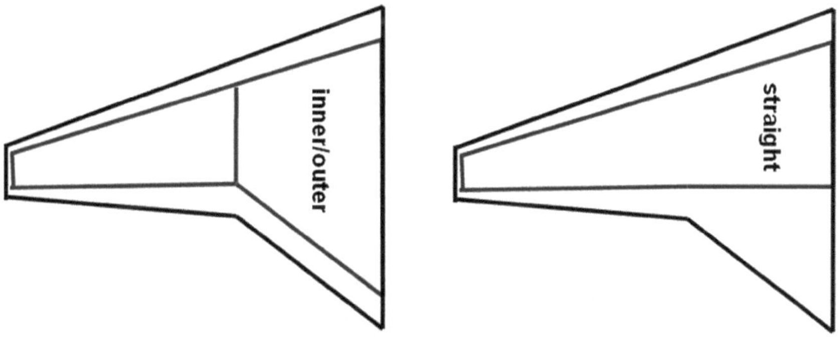

Fig. 18. Simplified wing box models

gathering some experience on the impact of different structural topologies. As before, the span, the leading edge sweep, the tip and mid chord as well as the wing twist were selected as independent variables. An airfoil optmization was added by the resultant shape of a mix of 4 characteristic profiles, its camber shape and profile thickness (Fig. 17).

The objective of the wing optimization was maximum range for a total aircraft weight, payload and engine conditions. Two different configurations demanding a light condition $C_L = 0.23$ and a heavy condition $C_L = 0.5$ at Mach 0.77. Upper and lower limits of the pitching moment were used as additional constraints.

The weight estimation was done as described for the evolutionary aircraft optimization. To check effects of wing box layouts two different approaches were

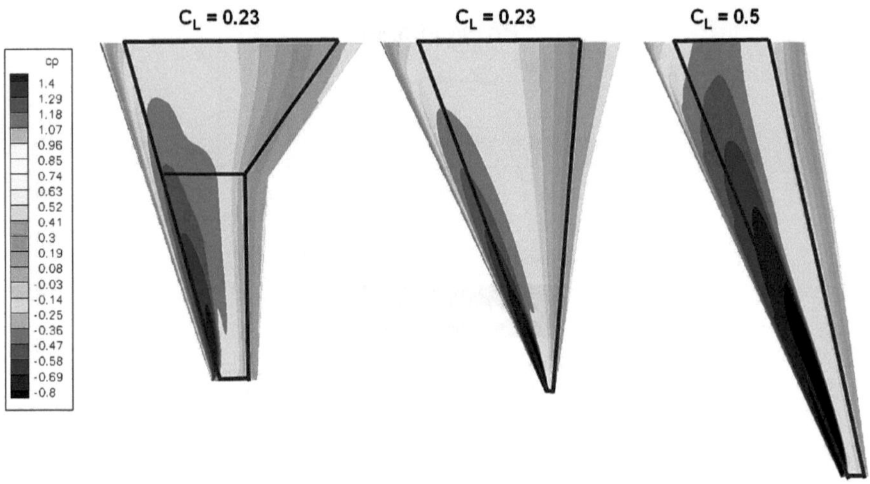

Fig. 19. Wing pressure distribution with different design conditions and wing box models

simulated. Again one simulated a straight wing box, while an other design divided the box into an inner and outer panel (Fig. 18). Optimizations have been performed for the combinations of the different load and wing-box type conditions. The objective of the optimization was a maximum range for a total aircraft with weight, payload and engine conditions. Two different configurations were considered at Mach 0.77. The first at $C_L = 0.23$ for a so called baseline version and the second much heavier aircraft at $C_L = 0.5$. Upper and lower limits of the pitching moment were used as additional constraints.

Fig. 19 shows optimized maximum range configurations and their surface pressure distribution of the low load case with both wing box models as well as a higher wing load with the straight wing box. As expected the lower lift demands reduced the aspect ratio to minimize the wing weight by a short span and rather thick and long root chord geometries. For high lift and aileron considerations the rather small taper ratio would need a revision – not included here. The development of the lambda wing shape with the inner/outer structural topology is depending on a simple averaging process of the inner and outer wing panel taper ratio to comply with the wing box weight formula used here. The higher lift, straight wing box case is dominated by aerodynamic drag which is reduced by a significant increase in span rspt. aspect ratio. Fig. 20 depicts the development of the range for the SIMPLEX approach for three different wings.

The $C_L = 0.5$, straight wing box design also shows a distinct development of twist to reduce drag by thriving for a more elliptic lift distribution. The best design of the optimizations shown here developed -1.75°of twist, while the span still increases (Fig. 21). Further design cycles would increase this tendency. The low lift cases showed a maximum twist of -0.6°only.

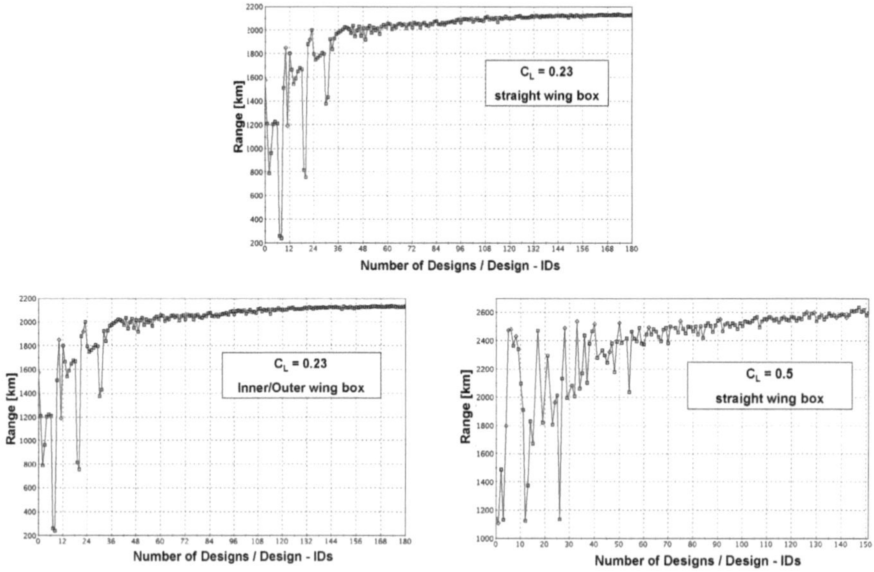

Fig. 20. SIMPLEX-optimizations for three wings at Mach = 0.77

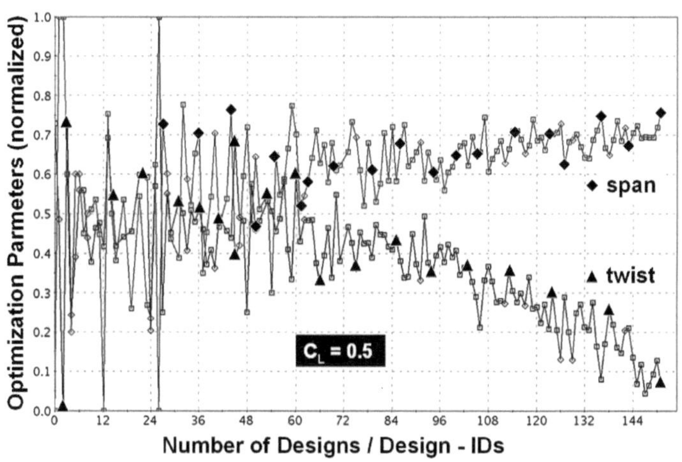

Fig. 21. Development of span and twist for the $C_L = 0.5$ wing max range design at Mach = 0.77

For the full aircraft the wing was developed along similar lines. For simplicity the airfoil was fixed to the RAE2822, while the planform parameters were the same as before. The angle of attack was chosen as an additional parameter to allow for a high aerodynamic efficiency via good L/D values. Range was chosen

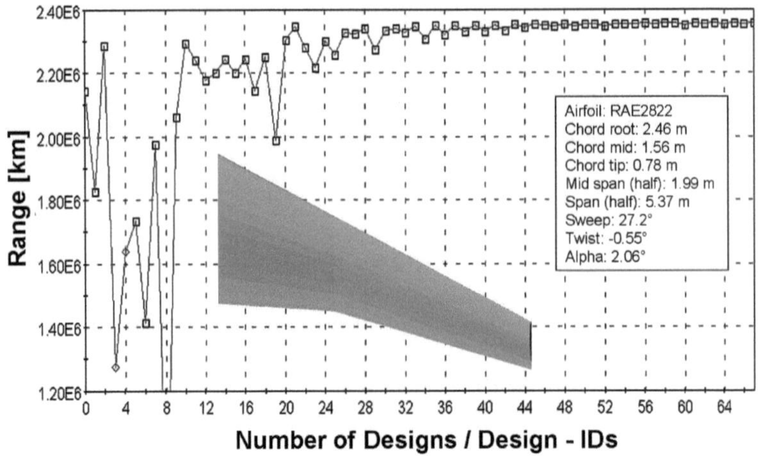

Fig. 22. Optimization for the full aircraft wing design at Mach = 0.77

as objective, however based on the Breguet-equation. The final wing shown in Fig. 22 was than introduced into the full aircraft model, where its position was optimized to reduce trim-drag. It should be noted that no optimization ever touched the geometrical limits set for a save CAD modelling.

6 Optimization of a Combat Aircraft by a MOGA Genetic Approach

For a full aircraft (Fig. 23) optimization the isolated wing was developed along the same lines as above. To allow for a high aerodynamic efficiency via good L/D values, α was chosen as an additional parameter, the objective range was evaluated via the Breguet-equation. The optimum angle of attack of the isolated wing (Fig. 22) was applied as its angle of incidence in the complete aircraft configuration(Fig. 18). The wing position along the center fuselage became a configurational parameter and was allowed to vary for more than 700 mm.

The location of the engine installation and the fuselage center section shape became additional configurational design parameters, while the incidence of the elevator as well as the angle-of-attack had to provide steady longitudinal flight conditions. The engine position could move in between 62% and 75% of the fuselage length. The upper part of the two mid fuselage cross section could be modified in between 350 mm to 650 mm in to accommodate more fuel volume.

Here, an autopilot functionality which guarantees level flight ($C_L = weight$, $C_m = 0.0$) operated on the basis of full RANS-simulations. Each autopilot trim requires at least three evaluations to obtain level flight angle-of-attack and elevator trim-incidence. The elevator trim- or autopilot evaluation-position was introduced via a chimera-mesh representation. For a typical autopilot trim the

Fig. 23. Full aircraft configuration

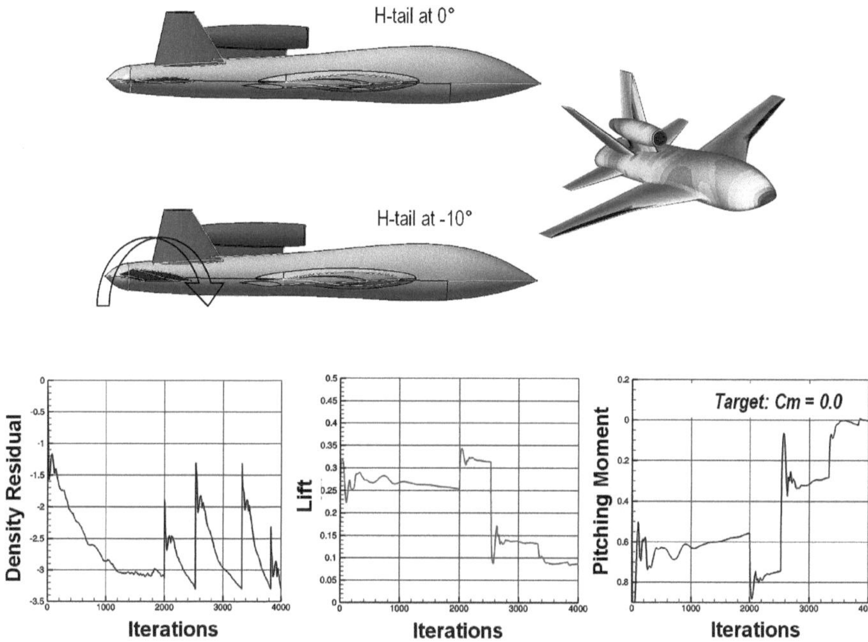

Fig. 24. Autopilot for level flight trim conditions

development of lift and pitching moment is shown in Fig. 24. While the begin of the convergence plots show the start of the first evaluation, the first interruptions in the convergence of C_L and C_m indicate the evaluation of the second evaluation to produce a lift and pitching moment derivative. The third break shows a second correction, while the last one finally results in the level flight conditions with $C_m = 0.0$.

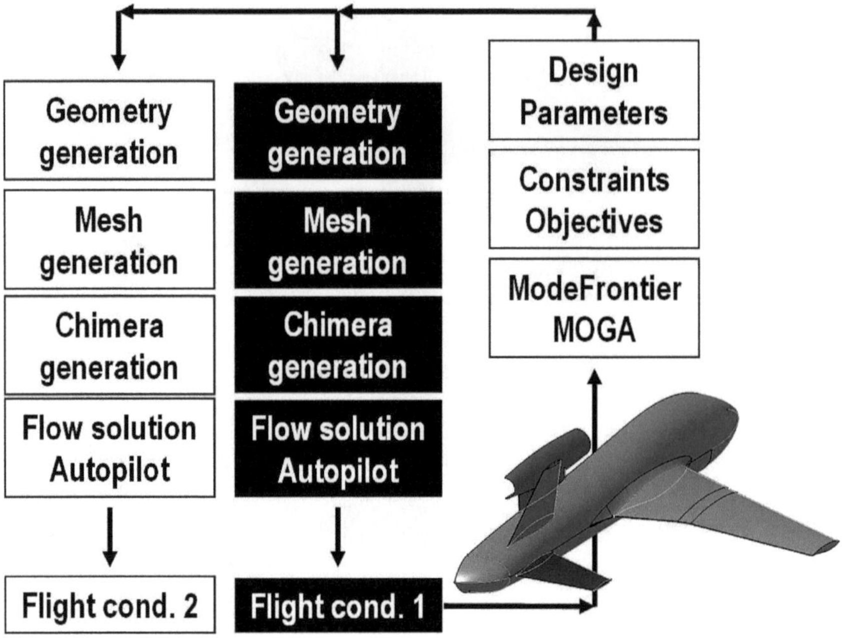

Fig. 25. ModeFrontier genetic algorithm MOGA optimization process

Fig. 25 depicts the optimization process as implemented in the ModeFrontier MOGA approach. Here the different flight conditions together with the autopilot trim functionality are shown.

By nature genetic algorithms do not show a development of objectives and parameters such gradient based or evolutionary methods. The recombination of parts of the most fit designs produces a random set of designs (Fig. 26), where the upper-most bound indicate the best designs. Fig. 26's blue and green stars mark the best and the worst design rspt.

A comparison of the worst and best designs surface pressure distribution is shown in Fig. 27. The forward position of the engine has multiple benefits for the design. Firstly it avoid the shock prone velocities in between the engine and the vertical tails, secondly it pushes the engine's centre of gravity closer to the overall c.g., improving the trim situation. Together with a smoother fuselage it also reduces the recompression effects ahead of the engine. The smooth and thicker fuselage provides some extra fuel and shows a much more favourable interference in between wing and fuselage. This allows for a much lower angle of attack which reduces the induced drag. An inspection of the elevator incidence shows a much reduced angle of incidence thereby reducing the trim-drag. The table gives a survey of the corresponding parameters and angles.

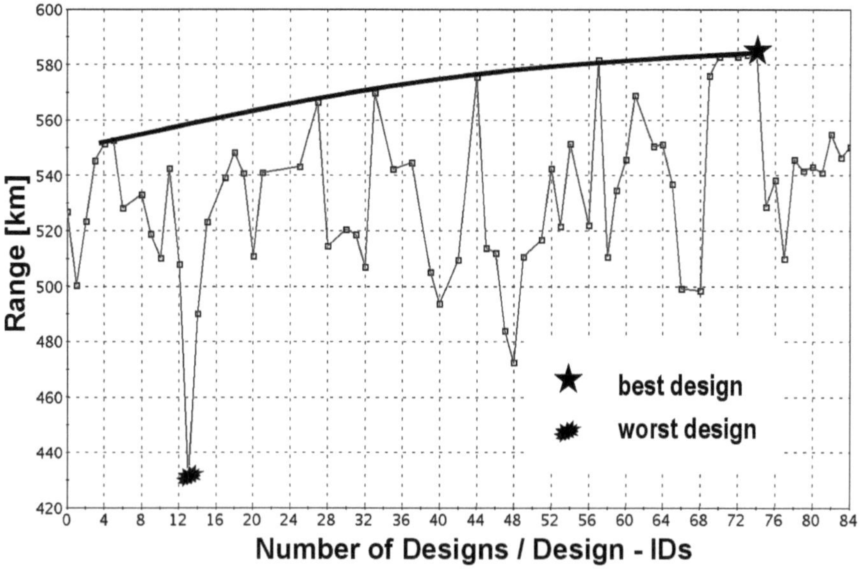

Fig. 26. Optimization of range by the genetic algorithm MOGA

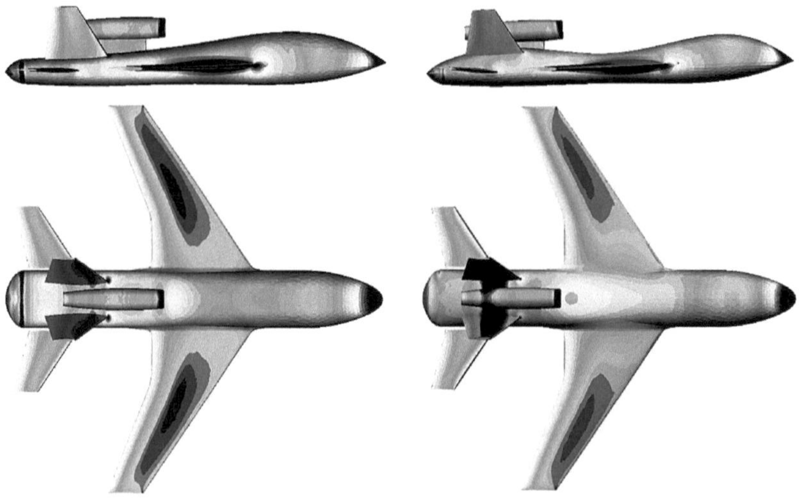

Fig. 27. Best and worst aircraft design at Mach = 0.77

Table 1. Design parameters, angles of attack and trim-angles

	best	worst
Wing position	2727.0 mm	2792.0 mm
Fuse thick 2	530.0 mm	350.5 mm
Fuse thick 3	401.5 mm	513.5 mm
Engine pylon	66.2%	71.3%
angle-of-attack	1.38°	-2.09°
elevator incidence	-2.25°	-6.70°

7 Summary

High fidelity simulation and optimization processes for aerodynamically driven optimization problems have been presented. The capability was demonstrated to set-up complex optimization processes involving advanced flow simulation tools for complete aircraft optimization problems by the successful performance of aircraft shape optimizations including plan form, wing and profile design issues for a mission driven optimization problem. Also taken into account are aircraft structural weights and engine performance issues. This process is open for the integration of flight dynamical as well as structural multidisciplinary considerations. The multidiscplinary capabilities are provided by an integrated simulation server environment. Evidence for a quick set-up of an optimization problem within the *modeFRONTIER* optimization system and process control environment has been shown.

It is obvious that most of the developments available will be used within an operational context after necessary adaptations and further validation exercises. In summary the capabilities and the competitiveness within flight physics has gained a lot from the MEGADESIGN project.

References

1. Rieger, H., Leicher, S., Fritz, W., Hitzel, S., Fornasier, L., Tremel, U., Weide, E.v.d.: Perspektiven der aerodynamischen Simulation in der Kampfflugzeugentwicklung. DGLR-2002-014, DGLR-Jahrestagung, Stuttgart (2002)
2. Rieger, H.: Experiences in Aerodynamic Shape Optimization at EADS-Military Aircraft Systems. In: 6th Int. Congress on Industrial and Applied Mathematics, Zurich, July 16-20 (2007)
3. Gerhold, T., Friedrich, O., Evans, J., Galle, M.: Calculation of Complex Three-Dimensional Configurations Employing the DLR TAU-Code. AIAA 97-0167 (1997)
4. modeFrontier – A multi-Objective Optimization and Design Environment, http://www.esteco.com

5. Tremel, U., Deister, F., Sørensen, K., Rieger, H., Weatherhill, N.: The SimServer – A Parallel Multidisciplinary Simulation Environment. DGLR-2003-099, München, Germany (2003)
6. Tremel, U., Hitzel, St., Sørensen, K., Weatherhill, N.: JDAM Store Separation from an F/A-18C – An Application of the Multidisciplinary SimServer System. In: AIAA 23rd Applied Aerodynamics Conference 2005, Toronto, Canada (2005)
7. Sørensen, K., Tremel, U., Rieger, H., Hitzel, St.: Simulation of a Manoeuvering Fighter Aircraft with the Unstructured Chimera Approach. AIAA – 2007 -123, Reno, NV (2007)
8. Raymer, D.: Aircraft Design: A Conceptual Approach. AIAA Education Series (1989)

Flexible Wing Optimisation Based on Shapes and Structures

Holger Barnewitz

Airbus, Airbus-Allee 1, D-28199 Bremen
holger.barnewitz@airbus.com

Summary. A multi-disciplinary optimisation (MDO) process chain for shape optimisation of a wing including the static deformation has been developed. The objective function, which should be minimized, is equivalent to the total aerodynamic drag force of an aircraft in stationary horizontal flight.

The CATIA V5 parametric model of the wing is controlled by the optimiser using an external CATIA-DesignTable. For four predefined wing sections there are parameters to control the thickness, camber, and twist independently giving in total 12 design parameters for the outer shape while maintaining a fixed wing planform. Additionally, two structure design parameters control the relative thickness change of the wing front and rear spars in combination with the upper and lower sheet thicknesses of the metallic wing box. The stiffness and the weight of the wing depend on these structural parameters.

An equivalent beam stick model is then automatically generated for any change of the wing box geometry. Iterative coupling between aerodynamic forces and weight forces (weight of wing- box, fuel, engines, payload, and engine thrust forces are taken into account) and equivalent beam stick bending and twisting is done until a steady state solution is obtained. Here, 12 coupling iterations are carried out using a Volume Spline technique to deform the CFD mesh according to the resulting equivalent beam stick deformation. TAU has been used as Navier-Stokes (RANS) solver on a structured mesh.

The optimiser chosen for this task is a Downhill Simplex method, which is very robust, does not require gradients, and performs well even with objective function evaluations that are subject to random noise (non-smooth). The reduction of the total drag force (objective function) and the total aircraft mass decreases due to wing box mass optimisation is shown while at the same time the aerodynamic performance L/D improves.

1 Introduction

Multi-Disciplinary Optimisation (MDO) looks to find an optimum for an engineering problem where several disciplines are interdependently coupled. The optimum of the simultaneous problem is superior to the design found by optimizing each discipline sequentially, since it can exploit the interactions and trades between the disciplines.

N. Kroll et al. (Eds.): MEGADESIGN and MegaOpt, NNFM 107, pp. 287–305.
springerlink.com

A typical application of MDO in the field of transport aircraft design is the simultaneous aerodynamic and structural optimisation of a wing. The wing is defined in terms of some geometric variables, and the effects on aerodynamics and structural behaviour are determined as the geometry is varied. Results are assessed versus some defined value of merit, and with respect to constraints which can be based on performance, safety, operability, or practicality.

The task given was to optimise a wing with respect to aerodynamics, structures, and performance under considerations of static aero-elastic effects as specified in [1]. In such a case there is an iterative loop that may take several iterations to converge before subsequent disciplines (in this case, performance) are executed. The study involves the sizing of the wing box skins and spars to minimum weight subject to static aero-elastic requirements. To characterize wing weight as a function of wing geometry and wing structure a shape parametrization strategy for geometry generation is a prerequisite for performing aerodynamic shape optimisation. Additionally the parametrization and the structure model for the wing must be available and re-generated depending on the parametrically generated outer wing shape and inner wing structure.

1.1 MDO Process Chain

The considered MDO process chain for shape optimisation of a wing including the static deformation is shown in Fig. 1. The objective function here is:

$$\mathrm{Obj} = W_{\mathrm{A/C}} \times C_{\mathrm{D}}/C_{\mathrm{L}}, \tag{1}$$

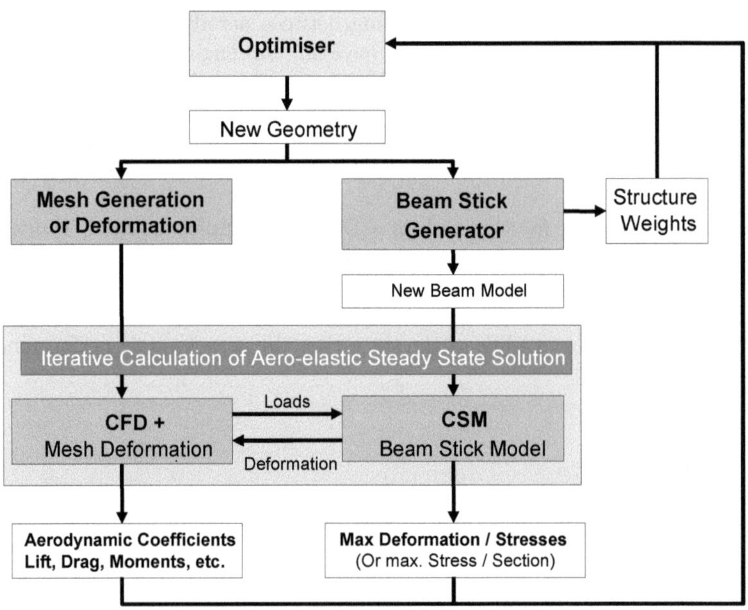

Fig. 1. Multi-disciplinary optimisation chain for wing shape and structure weight

where $W_{A/C}$ is the total weight of the aircraft, C_D is the overall aerodynamic drag coefficient, and C_L is the aerodynamic lift coefficient of the aircraft. The objective, thrust, is equivalent to the total aerodynamic drag force in stationary horizontal flight, which should be minimised.

2 Optimisation Process Chain Tools

2.1 CAD Geometry Handling

Starting point for the wing geometry is the pre-defined wind tunnel test wing SFB-401 given in CAD file format (IGES) [2], [3]. It has no parametric shape properties. According to the optimisation parameters (Sect. 3.3) and constraints to be used here a new parametric CATIA V5 model has been developed referring to the original geometry. It facilitates a fully automatic re-generation of the wing surface geometry. The wing planform is considered as fixed while airfoil twist, thickness, and camber can be changed parametrically at defined span-wise wing sections.

The CAD system CATIA V5 is run in batch mode on a Windows PC using a CATScript and a DesignTable to create a new geometry. The connection between the Unix compute cluster and Windows is handled by the Secure Shell (ssh) protocol to copy files and execute CATIA V5.

2.2 Wing Structural Model

Based on the outer shape of the generated wing geometry a simplified Finite Element Method (FEM) structural model for the wing box is generated with tools developed by RWTH-Aachen (FEAFA, ACM, etc.) [8].

Derived from this wing box structure an equivalent beam stick model is determined which quickly calculates the wing bending and torsion according to given forces and moments. Here, aerodynamic forces and moments, engine forces like thrust, and weight forces like fuel and engines as well as structure mass forces from the wing box and the fuselage (payload) are taken into account.

The beam stick model generator BEAMPREP calculates a FEM beam model from wing sections, from front and rear spars with equivalent metal sheet thicknesses. The number of ribs and spars is representative of an Airbus-type aircraft wing (see Fig. 3). Additionally, wing box stresses per unit force are calculated to respect stress constraints. Finally, the weight of the wing structure is calculated as part of the objective function.

FEAFA (Finite Element Analysis for Aero-elasticity, [8]) calculates FEM matrices which are used within the ACM (Aero-elastic Coupling Module) to iteratively find the static wing deformation.

2.3 Flow Solver

For the transonic aerodynamics the Reynolds Averaged Navier-Stokes, unstructured mesh flow solver Tau [6] is used [4], [5]. To ensure appropriate accuracy while using a low number of mesh points for the time consuming optimisation a structured mesh has been converted to the unstructured Tau format. For simple topologies like this wing without any complex attachments (pylon, nacelles, ailerons, flaps, etc.) this is proved to be highly efficient.

2.4 CFD/CSM Static Coupling

To calculate the static wing deformation during stationary horizontal flight a mesh deformation tool from DLR [7] is used which is fed by the beam stick deformation field. The strong interaction between aerodynamics (CFD) and structure (CSM) is controlled by the Aerodynamic Coupling Module (ACM) which also interpolates forces and moments between CFD surface mesh points and beam the stick model axis. For details refer to [8].

Iterative coupling between aerodynamic forces, weight forces, equivalent beam stick bending and twisting is done until a steady state solution is obtained.

Here, 12 coupling steps are carried out to achieve a converged solution. To initialize the CFD solution 200 CFD solver iterations are performed before starting the static CFD/CSM coupling steps where only 75 CFD iterations were found to be sufficient. This procedure has been adjusted manually for this test case before carrying out the real overall optimisation.

2.5 CAD Shape to Mesh Deformation Connector

The deformation tool is also applied to reflect the changed shape design generated with the parametric CATIA V5 model. For every varied CAD geometry the initial CFD mesh is deformed to reflect the changed geometry. The advantage of using mesh deformation is manifold:

- It avoids the problem of *numerical noise* for the calculated aerodynamic coefficients which might occur if new meshes are created for slightly changed geometries. The reason for noise is the change of mesh topology. Mesh deformation conserves the topology and small geometry variations produce small mesh deformations in a continuous way.
- A so called *restart* capability of the flow solver allows to start from a beforehand calculated flow solution to save computing time.
- Usually, deformation of an unstructured CFD mesh is faster than re-generate a new mesh, thus also saves computing time.

Extra effort has been undertaken by the author to develop a CAD to mesh deformation connector. The deformation field induced by the parametric CAD change is uniquely determined for the fixed topology by using an appropriate wire-frame representation of the surface. This wire-frame has the same number of curves for a fixed topology and offers the determination of a discrete deformation vector field. The idea is based on the comparison of uniquely defined points on discretised CAD curves. Varying CAD intersection curves can be handled with this technique as well.

Interpolations between the discrete deformation vector field and CFD mesh points are calculated using Radial Basis Functions methods [7].

3 Wing Optimisation Problem

3.1 Forces at Stationary Horizontal Flight

The optimisation problem consists of two disciplines, the aerodynamic performance described by the drag/lift forces ratio, and the structural weight of the wing. Both are coupled by the required lift coefficient C_L expressed in the following relations:

$$F_{\text{Thrust}}/F_{\text{Weight}} = F_{\text{Drag}}/F_{\text{Lift}} = C_D/C_L \tag{2}$$

In stationary horizontal flight lift equals weight and drag equals thrust with opposite signs, so the thrust/weight ratio equals the C_D/C_L ratio of aerodynamic coefficients.

To minimise the aircraft fuel consumption implies minimising the total thrust force F_{Thrust}. This yields the Objective Function:

$$\text{Obj} = F_{\text{Thrust}} = F_{\text{Weight}} \times C_D/C_L \qquad [\text{N}] \tag{3}$$

The complete aircraft mass $M_{\text{A/C}}$ is the sum of the wing structure mass M_w (considered as variable) and the rest mass M_0 (considered as fixed). For the current example M_0 is composed of 100 t structure mass (without wing), 40 t payload, 106 t fuel mass, and 20 t mass for four engines:

$$M_{\text{A/C}} = M_w + M_0 \tag{4}$$

$$M_0 = 100\,\text{t} + 40\,\text{t} + 106\,\text{t} + 20\,\text{t} = 266\,\text{t} \tag{5}$$

With these masses the target lift coefficient C_L for the complete aircraft and the wing lift coefficient $C_{L,\text{wing}}$ which depends on the wing mass M_w can be calculated:

$$C_L = F_{\text{Lift}}/(q * A) = n * g * (M_0 + M_w)/(q * A) \tag{6}$$

$$C_{L,\text{wing}} = (C_L - C_{L,\text{trim}})/f_{\text{rest}} \tag{7}$$

with:
- n : load factor, set to 1.0
- g : acceleration of gravity
- q : dynamic pressure
- f_{rest} : factor accounting for lift of fuselage, set to 1.14
- $C_{L,\text{trim}}$: change of lift due to HTP trimming, set to 0.01
- A : reference area for aerodynamic coefficients
- F_{Lift} : aerodynamic force (vertical component)

With every modification of the structure parameters a change of the wing structure mass requires a new target lift coefficient $C_{L,\text{wing}}$:

$$C_{L,\text{wing}}(M_w, \alpha) = \frac{1}{f_{\text{rest}}}\left(\frac{(M_0 + M_w) * n * g}{(q * A)} - C_{L,\text{trim}}\right) \tag{8}$$

The required angle of incidence α is determined iteratively within the flow solver Tau by using its target lift calculation mode. During optimisation the required lift coefficient $C_{L,\text{wing}}$ is adapted depending on the altered wing structure mass M_w.

3.2 Flow Conditions

For simplicity we assume that the aircraft flies at constant altitude of 11 km with design Mach number set to Ma = 0.82 and Reynold's length of 7 m. This yields for the flight conditions at Standard Atmosphere:

$$\begin{aligned} \text{Dynamic pressure} \quad & q = 10684\,\mathrm{N/m^2} \\ \text{Static pressure} \quad & p = 22700\,\mathrm{N/m^2} \\ \text{Density} \quad & \rho = 0.3648\,\mathrm{kg/m^3} \\ \text{Temperature} \quad & T = 216.74\,\mathrm{K} \\ \text{Reynolds number} \quad & Re = 43.45 \ * \ 10^6 \end{aligned}$$

With a wing reference area of 361.6 m^2 (wing span is 58 m) and half of maximum fuel mass of 53 t which results in total aircraft mass of 243.5 t the wing lift coefficient $C_{L,\mathrm{wing}}$ is approximately 0.534 for the initial wing structural model and its associated weights.

3.3 Design Parameters for Aerodynamic Shape and Wing Box Structure

A CATIA V5 parametric model of the wing is controlled by the optimiser using an external CATIA-DesignTable (ASCII format) , where all relevant shape parameters for the wing are listed. The shape of airfoils at four predefined wing sections (root, kink1, kink2, and tip section) can be changed parametrically to control the thickness, camber, and twist independently, giving in total 12 design parameters for the outer shape. In principle this can be extended to an arbitrary number of design variables. The wing planform is fixed, so the span, sweep angle, taper ratio, and wing area remain unchanged.

Additionally, two structure design parameters control the *relative thickness change* of the wing front and rear spars in combination with the upper and lower sheet thicknesses d_i of the wing box (see Fig. 2 and 3). The wing root section thicknesses are scaled with the first parameter and the wing tip thicknesses with the second structure parameter. A linear blending between the defining sections along the wing span varies these thickness factors. The stiffness and the weight of the wing are depending on these structure parameters. All components of the wing structure are made of isotropic material, e. g. aluminium.

3.4 Optimisation Algorithm

The Downhill Simplex Optimiser method from Nelder/Mead [9], [10] is used to find the minimum of the objective function (3). It is commonly used to solve nonlinear optimisation problems with a moderate number of design variables (ca. 10 to 50 variables).

The algorithm is gradient free – no derivative is required, the direction of search is determined using $n + 1$ objective function evaluations for n design variables. Experience proofs its robustness allowing 20% of failures of all analysis steps without affecting the optimiser convergence. Furthermore it performs

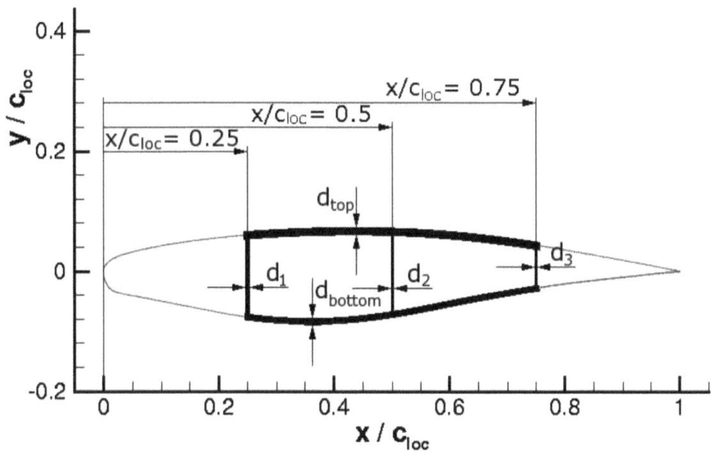

Fig. 2. Wing box section with equivalent thicknesses d_i for spars and upper/lower sides

well where objective function evaluations are subject to random noise (non-smooth) which is often the case for solutions of discretised RANS equations on unstructured meshes (refer to Sect. 2.5). However, the convergence rate is linear dependent on the number of design variables, e.g. for 10 design variables approximately 200 analysis steps are necessary to find the optimum.

3.5 Handling of Optimisation Constraints

Constraints for the optimisation are handled using penalty functions added to the objective function (3). The penalty method has a number of advantages: It is easy to use and it allows inexact constraints, i. e. constraints are not necessary to be fulfilled exactly. However, the factors determining the amount of penalty have to be chosen carefully, otherwise the optimiser may be driven by constraints only or does not fulfil constraints at all.

The following additional constraints have to be fulfilled to cover conditions at off-design points:

1. Wing bending maximum should be less than 1.3 m at the wing tip (half span is 29 m). This is to avoid a too flexible wing with unrealistic bending or possibly unstable behaviour like flutter. The maximum bending is calculated from deformation vectors calculated for the CFD surface mesh at the end of the iterative CFD/CSM coupling loops.
2. Wing twist change between root section and tip section should stay below 10 degrees.
3. Airfoil thickness should not exceed 15% at wing root section.

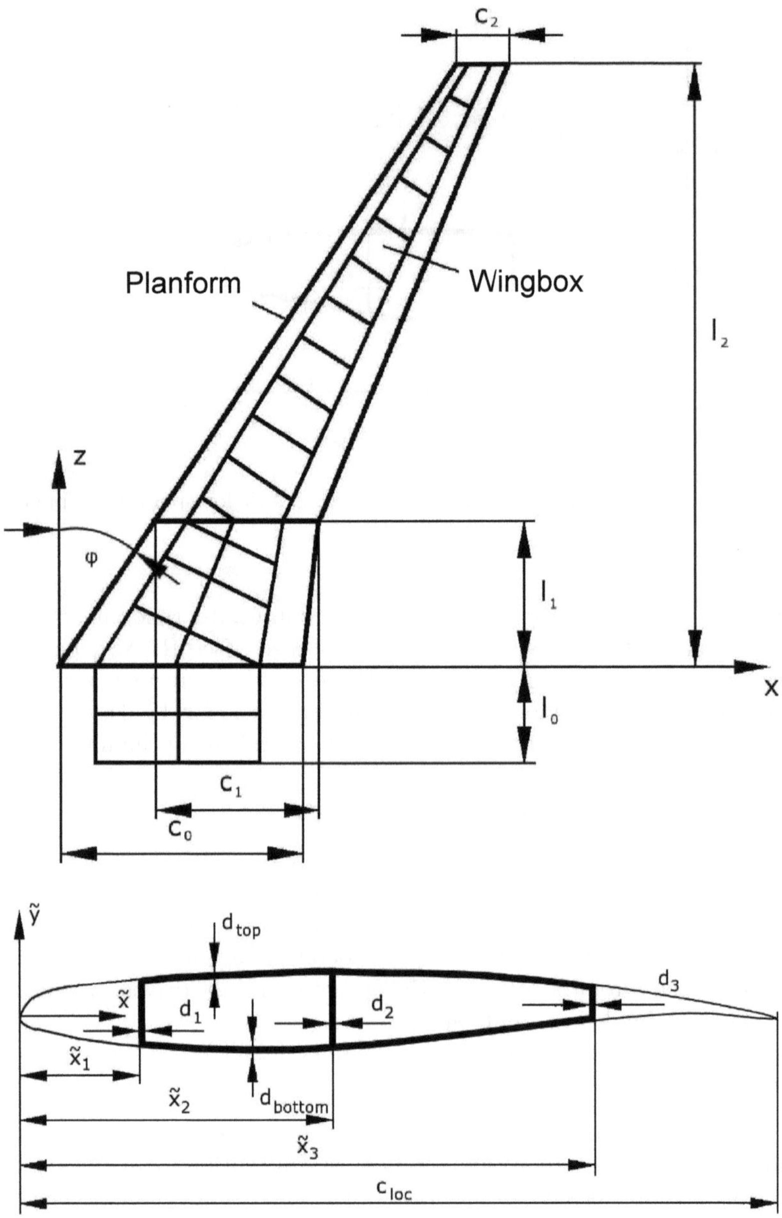

Fig. 3. Wing planform and wingbox section

4. The wing leading edge radius should be limited to avoid loss of low-speed performance. This is automatically fulfilled by the geometry parametrisation as there is no parameter to change to it.

It should be mentioned that such constraints might better be treated via a multi-point design optimisation approach. But this would have violated the test approach agreed in the MegaDesign project.

In order to accelerate and stabilise the convergence of the optimisation process the following techniques have been applied to overcome difficulties which were detected in a first test optimisation described in Sect. 4.1:

1. Aerodynamic coefficients are averaged during the last phase of the CFD solver iterations to minimize the effects of unstable or not fully converged CFD/CSM solver solutions and to smooth the objective function.
2. The standard deviation for this averaging for the aerodynamic coefficients is taken into account and added as a penalty to the objective function to avoid possibly unstable or oscillating solutions.

4 Optimisation Results

4.1 Test Optimisation

Some test optimisations have been performed with a reduced number of shape design variables to check the penalty factors β_i. According to a preliminary specification [1] the first test objective function was set to be (compare with (3)):

$$\text{Obj} = M_w * C_D/C_L + \sum_{i=1}^{m} \beta_i \, p_i \qquad \text{[kg]}. \tag{9}$$

The optimisation result was not satisfactory because too much emphasis is laid on the wing box mass M_w. The aerodynamic part C_D/C_L was not properly taken into account because the sensitivity for this part is too low, see Fig. 4. Additionally the constraint for maximum bending ($< 1.3\,\text{m}$ at wing tip) was not satisfied and lead to a very flexible wing with very low mass. A detailed sensitivity analysis should give more clarity and will be part of future investigations.

With the techniques described in Sect. 3.5 and improved factors β_i for the penalties p_i (maximum bending at tip, maximum root section thickness, maximum twist angle, variance of the aerodynamic coefficients during last part of the flow solver iteration, and smoothing/averaging of aerodynamic coefficients) the final optimisation is executed with the objective function:

$$\text{Obj} = F_{\text{Weight}} \times C_{\text{D}}/C_{\text{L}} + \sum_{i=1}^{k} \beta_i \, p_i \qquad \text{[N]}. \tag{10}$$

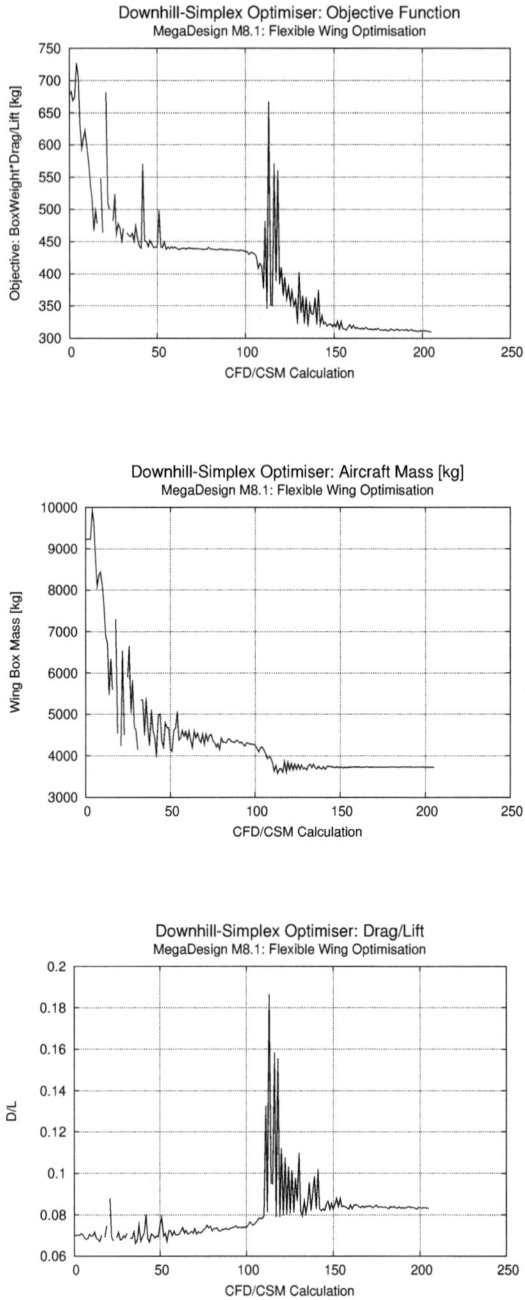

Fig. 4. Non-satisfactory optimisation run: Objective is improved by reduction of wing mass M_w, but aerodynamic performance D/L is degraded

4.2 Final Optimisation with 13 Design Variables

After adjustments of the penalty factors β_i the optimisation has been carried out successfully with 11 shape design parameters (twist, camber, thickness for 4 wing sections, twist fixed at wing root) and two structure design parameters (factors for thickness scaling), summing up to 13 design variables.

After 300 optimiser steps (this is equivalent to 300 coupled CFD/CSM calculations with 3600 beam stick model calculations and 300 Finite Element Model assemblies and 300 CATIA runs) the objective function nearly reaches its final value of 127000 N. This is the required thrust force of the considered four engines aircraft in stationary horizontal flight. However, we still did not do CFD calculations for the complete aircraft including fuselage, tails, etc.

One coupled CFD/CSM calculation took 20 min wall clock time on a 4 processor machine workstation (HP-UX-11.11, PA-RISC processor, model C8000). The complete optimisation lasts approximately four days.

Result Analysis

In Fig. 5 the reduction of the total drag force (objective function) is shown. Fig. 6 displays the decrease of total aircraft mass from 244 t to 238.4 t due to wing box mass reduction of 5.6 t. The initial equivalent metal sheet thickness at the root section has been reduced from 27 mm to 21.9 mm, and at the wing tip section from 20 mm to 18.7 mm. At the same time the aerodynamic performance L/D improves, see Fig. 7. The lift/drag ratio has been increased from initially 12 to

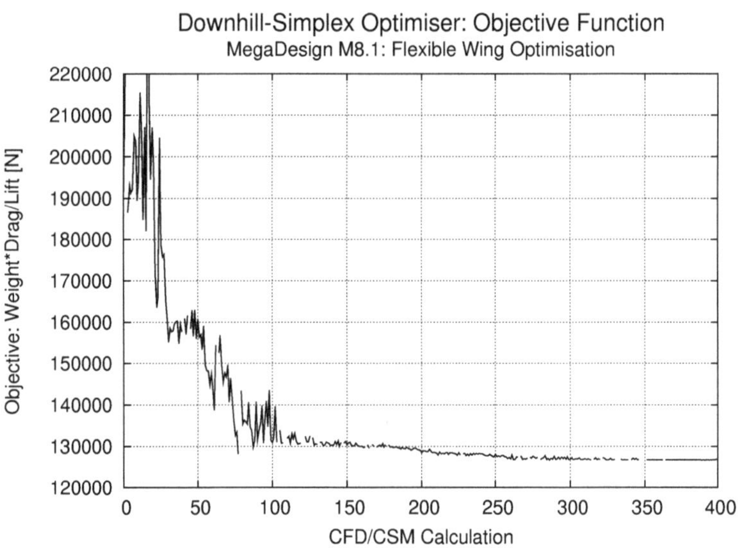

Fig. 5. Objective function development during optimisation with the gradient free Downhill Simplex optimiser

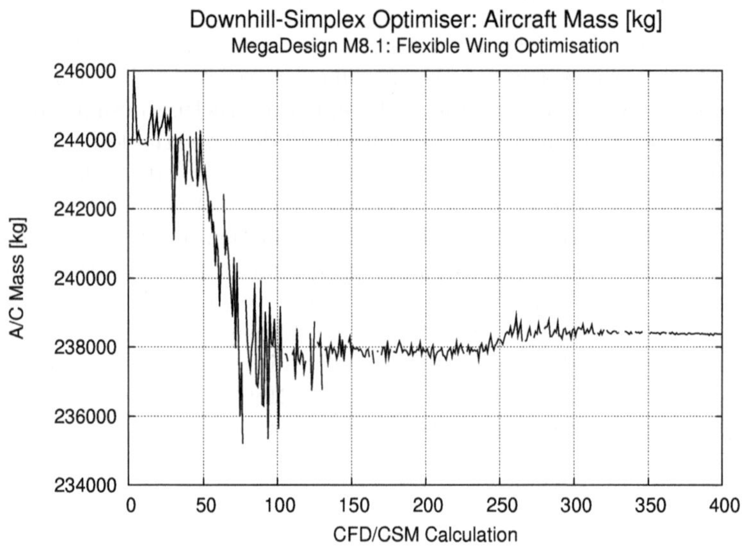

Fig. 6. Total aircraft mass reduction due to change of wing box structure mass during optimisation

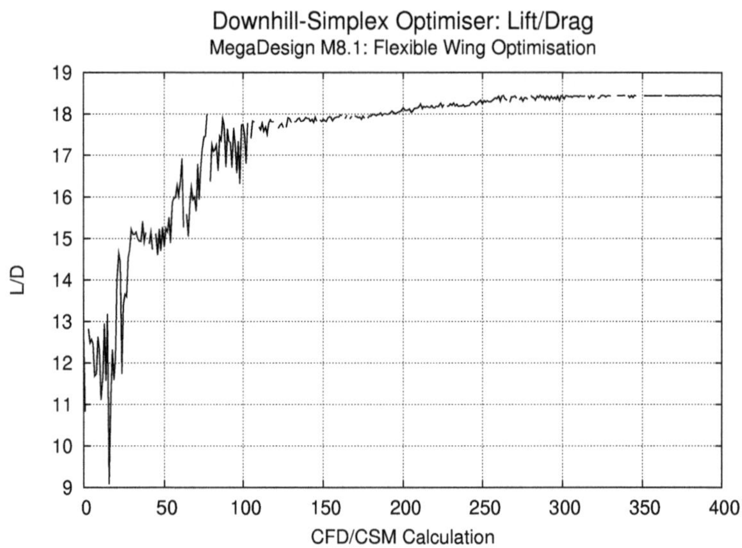

Fig. 7. Change of lift to drag ratio during optimisation run

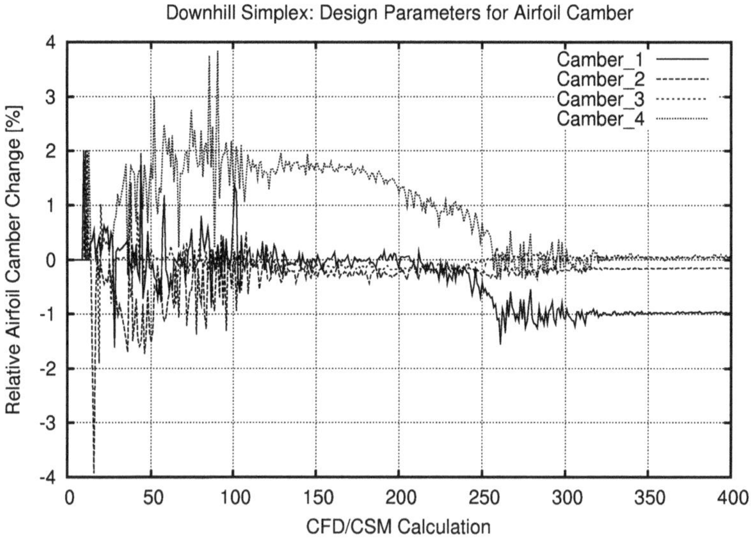

Fig. 8. Change of camber during optimisation

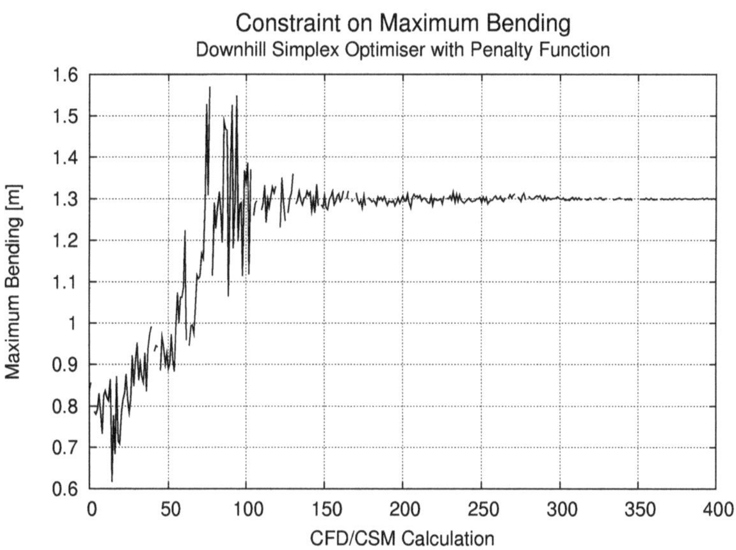

Fig. 9. Fulfilled wing tip bending constraint (maximum of 1.3 m) during optimisation, driven by penalty functions

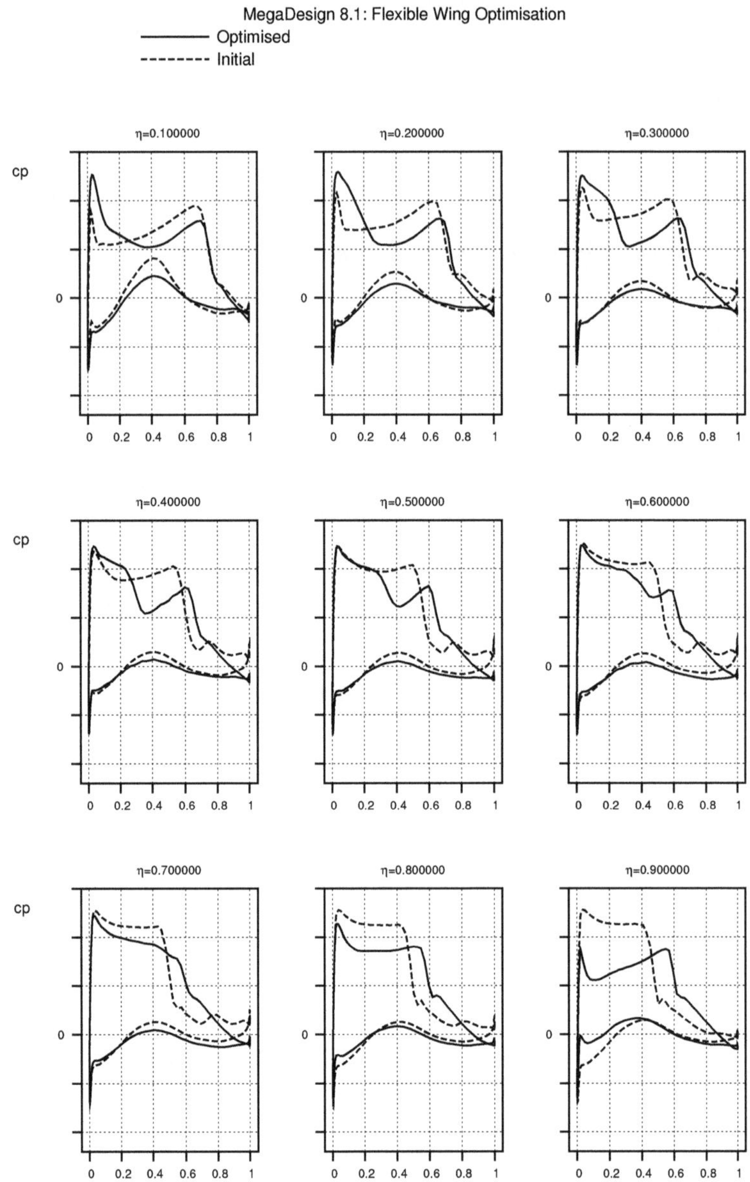

Fig. 10. Surface pressure coefficient distribution c_p at several wing sections for the optimised wing (solid curve) and for the initial wing (dashed)

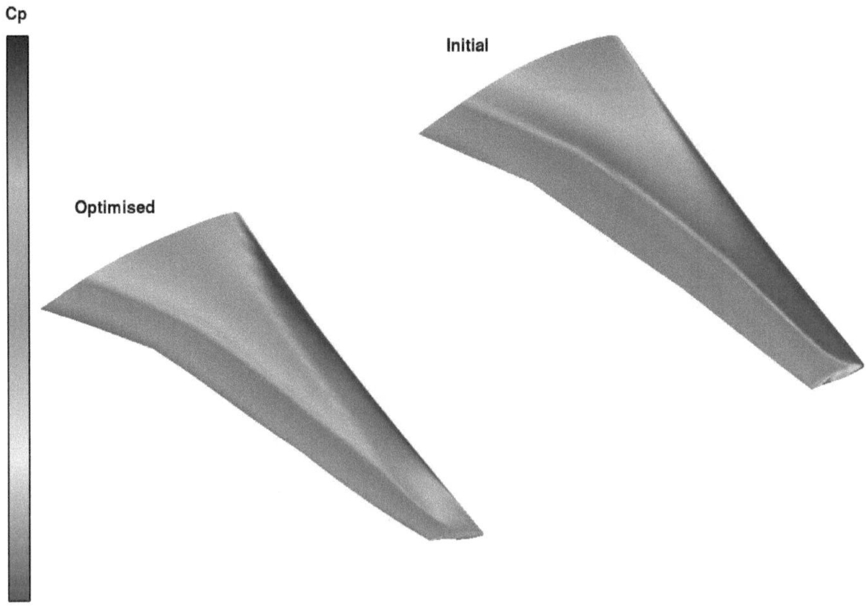

Fig. 11. Surface pressure coefficient c_p distribution in three-dimensional view for initial wing (right hand side) and for the optimised wing (left hand side)

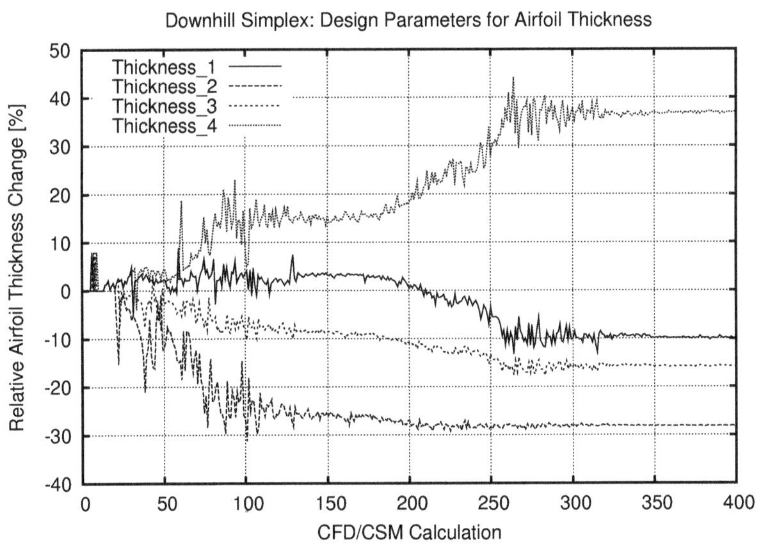

Fig. 12. Design parameter for relative wing section thicknesses during optimisation.

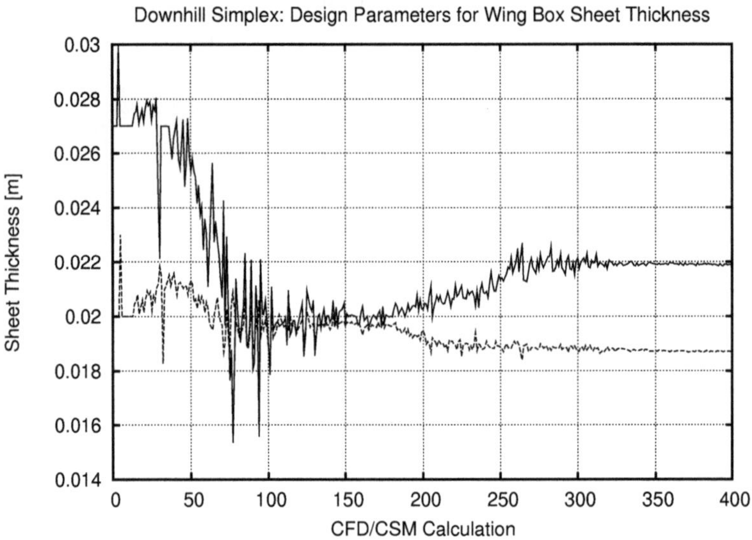

Fig. 13. Sheet thickness development during optimisation

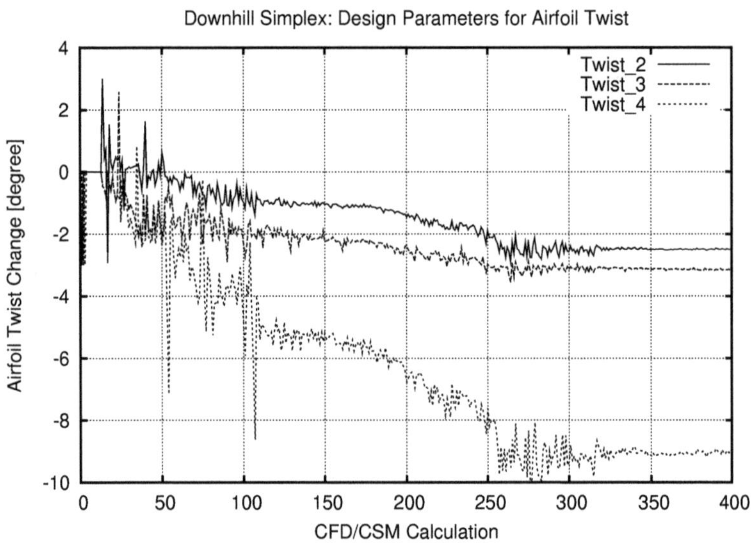

Fig. 14. Wing twist development during optimisation

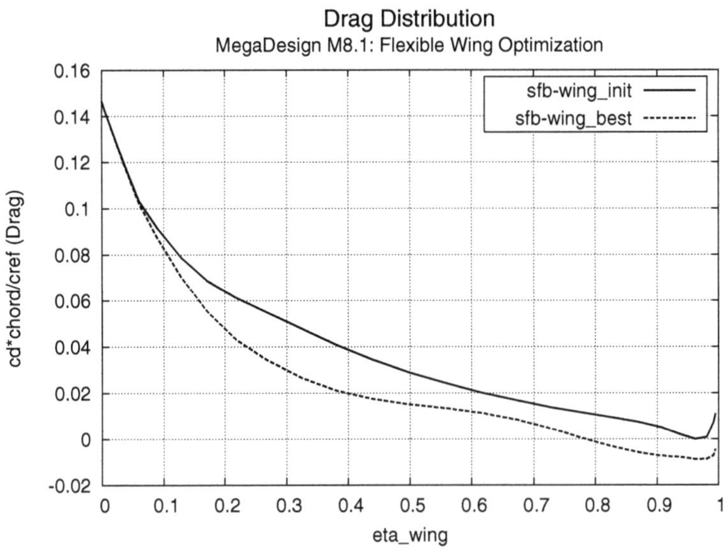

Fig. 15. Span-wise local drag distribution

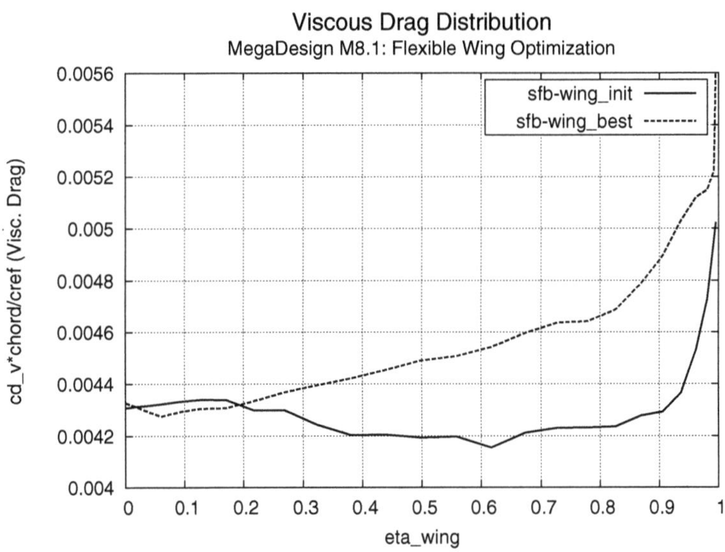

Fig. 16. Span-wise local viscous drag distribution

nearly 18.5. Also the constraint for maximum wing tip bending is maintained, see Fig. 9.

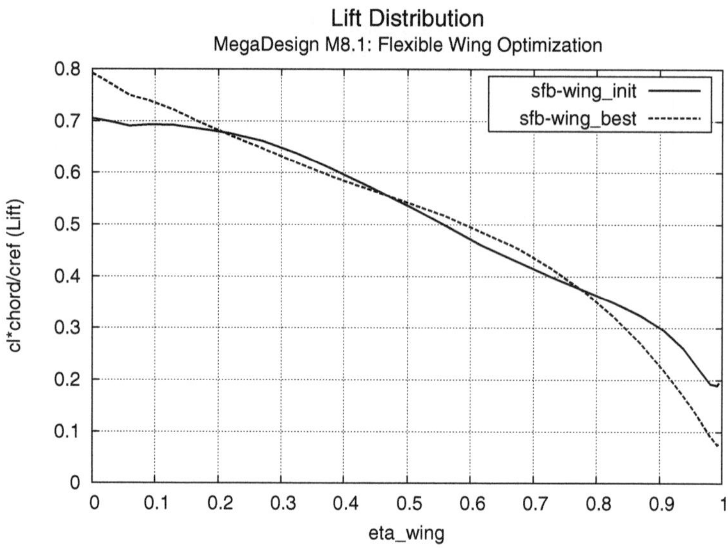

Fig. 17. Span-wise local lift distribution

In Fig. 10 the pressure coefficient distribution for several span-wise surface sections are displayed for the initial wing and for the optimised wing geometry.

The following effects lead to the reduction of aerodynamic drag:

1. Reduced shock strength decreases wave-drag with reduced affinity for flow separation occurring at relative wing section positions between $\eta = 0.4$ and $\eta = 0.7$ for the initial wing geometry.
2. Increased suction at the leading edge which decreases pressure drag
3. Higher pressure at the trailing edge.

Details for the span-wise drag distribution for the wing are shown in Fig. 15. The local sectional drag is decreased significantly in the range of 30 to 70% at the middle and outer region of the wing. For the wing root region the drag reduction is vanishing.

However, looking at the viscous part of the span-wise sectional drag (Fig. 16), we obtain a marginal increase of drag compared to the initial wing. But this is only a small fraction (one tenth) of the total local drag and the increase can be neglected.

The span-wise load distribution in Fig. 17 shows the change towards more inboard loading for the optimised wing. It reduces the wing root bending moment and the amount of wing tip deflection.

5 Conclusion

Several essential tools developed within the MegaDesign project were used and put together to build a multi-disciplinary optimisation chain. Using a mesh deformation tool based on Radial Basis Functions (RBF) allowing scattered data interpolation was the key technology to connect CAD based parametric shapc modifications with CFD mesh deformation. The idea to use an appropriate discretised wire-frame representation of the CAD geometry to calculate the discrete deformation field as input for the RBF deformation solved the problem of numerical noise.

The CFD/CSM coupling approach works robust and efficient. However, the correct problem formulation with adjusted factors for constraints and penalties decides between success and failure. Fuzzy Logic, automatic penalty factor determination and automatic sensitivity detection may help here.

Further improvement and development of tools would increase the use of numerical optimisation in industry. Availability of optimisers or optimisation systems is not the current bottleneck. Appropriate tools which form the building blocks of complex process chains need to be enhanced and equipped with compatible interfaces.

Finally, the use of adjoint solvers for gradient based optimisers seems to be possible for unstructured CFD meshes in combination with CAD based geometry parametrisation by making use of RBF based mesh deformation techniques.

References

1. Heinrich, R., Dargel, G., Kroll, N.: MegaDesign - Spezifikation des Testfalls für den Hauptmeilenstein M8.1 im Verbundvorhaben. DLR, Institut für Aerodynamik und Strömungsmechanik (2005)
2. Reimer, L., Braun, C., Chen, B.-H., Ballmann, J.: Computational Aeroelastic Design and Analysis of the HIRENASD Wind Tunnel Wing Model and Tests. In: International Forum on Aeroelasticity and Structural Dynamics (IFASD) 2007, Stockholm, paper IF071, June 17-20 (2007)
3. Ballmann, J., Dafnis, A., Braun, C., Korsch, H., Reimerdes, H.-G., Olivier, H.: Homepage of HIRENASD-Project, `https://heinrich.lufmech.rwth-aachen.de`
4. Tau-Code User Guide, Release 2006.1.0, DLR, Göttingen/Braunschweig - Germany (2006)
5. Technical Documentation of the DLR, TAU-Code, DLR, Göttingen/Braunschweig - Germany
6. Gerhold, T., Friedrich, O., Evans, J., Galle, M.: Calculation of Complex Three-Dimensional Configurations Employing the DLR Tau-Code. In: 35th Aerospace Science Meeting & Exhibit, Reno, NV (1997)
7. Heinrich, R.: Process chain description for MegaDesign task 8.1 including mesh deformation tool based on radial basis functions, DLR, private communication (2006)
8. Reimer, L., Braun, C.: Verbundvorhaben MegaDesign - Kurzdokumentation des Struktur-Präprozessors FEAFA (Finite Element Analysis for Aeroelasticity) zur Kopplung mit dem Aeroelastischen Kopplungsmodul (ACM). Lehr- und Forschungsgebiet für Mechanik, RWTH Aachen (2006)
9. Nelder, J.A., Mead, R.: Computer Journal 7, 308 (1965)
10. William, H., Flannery, B., Teukolsky, S., Vetterling, W.: Numerical Recipes in C – The Art of Scientific Computing. Cambridge University Press, Cambridge (1988)

Multidisciplinary Optimization of an UAV Combining CFD and CSM

S.M. Hitzel, L. Nardin, K. Sørensen, and H. Rieger

Aerodynamics and Methods, EADS Military Air Systems, Germany

Summary. Multidisciplinary Design Optimization (MDO) is a challenging goal for designers. The aim is to use optimization strategies to integrate a number of different disciplines simultaneously in the optimization process. The optimum of the multidisciplinary problem is better than the design found by optimizing each discipline sequentially, since it can exploit the interactions between the disciplines. However, including all disciplines simultaneously significantly increases the complexity of the problem.

At EADS Military Air Systems (EADS-MAS) the analysis of the multidisciplinary design optimization of an aircraft was carried out. While the whole aircraft has been optimized with regards to the aerodynamic part, structural concerns were taken into account for the wing alone. The whole process is not a real multidisciplinary optimization, because the structural part has not been integrated in a closed loop with the aerodynamic analysis. Instead, the structural code has been used to size the wing weight already optimized with regard to aerodynamic. The aim of this study was to prove that it was possible to eventually carry out such a multidisciplinary optimization with the available tools for future analyses.

1 Optimization Process

The optimization of both aerodynamic and structural characteristics of an unmanned aerial vehicle (Figure 1) was carried out. Actually the structural part regarded only the analysis of the wing alone.

Initially the weight of the wing was simply estimated using a very simple structural wing box model through an analytical formulation. Obviously this analysis was too simplistic: first of all because of the simplicity of the analytical model, and then because the forces resulting from the analytical analysis were not applied to the structure. Therefore a more complex methodology has been implemented, comprising an FEM structural analysis of the wing.

The complete optimization process is carried out as described in Figure 2: a CAD model is provided, the unstructured chimera mesh generated with the MAS Mesher [1] and used to compute the aerodynamic forces. The CATIA model provides also the FEM mesh which is used by the structural analysis system NASTRAN to compute the stresses of the wing structure. The optimization analysis is steered by modeFRONTIER [2].

N. Kroll et al. (Eds.): MEGADESIGN and MegaOpt, NNFM 107, pp. 307–312.
springerlink.com

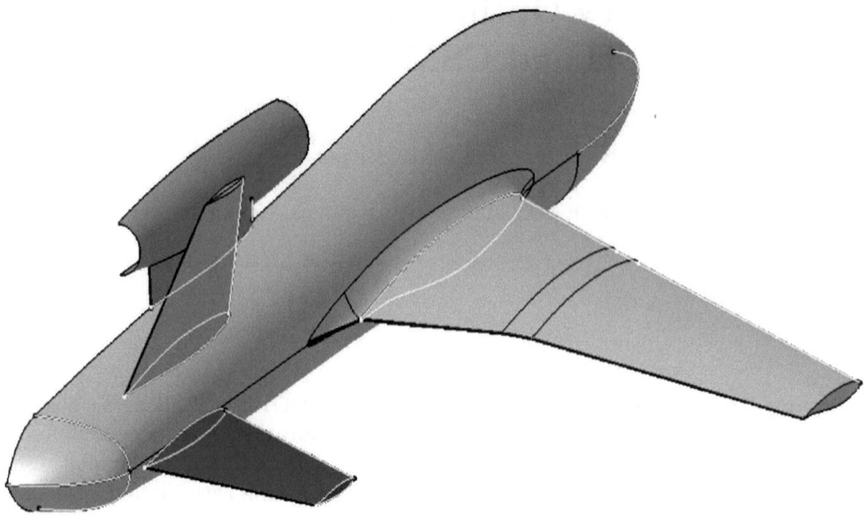

Fig. 1. Aircraft topology

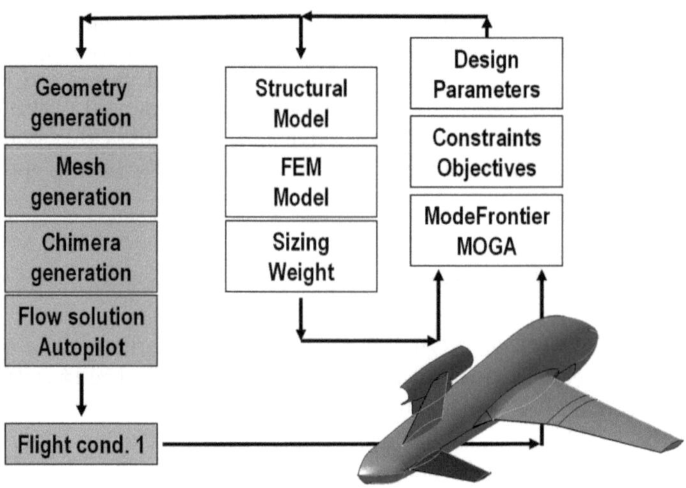

Fig. 2. Optimization process including structural weight evaluation via FEM

The optimization carried out is not a close-coupled one, because the results of the structural analysis are not feed up to the aerodynamic analysis. Furthermore, the design parameters steering the structural sizing are not the same driving the aerodynamic optimization: for the structural part the thicknesses of spars, ribs and of the skin were taken into account. None of these design variables have a direct influence on the aerodynamic design. Currently only the sizing procedure of the wing structure has been included.

The optimization process has been therefore steered by modeFRONTIER which changed the input design parameters accordingly to the aerodynamic simulations output. The structure of the wing was then sized through the analysis of the stresses calculated by Nastran.

2 Aerodynamic Analysis and Optimization

The aim of the aerodynamic optimization was to maximize a loiter phase as presented in a companion paper in this book. For simple stealth reasons the leading and trailing edge of the horizontal tail were kept parallel to the wing leading and trailing edges. Furthermore a single "rubber" engine was assumed and not modelled in detail in the CAD (Figure 1). The wing and tail profiles were kept unchanged. The mission weight was assumed to be constant for the loiter phase at an average of ZFW + 0.5 full fuel weight. The design variables of this aerodynamic part are wing parameters such as sweep, aspect ratio, taper ratio (inner, outer), inner wing kink span and possibly the position of the wing.

The fuel consumption was evaluated via the thrust equals drag condition, the drag being evaluated by the flow simulation. The max rated sea level thrust as a function of elevation and Mach number is computed using data for a generic high-bypass turbofan engine. Using generic engine data the specific fuel consumption for cruise and loiter fuel is found along the same lines. With the loiter thrust required assumed to depend on the glide ratio, and the glide ratio being determined via the flow simulations, the objective loiter time becomes:

$$t_{loiter} = \frac{W_{fuel} - W_{fuel,cruise}}{TSFC_{loiter} \cdot T_{loiter}}$$

In the formula W indicates the aircraft mass, $TSFC$ stands for the thrust specific fuel consumption and T means the corresponding thrust. Conducting the optimization through the use of the previous formula is better than optimizing the design for just lift or drag: in the above formulation the main characteristics steering the optimization are present, therefore maximizing the loiter time implicitly optimize lift and drag of the whole aircraft.

The aircraft had to be simulated at trimmed conditions. For a simplified version the assumption of linear aerodynamics at low angle-of-attack the angle-of-attack and angle-of-incidence for the horizontal tail trim-position where estimated according to Raymer [3].

3 FEM Integration

Provisions were made to design a realistic parameterized structural model geometry via CAD. The wing structure can be modelled with different numbers of profile-ribs and different parameterized positions of a front and rear wing spar. The modelling of the structure is integrated into the automatic optimization process and can be called during the optimization any time intended.

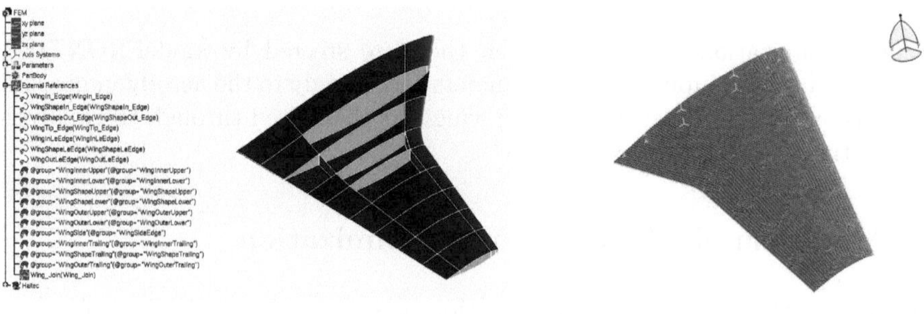

Fig. 3. Geometry of a structural wing model and FEM-model of the same wing

This geometry is automatically meshed into a complete FEM-model. Currently shell elements are used for each geometrical rib, spar and skin segment. Next, the evaluated aerodynamic pressure distribution is mapped onto the knots of the FEM-model in the CATIA environment. Together with the gravity loads and/or other inertia loads and external forces (e.g. from engines) a NASTRAN-deck is provided automatically, which then can be used to size the local thickness of the wing structure.

Figure 3 shows the internal wing structure of the full aircraft shown in Figure 1. The cut-away provides insight into the rib, spar, skin structural layout, while the surface of the FEM-model is depicted as well on the right side.

On the left side of Figure 4 the aerodynamic load is mapped automatically via a CATIA-V5 macro onto the FEM-knots. In the same picture the gravity forces are indicated by arrows. A first non-sized distribution of the von Mises stresses is shown on the right side of the picture. On the basis of these stresses the local parts of the structure, the individual spars, ribs and skins can be sized or optimized for minimum weight, respectively.

Initially the thickness of each element (spars and ribs) was sized on its entireness, but a high stress concentrated on a small area, usually at the junction between spars and ribs, resulted in a high thickness for the whole part. Actually the simplicity of the model didn't account for real structural components. Therefore each element of the FEM mesh was sized individually. This strategy allowed

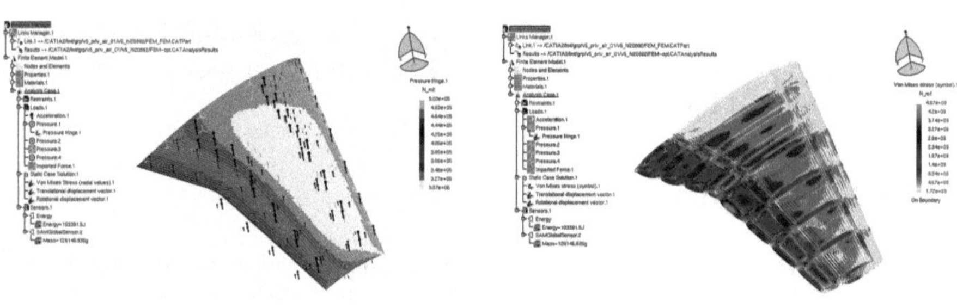

Fig. 4. Mapped pressure distribution and van Mises stress on the FEM-elements before sizing

defining different thicknesses not only for the different parts of the wing, but for each element. The mass was therefore more concentrated at the junctions, while the skin and the parts in which the stresses were not high resulted on having a reduced thickness.

4 Results

In Figure 5 each design is equipped with the same structural topology: no design change in the number of ribs was done. However the location of spars is different. Figure 5 indicates the force paths by a relatively high value for the local element thickness. It can be recognized clearly that in the Figure on the left the position of the front spar is not optimal because relatively high stress has to be carried by the skin. Also one can observe load paths normal to the elastic axis. The described deficit has been tried to improve by a topological change of the position of the front spar (fig. 5 right). Here the two load transfer points at the wing-fuselage interface are in coincidence with the front and rear spar attachment points. In principle the thickness distribution has been bettered, however, some disturbances are still visible in the mid wing region. The reason for that is unknown for the moment. Interestingly the wing structural weight of the improved spar locations has become lower. The structural analysis gives for the left wing topology a weight of 166 kg whereas the left wing has been lowered to 145 kg.

Further computational experiments are displayed in Figure 6. For the structural topology of Figure 5 (right) the FEM-mesh was refined by a factor of 2. The result is depicted on the left in Figure 6. It is noticed that the local high stresses in the mid wig region have disappeared and now are concentrated along the spars in span direction. Also disturbances are now visible mainly at the trailing edges. This could be explained by some insufficient FEM modelling in that area.

On the right in Figure 6 the topology of the ribs has been changed. The stress distribution has been very much improved. The maximum element thickness is

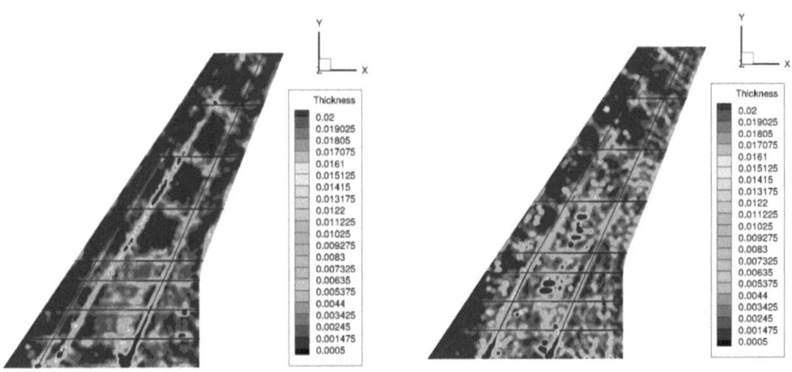

Fig. 5. Influence of spars location on element thickness distribution

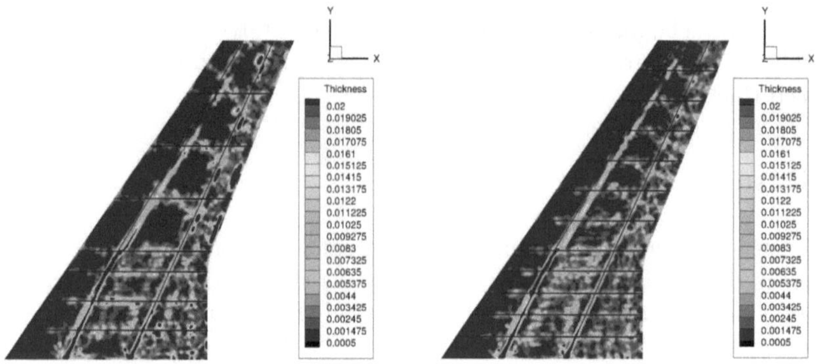

Fig. 6. Influence of FEM-mesh refinement (left) and number of ribs (right)

concentrated along the spars and the hot spots are almost gone. By inspection one can observe still area for improvements. This concerns the outer wing spar and ribs locations. Positioning of the ribs along the stress paths (normal to the elastic axis) are a matter of priority. Interestingly the weight of the left wing has improved to 135 kg.

5 Conclusions

From the results obtained it can be concluded that a stepwise optimization process has been successfully implemented. The process is starting with the aerodynamic configuration optimization based on a given mission scenario and encompasses the structural sizing optimization of the main structural components. The experience gained with this example brought us to determine a strategy for future optimizations, for example sizing each element separately and letting the positions of the spars and the number of ribs to be changed by the optimization analysis tools. In addition to that, a convergence study on the mesh size can help to obtain better solutions.

This proves that the tools to realize such a multidisciplinary design optimization are available. Future steps will be the close-coupled integration of the different disciplines, to really appreciate the interactions between them.

References

1. Tremel, U., Deister, F., Hassan, O., Weatherill, N.P.: Automatic Unstructured Surface Mesh Generation for Complex Configurations. Intern. Journal for Numerical Methods in Fluids (N.P. Weatherill, ed.) 44 (2004)
2. *modeFrontier – A multi-Objective Optimization and Design Environment*, `http://www.esteco.com`
3. Raymer, D.: Aircraft Design: A Conceptual Approach. AIAA Education Series (1989)

Author Index

Notes on Numerical Fluid Mechanics and Multidisciplinary Design

Available Volumes

Volume 107: Norbert Kroll, Dieter Schwamborn, Klaus Becker, Herbert Rieger, Frank Thiele (eds.): MEGADESIGN and MegaOpt – German Initiatives for Aerodynamic Simulation and Optimization in Aircraft Design. ISBN 978-3-642-04092-4

Volume 106: Wolfgang Nitsche, Christoph Dobriloff (eds.): Imaging Measurement Methods for Flow Analysis - Results of the DFG Priority Programme 1147 "Imaging Measurement Methods for Flow Analysis" 2003–2009. ISBN 978-3-642-01105-4

Volume 105: Michel Deville, Thien-Hiep Lê, Pierre Sagaut (eds.): Turbulence and Interactions - Keynote Lectures of the TI 2006 Conference. ISBN 978-3-642-00261-8

Volume 104: Christophe Brun, Daniel Juvé, Michael Manhart, Claus-Dieter Munz: Numerical Simulation of Turbulent Flows and Noise Generation - Results of the DFG/CNRS Research Groups FOR 507 and FOR 508. ISBN 978-3-540-89955-6

Volume 103: Werner Haase, Marianna Braza, Alistair Revell (eds.): DESider – A European Effort on Hybrid RANS-LES Modelling - Results of the European-Union Funded Project, 2004–2007. ISBN 978-3-540-92772-3

Volume 102: Rolf Radespiel, Cord-Christian Rossow, Benjamin Winfried Brinkmann (eds.): Hermann Schlichting – 100 Years - Scientific Colloquium Celebrating the Anniversary of His Birthday, Braunschweig, Germany 2007. ISBN 978-3-540-95997-7

Volume 101: Egon Krause, Yurii I. Shokin, Michael Resch, Nina Shokina (eds.): Computational Science and High Performance Computing III - The 3rd Russian-German Advanced Research Workshop, Novosibirsk, Russia, 23–27 July 2007. ISBN 978-3-540-69008-5

Volume 100: Ernst Heinrich Hirschel, Egon Krause (eds.): 100 Volumes of 'Notes on Numerical Fluid Mechanics' - 40 Years of Numerical Fluid Mechanics and Aerodynamics in Retrospect. ISBN 978-3-540-70804-9

Volume 99: Burkhard Schulte-Werning, David Thompson, Pierre-Etienne Gautier, Carl Hanson, Brian Hemsworth, James Nelson, Tatsuo Maeda, Paul de Vos (eds.): Noise and Vibration Mitigation for Rail Transportation Systems - Proceedings of the 9th International Workshop on Railway Noise, Munich, Germany, 4–8 September 2007. ISBN 978-3-540-74892-2

Volume 98: Ali Gülhan (ed.): RESPACE – Key Technologies for Reusable Space Systems - Results of a Virtual Institute Programme of the German Helmholtz-Association, 2003–2007. ISBN 978-3-540-77818-9

Volume 97: Shia-Hui Peng, Werner Haase (eds.): Advances in Hybrid RANS-LES Modelling - Papers contributed to the 2007 Symposium of Hybrid RANS-LES Methods, Corfu, Greece, 17–18 June 2007. ISBN 978-3-540-77813-4

Volume 96: C. Tropea, S. Jakirlic, H.-J. Heinemann, R. Henke, H. Hönlinger (eds.): New Results in Numerical and Experimental Fluid Mechanics VI - Contributions to the 15th STAB/DGLR Symposium Darmstadt, Germany, 2006. ISBN 978-3-540-74458-0

Volume 95: R. King (ed.): Active Flow Control - Papers contributed to the Conference "Active Flow Control 2006", Berlin, Germany, September 27 to 29, 2006. ISBN 978-3-540-71438-5

Volume 94: W. Haase, B. Aupoix, U. Bunge, D. Schwamborn (eds.): FLOMANIA - A European Initiative on Flow Physics Modelling - Results of the European-Union funded project 2002 - 2004. ISBN 978-3-540-28786-5

Volume 93: Yu. Shokin, M. Resch, N. Danaev, M. Orunkhanov, N. Shokina (eds.): Advances in High Performance Computing and Computational Sciences - The Ith Khazakh-German Advanced Research Workshop, Almaty, Kazakhstan, September 25 to October 1, 2005. ISBN 978-3-540-33864-2

Volume 92: H.J. Rath, C. Holze, H.-J. Heinemann, R. Henke, H. Hönlinger (eds.): New Results in Numerical and Experimental Fluid Mechanics V - Contributions to the 14th STAB/DGLR Symposium Bremen, Germany 2004. ISBN 978-3-540-33286-2

Volume 91: E. Krause, Yu. Shokin, M. Resch, N. Shokina (eds.): Computational Science and High Performance Computing II - The 2nd Russian-German Advanced Research Workshop, Stuttgart, Germany, March 14 to 16, 2005. ISBN 978-3-540-31767-8

Volume 87: Ch. Breitsamter, B. Laschka, H.-J. Heinemann, R. Hilbig (eds.): New Results in Numerical and Experimental Fluid Mechanics IV. ISBN 978-3-540-20258-5

Volume 86: S. Wagner, M. Kloker, U. Rist (eds.): Recent Results in Laminar-Turbulent Transition - Selected numerical and experimental contributions from the DFG priority programme 'Transition' in Germany. ISBN 978-3-540-40490-3

Volume 85: N.G. Barton, J. Periaux (eds.): Coupling of Fluids, Structures and Waves in Aeronautics - Proceedings of a French-Australian Workshop in Melbourne, Australia 3-6 December 2001. ISBN 978-3-540-40222-0

Volume 83: L. Davidson, D. Cokljat, J. Fröhlich, M.A. Leschziner, C. Mellen, W. Rodi (eds.): LESFOIL: Large Eddy Simulation of Flow around a High Lift Airfoil - Results of the Project LESFOIL supported by the European Union 1998 - 2001. ISBN 978-3-540-00533-9

Volume 82: E.H. Hirschel (ed.): Numerical Flow Simulation III - CNRS-DFG Collaborative Research Programme, Results 2000-2002. ISBN 978-3-540-44130-4

Volume 81: W. Haase, V. Selmin, B. Winzell (eds.): Progress in Computational Flow Structure Interaction - Results of the Project UNSI, supported by the European Union 1998-2000. ISBN 978-3-540-43902-8

Volume 80: E. Stanewsky, J. Delery, J. Fulker, P. de Matteis (eds.): Drag Reduction by Shock and Boundary Layer Control - Results of the Project EUROSHOCK II, supported by the European Union 1996-1999. ISBN 978-3-540-43317-0

Volume 79: B. Schulte-Werning, R. Gregoire, A. Malfatti, G. Matschke (eds.): TRANSAERO - A European Initiative on Transient Aerodynamics for Railway System Optimisation. ISBN 978-3-540-43316-3

Volume 78: M. Hafez, K. Morinishi, J. Periaux (eds.): Computational Fluid Dynamics for the 21st Century. Proceedings of a Symposium Honoring Prof. Satofuka on the Occasion of his 60th Birthday, Kyoto, Japan, 15-17 July 2000. ISBN 978-3-540-42053-8

Volume 77: S. Wagner, U. Rist, H.-J. Heinemann, R. Hilbig (eds.): New Results in Numerical and Experimental Fluid Mechanics III. Contributions to the 12th STAB/DGLR Symposium, Stuttgart, Germany 2000. ISBN 978-3-540-42696-7

Volume 76: P. Thiede (ed.): Aerodynamic Drag Reduction Technologies. Proceedings of the CEAS/DragNet European Drag Reduction Conference, 19-21 June 2000, Potsdam, Germany. ISBN 978-3-540-41911-2

Volume 75: E.H. Hirschel (ed.): Numerical Flow Simulation II. CNRS-DFG Collaborative Research Programme, Results 1998-2000. ISBN 978-3-540-41608-1

Volume 66: E.H. Hirschel (ed.): Numerical Flow Simulation I. CNRS-DFG Collaborative Research Programme. Results 1996-1998. ISBN 978-3-540-41540-4

MIX
Papier aus verantwortungsvollen Quellen
Paper from responsible sources
FSC® C105338

If you have any concerns about our products,
you can contact us on
ProductSafety@springernature.com

In case Publisher is established outside the EU,
the EU authorized representative is:
Springer Nature Customer Service Center GmbH
Europaplatz 3, 69115 Heidelberg, Germany

Printed by Libri Plureos GmbH
in Hamburg, Germany